Alkali-Activated Materials: Advances and Novel Applications

Alkali-Activated Materials: Advances and Novel Applications

Guest Editors

Guilherme Ascensão
Sandra Lucas

MDPI

Basel • Beijing • Wuhan • Barcelona • Belgrade • Novi Sad • Cluj • Manchester

Guest Editors
Guilherme Ascensão
Department of Civil Engineering
University of Aveiro
Aveiro
Portugal

Sandra Lucas
Department of the Built Environment
Eindhoven University of Technology
Eindhoven
The Netherlands

Editorial Office
MDPI AG
Grosspeteranlage 5
4052 Basel, Switzerland

This is a reprint of the Special Issue, published open access by the journal *Applied Sciences* (ISSN 2076-3417), freely accessible at: https://www.mdpi.com/journal/applsci/special_issues/Alkali_Activated_Materials_Advances_Applications.

For citation purposes, cite each article independently as indicated on the article page online and as indicated below:

Lastname, A.A.; Lastname, B.B. Article Title. *Journal Name* **Year**, *Volume Number*, Page Range.

ISBN 978-3-7258-5761-6 (Hbk)
ISBN 978-3-7258-5762-3 (PDF)
https://doi.org/10.3390/books978-3-7258-5762-3

Contents

About the Editors

Guilherme Ascensão

Guilherme Ascensão is a Researcher in sustainable building materials and circular construction in the CERIS research unit, Department of Civil Engineering, at the University of Aveiro. His expertise includes the development and characterization of low-clinker cements, clay-based materials, and geopolymers for construction and building heritage rehabilitation. Guilherme Ascensão obtained a BEng in Civil Engineering in 2012, and an MSc in Civil Engineering in 2015 from the University of Aveiro (Portugal). He then joined the CICECO Aveiro Institute of Materials as a junior researcher. In 2016, he became a Marie Curie Doctoral researcher within the NEW-MINE project, working with the Heidelberg Cement Group (Bergamo, Italy), the University of Padova, Italy, and KU Leuven (Belgium). In 2021, Guilherme Ascensão joined the CECOLAB Association—Collaborative Laboratory Towards Circular Economy. In 2022, he joined the CERIS research group, Civil Engineering Research and Innovation for Sustainability, at the Department of Civil Engineering, University of Aveiro. With both international and industry experience, Ascensão's research has been dedicated to developing sustainable building materials and innovative construction solutions.

Sandra Lucas

Sandra Lucas is an Associate Professor in autonomous manufacturing of construction materials in the research group Concrete Structures at the TU/e Department of the Built Environment (unit Structural Design). Sandra's research goal is to understand how machine learning and data science accelerate the discovery and development of new construction materials that are functionally advanced and environmentally sustainable. This involves synthesizing materials with embedded functionalities that adapt to environmental conditions or mitigate impacts. She is the project leader of two NWO projects to develop smart and sustainable 3D-printed concrete. Sandra Lucas obtained a BEng in ceramic and glass engineering, an MSc in environmental management and materials and, in 2011, a PhD in Civil Engineering from the University of Aveiro (Portugal). She then became a Marie Curie Postdoctoral researcher at the Fraunhofer Institute UMSICHT (Oberhausen, Germany) for the SHeMat project. As part of this project, she was a visiting researcher at EPSI ParisTech (Paris, France) and Delft University of Technology (The Netherlands). In 2014, she joined the University of Greenwich (London, UK) as a senior lecturer in construction materials and sustainability. In 2018, Sandra Lucas joined the research group Concrete Structures at the Department of the Built Environment of the Eindhoven University of Technology (TU/e, the Netherlands).

applied sciences

MDPI

Article

An Investigation on the Synthesis of Alkali Activated Materials from Thermally Modified Clays

Guilherme Ascensão [1,*], Enrico Bernardo [2] and Victor M. Ferreira [1]

[1] RISCO, Department of Civil Engineering, University of Aveiro, 3810-193 Aveiro, Portugal
[2] Department of Industrial Engineering, University of Padova, 35131 Padova, Italy
* Correspondence: guilhermeascensao@ua.pt; Tel.: +351-966944248

Abstract: The sustainability and economic competitiveness of alkali activation technology greatly depends on expanding the raw materials database with locally available resources. Therefore, a notable trend has been witnessed toward the exploitation of common clays as alternatives to well-established solid aluminosilicate precursors due to their availability and wide geographical distribution. However, common clays are complex and dedicated research is needed to tailor synthesis procedures and mix designs for different clay resources. This paper describes the outcomes of a study conducted to investigate the influence of several synthesis parameters (solid-to-liquid ratio, NaOH molarity, Si availability, and curing conditions) on the properties of alkali activated binders produced from different thermally modified clays. Optimal synthesis conditions for benchmark metakaolin systems have been identified and binders were produced with progressive dosages of metakaolin replacement by common local clays. Fundamental physical and mechanical properties such as apparent density, open porosity, water absorption, and compressive strength were examined at different curing ages, and X-ray diffraction (XRD) was used to provide complementary mineralogical insights. By combining the effects of the parameters studied, mortar specimens were produced with the developed binders, reaching compressive strength values exceeding 28.2 ± 0.1 MPa, a bulk density as low as 1.78 ± 0.0 g/cm^3, and open porosity and water absorption values lower than 15% and 8%, respectively. These properties are comparable to those of conventional hydraulic products, which presents them as interesting candidates for construction. Ultimately, this work aims to contribute with valuable insights toward the valorization of a large group of unexploited clay precursors by demonstrating the feasibility of producing technologically competitive alkali activated materials with little or no use of the prime precursors, thus adding to the extant knowledge and contributing to future scientific and industrial developments in this field.

Keywords: alternative binders; alkali activated materials; secondary resources; valorization of common clays

1. Introduction

In the coming decades, global tensions and conflicts over natural resources, social inequality, forced migration, and global economic instability will be exacerbated by rampant climate changes if profound actions are not immediately taken to modify our current development models. Global warming is among the most pressing environmental problems that should be prioritized when tackling climate change, and anthropologic greenhouse gas (GHG) emissions must be drastically reduced. GHGs are mostly driven by CO_2 emissions, which account for approximately 80% of the total GHG [1]. Carbon capture and storage technologies are expected to deliver massive CO_2 uptakes [2,3], but these technologies still require massive capital investments and elevated operating costs that throttle their economic feasibility and prevent their worldwide spread with meaningful global impacts. Therefore, redefining our current industrial and manufacturing technologies has become urgent to reduce GHG emissions, limit the rate of climate change progression, and meet the

UN Sustainable Development Goals. The importance of the built environment in this matter can be gleaned from the numbers: the construction and operation of buildings consumes approximately 36% of all energy produced worldwide, and accounts for about 39% of global CO_2 emissions [4]. The world's growing population will drive further urbanization needs and concomitant increases in the consumption of building materials. Steel and cement are the most significant building materials and their manufacturing jointly accounts for 12–15% of the global annual CO_2 emissions [5]. Cement production alone represents 20% of industrial CO_2 emissions [2] and global consumption patterns forecast a significant soar in cement demand until 2050 [6]. Therefore, the development of low carbon binders able of replace ordinary Portland cement (OPC) is crucial to address the growing housing needs and decouple social development and well-being from climate change. Emergent binders with environmentally favorable profiles that are being intensely investigated include, among others, belite-rich Portland and belite ye'elimite-ferrite cements [7–11], magnesium-based cements [7], carbonation-hardening cements [7,12], calcined clay limestone cements [13–16], geopolymers, and alkali-activated materials [17–19].

Alkali-activated materials (AAMs) are one of most promising candidates to replace OPC-based products. Initial research on AAMs have been driven by the economic benefits delivered by using industrial by-products as alternative feedstock raw materials, but AAMs have proven to offer excellent technical and environmental performances if some conditions and predicates are safeguarded in the mixture design stage (i.e., thoughtful definition of precursors and alkaline solutions) [17,19–24]. AAMs have also shown enhanced resistance to chemical attack and harsh environments, which can reduce the maintenance costs and recurrent environmental impacts, further increasing the body of arguments supporting AAMs as alternative green binders suitable for the construction sector. In addition, as AAMs are the product of the reaction between an alkali metal source and amorphous or semi-vitreous aluminosilicate precursors, virtually any powdered material with sufficient alumina and silica in a reactive form—meaning sufficiently soluble in alkaline environments—can be used as a precursor, if adequate synthesis conditions are provided. This allows for resilient industrial manufacturing processes and extreme design versatility that allows for the use of locally available resources to address the specific needs and opportunities of regional markets while by-passing the external constraints of global value chains.

Natural and artificial pozzolans (e.g., volcanic ashes [25], pumice [26], and metakaolin [27]) and industrial by-products such as waste glass [28], bauxite residues [29,30], non-ferrous metallurgical slags [17,31,32], and coal and biomass fly ashes [33–36] have been examined as potential precursors, while silicates, hydroxides, carbonates, sulfates, and mixtures thereof can be used as activators. Metakaolin is the most well-established and technologically mature precursor used for AAM production. Metakaolin has attracted attention as a precursor due to its purity and consistent composition as well as its outstanding mechanical properties and the thermal stability of metakaolin-based binders. Technical challenges include the high specific surface area of metakaolin, which often results in poor workability and demand for high water dosages that later result in considerable shrinkage and cracking [37]. Several shrinkage mitigation strategies have been recently described (e.g., heat curing, shrinkage reducing admixtures) and can be used to counteract such deleterious phenomena [32,38].

However, the commercial exploitation of metakaolin-based AAMs can be limited by local shortages and the excessive cost of high-grade kaolinitic clays. The sustainability profiles and green credentials of building materials will be compromised if long transport distances are considered, and the use of endogenous materials as precursors should be favored to further future-proof alkali activation as a scalable, environmentally friendly, and cost-competitive technology. In this view, the interest in using low- and medium-grade clays (kaolinitic and otherwise) abundant in many locations around the globe as metakaolin surrogates is increasing [39–43].

Portugal has plenteous clay resources including kaolinitic clays. Kaolin deposits are distributed mostly in Portugal's northwest region comprising the Lusitanian Meso-

Cenozoic basin and a narrow northwest strip located between Viana do Castelo and Aveiro districts [44]. In 2016, Portugal ranked 12th in kaolin production in Europe [45], and although abundant, most kaolin resources are consumed by the local ceramic industry and other high-end markets such as paper, paint, rubber, and agrochemical production. Therefore, highly pure kaolinic clays (i.e., without significant levels of "industrial contaminants" such as quartz, mica, montmorillonite, iron, and titanium oxides) are highly disputed resources that are unlikely to be available for the construction market. Affordable alternatives for AAM manufacturing are needed, and several common clays might have the potential to be used in (partial) lieu of metakaolin. Many of these clays contain kaolinite minerals, but not in the desired quantities for metakaolin commercialization. High iron contents can also be allowed in AAM production and will not constitute a limiting factor as previously demonstrated [17,46,47]. Common clays are considerably cheaper, are available in significant volumes in Portugal, and are even often discharged as industrial waste in activities such as dredging, sand-washing, and kaolin extraction. Therefore, this research focused on the use of a common clay found in central Portugal as precursors, taken as a representative case study of a broad group of currently underutilized clay resources that could potentially be valorized via alkali activation. The influence of various synthesis factors on the fundamental properties of metakaolin systems was examined and taken as the baseline to describe the effects of replacing metakaolin by calcined common clays. Mortar specimens were also produced with progressive contents of common clay to investigate the key fresh and hardened-state properties.

The aim of this groundwork research was to establish the first principles of design for this novel valorization pathway and contribute to the development of innovative building materials that minimize the environmental impacts of construction while enforcing upcycling practices and sustainable development models in Portugal and elsewhere.

2. Experimental Conditions

2.1. Materials

A commercial metakaolin (hereafter MK) was purchased under the name of Argical™ from AGS-mineraux®, Clérac, France. According to the producer, MK is mostly amorphous and almost exclusively composed of SiO_2 and Al_2O_3 at 55.0 and 40.0 wt.%, respectively. A common local red clay (hereafter RC) was provided by a local supplier, Mota Ceramic Solutions®, Oiã, Portugal. RC was calcinated in a laboratory oven at 800 °C with an isothermal step of 2 h (heating rate of 3 °C/min up to 500 °C, and 10 °C/min up to 800 °C) followed by natural cooling. Similar thermal activation procedures have been adopted elsewhere [40,43]. The calcined clay (hereafter CRC) was used to gradually replace MK in the alkali activated binders and mortars (the mix proportions are further detailed in Section 2.2). Densified silica fume (hereafter SF; Elkem Microsilica® Grade 940, ELKEM, Norway) was used as a secondary source of reactive SiO_2. SF has been characterized in previous works and determined to be composed of 95.0 wt.% SiO_2 with minor elements such as CaO, Fe_2O_3, and Al_2O_3 being considered negligible (<0.5 wt.%). The BET specific surface area was determined to be approximately 22,200 m^2/kg [17,48].

Activating solutions were prepared by mixing sodium hydroxide and sodium silicate solutions. Sodium hydroxide solutions (8, 10 and 12 M) were prepared in the laboratory by dissolving sodium hydroxide beads (M = 40.00 g/mol, 97% pure NaOH pellets, CAS [1310-73-2], Scharlab, Sentmenat, Spain) in distilled water. A commercial sodium silicate solution (CAS [338443-3L], 10.6 wt.% Na_2O, 26.5 wt.% SiO_2, and 62.9 wt.% H_2O, Sigma Aldrich, Burlington, EUA) was employed without further elaboration. River sand ($\leq$2 mm) was used as the fine aggregate to prepare the mortar formulations.

2.2. Binder and Mortar Preparation

MK was used as the benchmark precursor to first examine the influence of several compositional factors and establish adequate synthesis conditions for the baseline system. To evaluate the feasibility of using CRC as an alternative precursor and identify optimal

MK replacement dosages, different binder compositions were produced with CRC steps of 25.0% up to total MK substitution. The details of the mixture proportions and synthesis factors examined are presented in Table 1. All binder formulations were designed to respect the following molar ratios: (i) $SiO_2/(Al_2O_3 + Fe_2O_3)$ = [3.75–5.15]; (ii) $(Na_2O + K_2O)/(Al_2O_3 + Fe_2O_3)$ = [1.00–1.30]; (iii) $(Na_2O + K_2O)/SiO_2$ = [0.20–0.30]; and (iv) $H_2O/(Na_2O + K_2O)$ = [10.0–15.0] and $SiO_2/(Na_2O + K_2O)$ = [0.50–0.80] in the solution; calculated assuming the precursors complete dissolution and participation in the polymerization reactions. The binder specimens were produced following the procedure described elsewhere by Ascensão et al. [17,29]. In short, it involved: (a) homogenization of sodium silicate and sodium hydroxide solutions at 60 rpm for 300 s; (b) manually mixing the powdered precursors in a plastic container; and (c) mixing the binders' solid components and the alkaline solutions at 60 rpm for 600 s in a mortar mixer. Then, the resulting slurries were transferred to plastic molds (Ø ≈ 20 mm; height ≈ 40 mm), placed inside sealed plastic containers, and left to cure in the mold for 24 h according to the conditions described in Table 2. Alkali activated mortars were prepared following the same procedure but with two additional mixing steps: (i) 30 s mixing step at 60 rpm for introduce the aggregates; and (ii) 60 s mixing step at 95 rpm to further guarantee a proper homogenization. Mortar specimens (4 × 4 × 16 cm^3) were produced with a binder-to-aggregate ratio of 1.0 and cast as prescribed by EN 1015-11 [49]. Mortars were coded after their parent binder using M to mark the difference between them (e.g., R50 represents a binder produced as per Table 1, while MR50 will be used to refer to mortar specimens produced with R50 binders).

Table 1. Experimental plan: Binders' mix design and variables assessed.

	Mixture Proportion (wt.%)					S/L *	C/S *	SH/SS *	NaOH	Curing	Factor
	MK	SF	CRC	NaOH	Na_2SiO_3	wt.%	wt.%	wt.%	mol/dm³	Cond.	
SL0.90	37.8	9.5	0.0	31.7	21.1	**0.9**	4.0	1.50	10.0	C1	S/L ratio
SL1.00	40.0	10.0	0.0	30.0	20.0	**1.0**	4.0	1.50	10.0	C1	
SL1.10	42.0	10.5	0.0	28.5	19.0	**1.1**	4.0	1.50	10.0	C1	
CS4.00	40.0	10.0	0.0	30.0	20.0	1.0	**4.0**	1.50	10.0	C1	C/S ratio
CS3.50	38.9	11.1	0.0	30.0	20.0	1.0	**3.5**	1.50	10.0	C1	
CS3.00	37.5	12.5	0.0	30.0	20.0	1.0	**3.0**	1.50	10.0	C1	
HS1.50	40.0	10.0	0.0	30.0	20.0	1.0	4.0	**1.50**	10.0	C1	SH/SS ratio
HS1.25	40.0	10.0	0.0	27.8	22.2	1.0	4.0	**1.25**	10.0	C1	
HS1.00	40.0	10.0	0.0	25.0	25.0	1.0	4.0	**1.00**	10.0	C1	
M8	40.0	10.0	0.0	30.0	20.0	1.0	4.0	1.50	**8.0**	C1	NaOH M
M10	40.0	10.0	0.0	30.0	20.0	1.0	4.0	1.50	**10.0**	C1	
M12	40.0	10.0	0.0	30.0	20.0	1.0	4.0	1.50	**12.0**	C1	
C1	40.0	10.0	0.0	30.0	20.0	1.0	4.0	1.50	10.0	**C1**	Curing cond.
C2	40.0	10.0	0.0	30.0	20.0	1.0	4.0	1.50	10.0	**C2**	
C3	40.0	10.0	0.0	30.0	20.0	1.0	4.0	1.50	10.0	**C3**	
R0	37.5	12.5	**0.0**	25.0	25.0	1.0	3.0	1.00	12.0	C3	CRC dosage
R25	28.1	12.5	**9.4**	25.0	25.0	1.0	3.0	1.00	12.0	C3	
R50	18.8	12.5	**18.8**	25.0	25.0	1.0	3.0	1.00	12.0	C3	
R75	9.4	12.5	**28.1**	25.0	25.0	1.0	3.0	1.00	12.0	C3	
R100	0.0	12.5	**37.5**	25.0	25.0	1.0	3.0	1.00	12.0	C3	

* Mass ratios are abbreviated as: S/L—solid to liquid ratio; C/S—calcined clay (MK, CRC, or MK + CRC) to silica fume ratio; SH/SS—sodium hydroxide to sodium silicate ratio.

2.3. Characterization Methods

The precursors' particle size distribution was determined by laser diffraction (Coulter LS230 analyzer) combining the laser diffraction technique and polarization intensity differential scattering (PIDS). The chemical composition of the precursors was obtained using X-ray fluorescence (Philips X'Pert PRO MPD spectrometer), with loss on ignition (LOI) at 1000 °C also determined. The crystallinity of the starting precursors and binders was assessed by X-ray diffraction (XRD). Experiments were carried out on a Bruker D8 Advance (Bruker AXS, Karlsruhe, Germany) (CuKa radiation, 5–70°, 0.05° step, and 1 s/step, with the support of a position sensitive detector) and on a Rigaku Geigerflex D/max-Series instrument (Rgaku Corporation, Tokyo, Japan) (CuKa radiation, 10–80°, 0.02° step and 10 s/step). The phase identification was performed by means of the Match! program package (Crystal Impact GbR, Bonn, Germany), supported by the database ICDD-PDF-2. Binder samples were collected from the mechanical tests after 28 day of curing, ground, sieved (≤63 μm), and kept sealed prior to testing. The bulk density of the binders and mortars was determined by the relation between the weight and volume of each sample. The weight (0.1 g accuracy) and dimensions (0.01 mm accuracy) of each individual specimen were measured after 7 and 28 days of curing. The Archimedes method (using water as the immersion fluid) was employed to evaluate the water absorption and open porosity after 28 days of curing. The compressive strength of the binders was determined using a Universal Testing Machine (Shimadzu model AG-25 TA) running at a displacement rate of 0.5 mm min^{-1}. A minimum of four cylindrical specimens were tested per formulation and curing age (7 and 28 days), and the average strength values are reported here after computing the correction factors for cylindrical specimens as prescribed by ASTM C42/C42M [50]. The compressive strength of the mortar specimens was determined as per EN 1015-11 [49]. To further complement the characterization of the CRC mortars, consistence (flow table test), fresh density, dynamic elasticity modulus, and capillary index were determined as per EN1015-3 [51], EN 1015-6 [52], EN 12504-4 [53], and EN 1015-18 [54], respectively.

Table 2. The examined curing conditions.

C1	40 °C ± 95 RH **1 day**	20 °C ± 95% RH **6 days**		20 °C ± 65 RH **21 days**
C2	20 °C ± 95 RH **1 day**	40 °C ± 95 RH **1 day**	20 °C ± 95 RH **5 days**	20 °C ± 65 RH **21 days**
C3	20 °C ± 95 RH **1 day**	40 °C ± 95 RH **5 days**	20 °C ± 95 RH **1 day**	20 °C ± 65 RH **21 days**
	Time (days)			

3. Results and Discussion

3.1. Precursor Characterization

Figure 1a reports the diffraction analysis of the local common clay, before and after calcination at 800 °C. Considering the intensity of the characteristic diffraction maxima, the content of quartz (PDF#87-0743) appears to be substantial. Accessory mineral phases consisted of feldspar (microcline, PDF#77-0135) and mica (muscovite, PDF#80-0743). Signals consistent with clay minerals such as kaolinite ($Al_2Si_2O_5(OH)_4$, PDF#89-6538) were visible only in the as-received clays, in good agreement with dehydroxylation and transformation into amorphous metakaolin upon calcination. The formation of metakaolin provided initial evidence in favor of the CRC potential as an alternative precursor in alkali activated systems. The enhanced peak at 2θ~23° is consistent with the formation of anhydrous aluminum silicate ($Al_{0.5}Si_{0.75}O_{2.25}$, PDF#37-1480), likely formed by the same dehydroxylation process. Increases in the intensity of the quartz-related peaks are most likely attributed to the 'concentration' mechanisms, as hydrated minerals undergo obvious mass reductions with dehydroxylation. Thermal decomposition probably did not solely involve clay minerals, but also (partially) muscovite. Similar observations were made by Merabtene et al. [55], who examined the decomposition of muscovite and reported the formation of additional quartz and leucite. The formation of leucite cannot be proved here considering the overlap with signals from microcline.

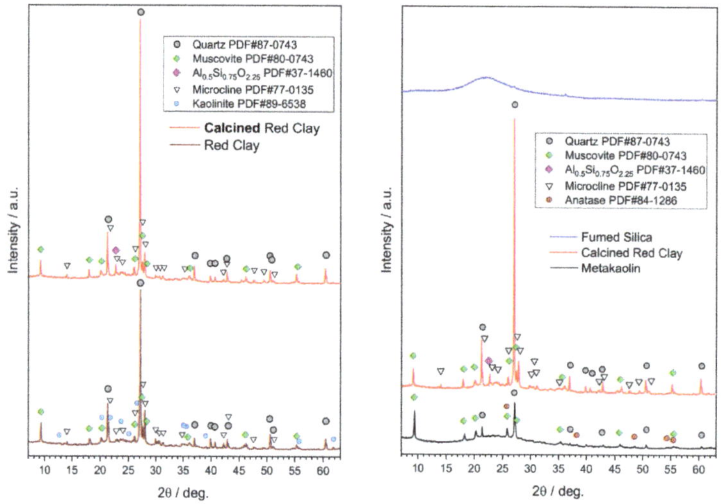

(**a**) (**b**)

Figure 1. Diffraction analysis of the starting raw materials used to synthesize AAM binders: (**a**) comparison between the local red clay before and after calcination; (**b**) comparison between the commercial metakaolin (MK), local common clay calcined at 800 °C (CRC), and fumed silica (SF).

Figure 1b shows a comparison between the mineralogical assemblage of precursors. MK presents muscovite and quartz as predominant minerals, accompanied by traces of anatase (TiO2, PDF#84-1286). As can be seen, the most significant differences between MK and CRC concern the contents of quartz (much reduced in MK) and amorphous material (much enhanced in MK considering the broad convexity located at 2θ~18–32°). SF was found to be completely amorphous, as already reported in previous works [17,48]. The high content of highly amorphous Si and the absence of major crystalline peaks justify the selection of SF as a secondary source of reactive silica in activated alkaline systems.

The chemical bulk compositions of MK, SF, and CRC are shown in Table 3. The most abundant oxides observed in the MK composition were silicon and aluminum oxides, related to the metakaolin content. These results agreed well with the product data sheet, albeit with a slightly lower content of Al_2O_3. Some variability in chemical composition can be easily attributed to the MK geological origin, and similar Al_2O_3 contents have been previously reported elsewhere for this commercial MK [56]. CRC was composed of similar SiO_2 and Al_2O_3 contents whereas major differences were found in alkaline oxides ($Na_2O + K_2O$), TiO_2, and Fe_2O_3. Such differences are in good agreement with the presence of alkali feldspars and anatase in the CRC mineralogical assemblage. The loss of ignition values in both calcined clays was inferior to 2.0 wt.%, which in part can be attributed to the thermal decomposition of muscovite.

Table 3. The chemical composition of the raw materials (% by mass).

	LOI *	SiO_2	Al_2O_3	CaO	Fe_2O_3	MgO	TiO_2	K_2O	Na_2O	Others
MK	1.9	57.0	34.5	0.2	1.2	0.2	1.8	0.4	nd *	2.9
CRC	1.2	56.0	33.9	0.2	4.2	0.5	0.4	3.4	0.1	0.3
SF	1.3	97.4	0.5	0.2	nd *	0.1	nd *	0.2	0.2	0.2

* LOI: loss on ignition, nd: less than quantification limit.

The particle size distribution of the precursors is provided in Figure 2. It can be seen that SF showed a coarser particle size distribution when compared with the remnant precursors. The calcined clays presented finer grain size distributions, with a smaller d10, d50, and d90 being found in CRC (Table 4). Among other factors, the precursors' particle size distribution determines the surface exposed to the alkaline medium, and thus influences the precursors' reactivity, the reaction kinetics, and the products formed. The finer particles of CRC precursors can possibly compensate to some extent for the lower degree of structural disorder and amorphous phases present in CRC when compared to MK (Figure 1b), thus resulting in precursors with potential akin to alkali activation technology. This mechanism seems to be corroborated by the strength and porosity results discussed in Section 3.2.6. In Figure 2 and Table 4, it can also be seen that the RC and CRC data points largely overlap with no significant differences being observed in the particle size distribution. Such similarity between the CRC and RC samples shows that the calcination process occurred without significant agglomeration of the clay particles, and thus without showing signs of sintering. It should be mentioned here that no grinding process was conducted after calcination to avoid mechanically imposed changes in the particle size distribution and reactivity, which would be intermingled in the interpretation of results. Overall, and despite the chemical, mineralogical, and physical differences regarding the benchmark precursor (MK), the results showed auspicious characteristics for the alkali activation of the examined local clays after calcification.

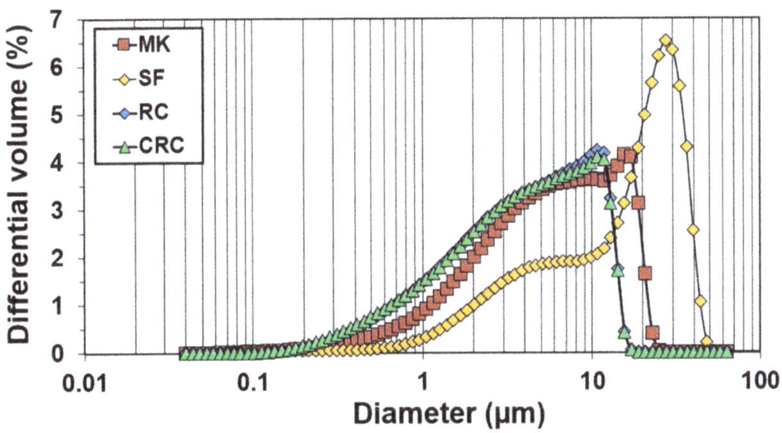

Figure 2. Particle size distribution of the precursors used to manufacture alkali activated binders and mortars: MK—commercial metakaolin; SF—silica fume; RC—as-received clay; CRC—calcined clay.

Table 4. The physical characteristics of the precursors.

Material	Mean Particle Size (µm)	D10 (µm)	D50 (µm)	D90 (µm)
MK	7.73	1.41	6.22	16.92
SF	18.11	2.70	17.77	35.16
RC	5.43	0.94	4.37	11.74
CRC	5.33	0.88	4.25	11.66

3.2. *Alkali Activated Binders Characterization*

3.2.1. Influence of Solid-to-Liquid Ratio

Compressive strength is among the most important properties of building materials as a measure of their load-bearing capacity, while density and porosity can be used as a proxy for durability. Figure 3 shows the apparent density (AD) and compressive strength

(CS) of MK-binders produced with varying solid-to-liquid ratios (S/L ratio) while keeping other synthesis parameters constant (Table 1). After 7 days of curing, the SL0.90 samples recorded the highest CS values (9.8 ± 0.7 MPa) whereas SL1.00 showed the lowest CS (7.1 ± 1.2 MPa). The SL1.10 binders showed intermediate CS values of 9.2 ± 2.0 MPa but considerable standard deviations were recorded in all binders, which can be attributed to the specimens' early stage of maturation and ongoing polymerization reactions [17]. Later strength gains were observed in all binders, with SL1.00 being the most performant in terms of CS at 28 days. AP showed the opposite trend, with significant reductions being observed as curing progressed. Such reductions in AD can be mostly related to free water egress, as suggested by the linear relationship between solution load and AD at 28 days. Conversely, CS cannot be described as a linear function of the S/L ratio (nor of AD), as an inflection point appears to exist around a S/L ratio of 1.0. At a higher S/L ratio, AAM pastes presented a less workable consistency and increases in viscosity that could result in air entrapment within the structure in the samples. In fact, some air bubbles and minor voids were visible on the surface of the SL1.10 specimens, along with increases in the open porosity (OP) and water absorption (WA) values (Table 5). Increases in porosity can compromise SL1.10 stability and partially explain the reduced CS values observed. In addition, alkaline hydrolysis occurs at the interface between the alkaline media and solid particles (kaolinite minerals), and high S/L ratios can reduce diffusion, limit the degree of reaction, and ultimately reduce the binder formation. Reductions in CS were not, however, accompanied by reductions in AD, which were increased by the higher solid load existing in the SL1.10 systems. Conversely, lowering the S/L ratio to 0.90 yielded more workable pastes but resulted in less dense and mechanically resistant binders. In addition to the higher amount of free water in these systems, increased dosages of activating solution and Na^+ content that assist in dissolution may also hinder the polycondensation rate as the solid load available for reaction is limited and dispersed in a very fluid medium. The SL0.90 binders displayed limited CS increments from 7 to 28 days, not being able to outperform the SL1.0 samples. Therefore, the CS results suggest that SL1.00 binders allow for higher dissolution rates of the starting precursors while not hindering the polymerization reactions. Overall, all binders displayed sufficient CS for civil engineering applications and low AD values at 28 days that ranged from 1.28 ± 0.03 g/cm^3 to 1.36 ± 0.01 g/cm^3, whereas OPC pastes typically had AD greater than 1.80 g/cm^3. Considering these results, 1.00 was selected as the S/L ratio for further examination of the MK and CRC binders.

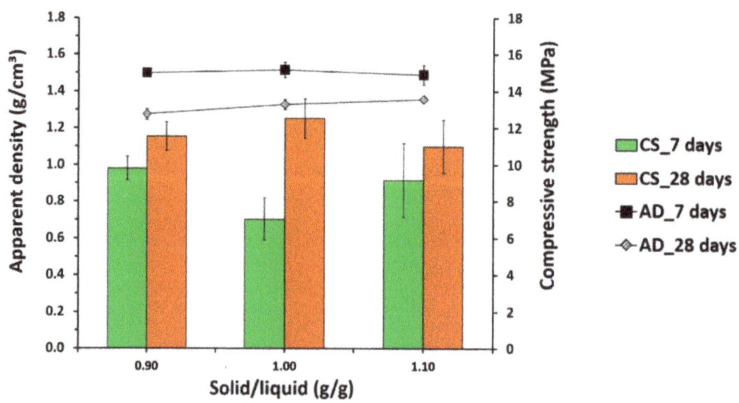

Figure 3. The apparent density and compressive strength of the alkali activated MK-binders at 7 and 28 days of curing as a function of the solid-to-liquid mass ratio.

Table 5. The water absorption and open porosity of the alkali activated binders after 28 days of curing.

	Water Absorption (WA)	Open Porosity (OP)
	%	%
SL0.90	15.68 ± 0.16	24.18 ± 0.17
SL1.00	14.34 ± 0.07	22.48 ± 0.15
SL1.10	14.96 ± 0.66	24.23 ± 1.09
CS4.00	14.34 ± 0.07	22.48 ± 0.15
CS3.50	14.35 ± 0.20	22.23 ± 0.37
CS3.00	14.45 ± 0.73	22.09 ± 0.91
HS1.50	14.34 ± 0.07	22.48 ± 0.15
HS1.25	15.47 ± 0.12	24.38 ± 0.16
HS1.00	15.51 ± 0.17	24.50 ± 0.38
M8	19.02 ± 0.27	29.36 ± 0.47
M10	14.34 ± 0.07	22.48 ± 0.15
M12	11.82 ± 0.28	18.56 ± 0.61
C1	14.34 ± 0.07	22.48 ± 0.15
C2	16.98 ± 0.22	27.90 ± 0.45
C3	15.16 ± 0.07	24.78 ± 0.24
R0	13.58 ± 0.72	21.81 ± 1.79
R25	14.27 ± 0.14	23.90 ± 0.15
R50	14.71 ± 0.82	24.08 ± 1.39
R75	13.00 ± 0.11	21.07 ± 0.20
R100	10.77 ± 0.16	17.54 ± 0.27

3.2.2. Influence of SF Incorporation

Given the insufficient availability of Si in the composition of some clay precursors, the addition of secondary sources of silica is often reported as necessary to deliver competent alkali-activated materials. The provision of extra silicon can be delivered by modifying the concentration and dosage of silicate solutions (discussed in Section 3.2.3) and/or by introducing amorphous Si-rich powdered materials (e.g., silica fume) in partial lieu of clay precursors. However, optimal replacement levels vary according to the characteristics of the systems examined, thus requiring dedicated investigations as performed here.

Figure 4 shows the effects of MK replacement by SF on CS and AD. The results show that all binder formulations increased in CS and reduced in AD with curing, regardless of the MK to SF mass ratio (CC/SF ratio). After 28 days of curing, binders with the lowest CC/SF ratio (CS3.00) recorded the lowest CS values (10.1 ± 0.9 MPa), whereas the CS4.00 formulations showed the highest CS (12.5 ± 1.1 MPa). Oshani et al. [57] reported increases in CS up to 18 wt.% SF, but decreases for higher dosages. Previous studies have shown that excessive SF can be deleterious for CS development due to its volume expansion [58,59]. The authors reported that SF contents over 10 wt.% led to the formation of large voids and concomitant reductions in CS. However, residual modifications on AD, OP, and WA were imposed by the CC/SF ratio (Table 5), suggesting similar porous structures being formed in all binders. The highest early strength of the CS3.00 samples (8.2 ± 1.5 MPa; CS3.50 = 5.8 ± 0.7 MPa; CS4.00 = 7.1 ± 1.1 MPa) may indicate that SF promotes early age polymerization but hinders the later dissolution of clay precursors and/or the diffusion of dissolved species. However, increases in early CS can be an interesting technological feature if one is prepared to partially sacrifice strength development at later ages. Economic and environmental benefits can also accrue from the partial replacement of calcined clay by other Si-rich waste-based materials such as among others, rice husk ash, and the versatility and resilience introduced into the manufacturing process should also not be neglected here. Therefore, 3.00 was selected as the CC/SF ratio for the further examination of the MK and CRC binders.

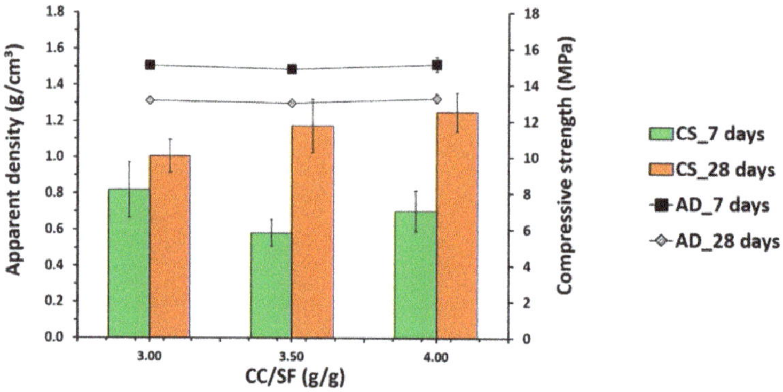

Figure 4. The apparent density and compressive strength of the alkali activated binders at 7 and 28 days of curing as a function of the calcined clay-to-silica fume mass ratio (CC/SF ratio).

3.2.3. Influence of NaOH-to-Na_2SiO_3 Ratio

Figure 5 shows the AD and CS of the binders produced with various NaOH-to-Na_2SiO_3 ratios by mass (SH/SS ratio), as per Table 1. The results showed that CS increased in all formulations with curing. AD showed the opposite trend but no noticeable differences were observed after 28 days of curing ($\approx$1.32 g/cm^3). Differences in WA and OP were also found to be residual, particularly between HS1.00 and HS1.25 binders (Table 5). However, the SH/SS ratio showed a considerable influence on strength development. After 7 days of curing, HS1.00 and HS1.25 recorded similar CS ($\approx$10 MPa), whereas HS1.50 showed significantly lower values (7.1 $\pm$ 1.2 MPa). At later ages, CS was maximized at the SH/SS ratio of 1.00 (14.3 $\pm$ 1.6 MPa), with HS1.25 showing a CS similar to those of HS1.50 of 12.5 $\pm$ 2.0 MPa and 12.5 $\pm$ 1.0 MPa, respectively. These findings are in agreement with previous results reported by, among others, Albidah et al. [60], Supit et al. [61], and Zhao et al. [62], who found increases in CS when reducing the SH/SS ratio. The authors also suggested the existence of the optimum SH/SS ratio below which further reductions will not deliver increases in strength. However, the existence of such a threshold limit cannot be inferred from the results discussed here. Increasing Na_2SiO_3 promoted the formation of more Si-rich structures, which contributed to the development of the highest CS in HS1.0. In addition, reducing the SH/SS ratio also decreased the activating solution pH, leading to more progressive dissolution and polycondensation, but ultimately allowing for higher reaction degrees [20]. It is interesting to point out the differences in the effects of introducing soluble silicates in the activating solution rather than as secondary powdered precursors (Section 3.2.2). In previous works [17], the authors demonstrated that using soluble silicates (K_2SiO_3) resulted in more densified structures and enhanced the mechanical performances. Although these effects can be largely determined by the precursor's composition and granulometric distribution, and by the concentration and nature of the silicate solutions (Na^+ vs. K^+), the results in Figure 5 recommend an SH/SS ratio of 1.00 to achieve the maximum CS values when synthesizing AAMs from CRC.

It should be mentioned that silicate solutions entail significant production costs and environmental burdens [20,21,63] and restrictions in the Na_2SiO_3 dosage should be considered in the following optimization studies, if technologically possible.

3.2.4. Influence of NaOH Molarity

Figure 6 presents the CS and AD of the MK-binders activated with the Na_2SiO_3 and NaOH solutions, in which the molarity of the latter was 8 M, 10 M, or 12 M (Table 1). The influence of NaOH molarity on CS development is notorious. Yearly age CS ranged between 2.7 $\pm$ 0.7 MPa (M8) and 12.2 $\pm$ 1.5 MPa (M12), being greatly proportional to the NaOH molarity. At 28 days, NaOH 8 M solutions delivering binders with CS did not exceed 3.4 ± 0.7 MPa, whereas the M10 and M12 binders displayed CS values around 12.5 MPa. The results suggest that in these systems, the optimal alkali environment for MK activation requires NaOH solutions with a minimum molarity of 10 M, if CS values higher than 10 MPa are to be attained. If sufficient alkalinity is not provided, MK dissolution may be partially impaired and the degree of polymerization of the final products may be weakened. Conversely to what was previously suggested [60–62], no inflection point was observed in CS when increasing the Na^+ availability. An excess of hydroxide ions and the deleterious formation of Na_2CO_3 did not seem to exist to a large extent as no efflorescence was visible on the surfaces of the M10 and M12 samples. AD was reduced in all binders with curing, but after 28 days, the M12 binders were slightly denser. It is also interesting to note that the WA and OP values increased as the NaOH molarity decreased, even at similar AD values (Table 5). This may suggest modifications in the pore connectivity and tortuosity favored by higher Na^+ availability. For better understanding of the morphology, distribution, and connectivity of pores, microstructural analysis and BET analysis can be suggested and will pursued in future works. Reducing the NaOH molarity, and thus the $(Na_2O + K_2O)/SiO_2$ molar ratio, also slightly enhanced the binders' viscosity and workability, but 12 M solutions were selected for further experimentation to maximize the yearly CS in the CRC binders.

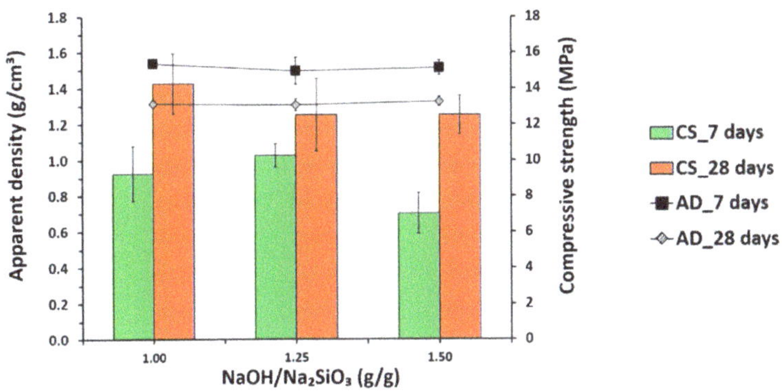

Figure 5. The apparent density and compressive strength of the alkali activated binders at 7 and 28 days of curing as a function of the sodium hydroxide-to-sodium silicate mass ratio.

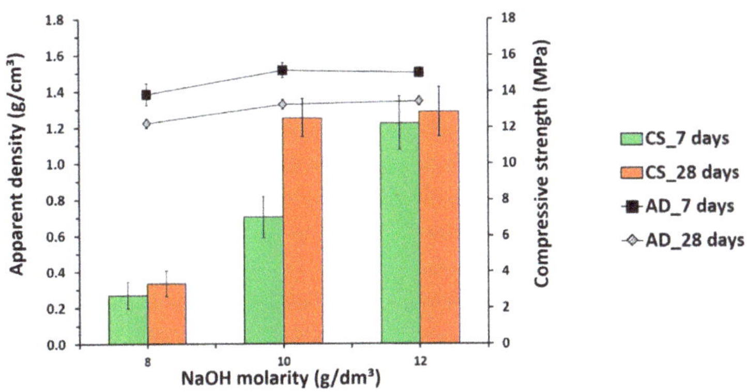

Figure 6. The apparent density and compressive strength of the alkali activated binders at 7 and 28 days of curing as a function of sodium hydroxide molarity.

3.2.5. Effect of Curing Conditions

Different curing conditions were evaluated and their effects on the binder properties were examined to define the optimum mix designs. At curing conditions 1 and 2, the specimens were cured at slightly elevated temperatures (40 °C) for a 24 h period, albeit starting immediately after mixing or 24 h later, C1 and C2, respectively. When employing curing condition 3 (C3), heat curing was initiated as in C2, but maintained for a 5 day period. In short, C1–C2 served to investigate the effects associated with heat curing location in time, while C2–C3 determined its most adequate duration for the systems examined here. Figure 7 describes the AD and CS of the binders cured with different curing regimes. After 28 days, the C3 samples recorded the highest AD (1.39 $\pm$ 0.02 g/cm^3) and CS values (13.6 $\pm$ 1.2 MPa). C1 and C2 showed a CS within the experimental error range (12.5 $\pm$ 1.1 MPa and 12.1 $\pm$ 1.9 MPa, respectively), and a slightly higher AD was observed in the C2 samples of 1.35 $\pm$ 0.02 g/cm^3 (C1 = 1.33 $\pm$ 0.03 g/cm^3). In addition, the yearly age CS was considerably ameliorated with the increment of the curing period (C2–C3) and benefitted from a dormant period in the initial stage of the reaction (C1–C2). Increases in AD and CS indicate highly polymerized and resistant structures being formed in C3, which, however, cannot be related to the densification and reduction in porosity (Table 5). Similar findings have previously been reported on slag-based systems where the benefits of curing at a slightly elevated temperature with an initial dormant period were demonstrated [17,32,38]. The volumetric stability of the specimens was not determined, but pronounced shrinkage phenomena able to provoke visible cracks and severely compromise CS did not seem to exist. The higher CS and AD values found in the C3 binders justify the selection of the C3 conditions for further experimentation when replacing MK with the CRC precursors.

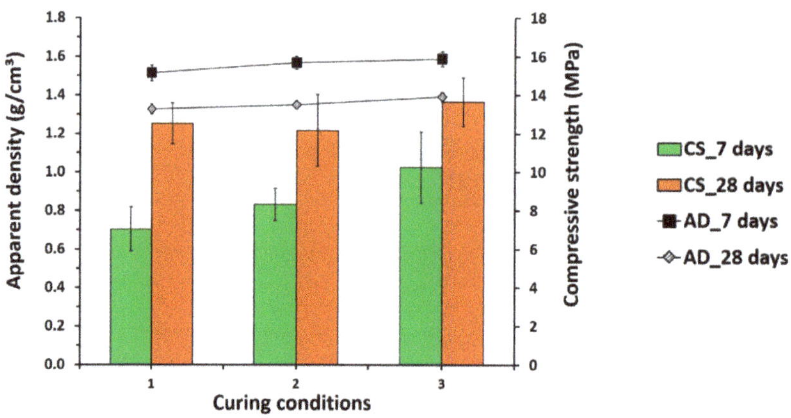

Figure 7. The apparent density and compressive strength of the alkali activated binders at 7 and 28 days of curing as a function of the curing conditions used as per Table 2.

3.2.6. Influence of CRC Incorporation

In Sections 3.2.1–3.2.5, the baseline MK system was investigated and the optimal synthesis conditions were selected. A binder synthesized with the S/L = 1.0, C/S = 3.0, SH/SS = 1.0, NaOH molarity = 12 M, and cured in C3 conditions (Table 1) was defined as the reference binder (R0; 0.0 wt.% CRC) to investigate the effects of MK replacement by CRC. Figure 8 presents the representative photographs of the alkali activated binders produced with the progressive replacement of MK by CRC (0, 25, 50, 75, and 100 wt.%).

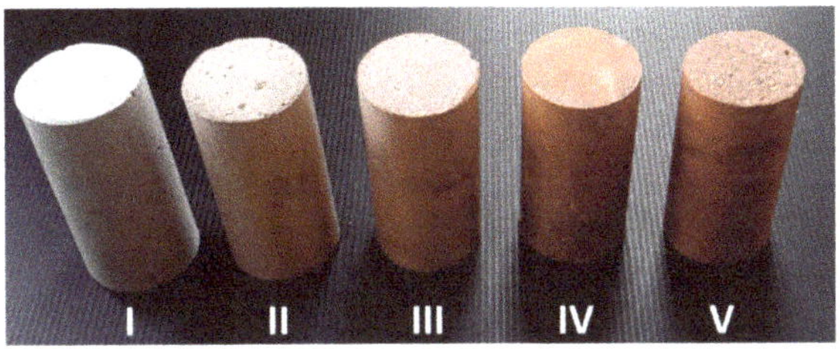

Figure 8. Representative photographs of the alkali activated binders produced with the progressive replacement of MK by CRC. From left to right: (i) 0.0%, (ii) 25.0%, (iii) 50.0%, (iv) 75.0%, and (v) 100.0% MK replacement.

Figure 9 shows the AD and CS of the binders with varying CRC dosages after 7 and 28 days of curing. Residual increases in AD could be seen as the CRC content rose (AD_{28dmin}, R0 = 1.37 ± 0.02 g/cm^3; AD_{28dmax}, R100 = 1.45 ± 0.01 g/cm^3), which can be attributed to the presence of iron in the CRC composition. CRC incorporation levels up to 75 wt.% resulted in slight increases in the yearly age CS. An optimal replacement level seemed to exist around 50 wt.% CRC, where the R50 samples recorded the highest CS values after 7 days of curing (12.5 ± 0.7 MPa), representing a 32.6% increase in comparison to the pure MK-binders (Figure 9). MK total replacement by CRC seemed to have no deleterious effects on early CS, but considerably increased the mechanical performance after 28 days of curing. Curiously, as the CRC content rises, the theoretical $SiO_2/(Al_2O_3 + Fe_2O_3)$ molar ratio of the mixtures decreases, thus less polymerized structures would be expected. Differences in the reaction rates arising from the CRC particle size distribution and composition (as discussed in Section 3.1) can explain to some extent the late strength developments in the R100 binders, but do not provide a general definition for the trends observed in the CS nor the WA and OP results.

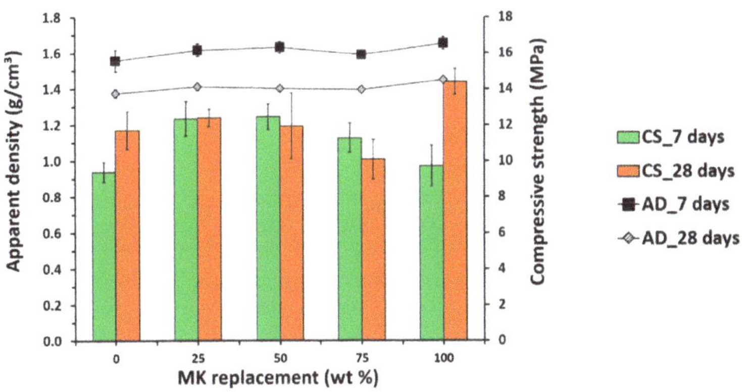

Figure 9. The apparent density and compressive strength of the alkali activated binders at 7 and 28 days of curing as a function of metakaolin replacement by calcined common clay.

Therefore, the mineralogical differences were investigated by means of XRD and the diffraction analysis of the binders prepared with the progressive replacement of MK by CRC (0, 50 and 100%) is shown in Figure 10. Amorphous material (broad convexity located at 2θ~22–35°) could be appreciated in all cases, which confirmed polymerization, even when totally replacing MK with CRC. The differences in the intensity of the quartz peaks found in the starting materials remained after alkali activation, but they appeared as much less remarkable. The dissolution of quartz from CRC is consistent with the increase in the relative intensity between the main peaks of microcline (PDF#77-0135) and quartz (PDF#87-0743). However, some crystalline quartz persisted, showing that some Si units did not partake in the reactions. Muscovite (PDF#80-0743) was detected in both the MK and CRC-containing binders, and anhydrous aluminum silicate phases ($Al_{0.5}Si_{0.75}O_{2.25}$, PDF#37-1480) identified in CRC precursors could not be detected, suggesting their dissolution and participation in polymerization reactions. The contribution of these phases can explain the porosity values and CS increases to some extent, but dedicated studies are still needed to better understand the composition and morphology of the reaction products that formed to provide a discretionary explanation of the reaction mechanisms involved.

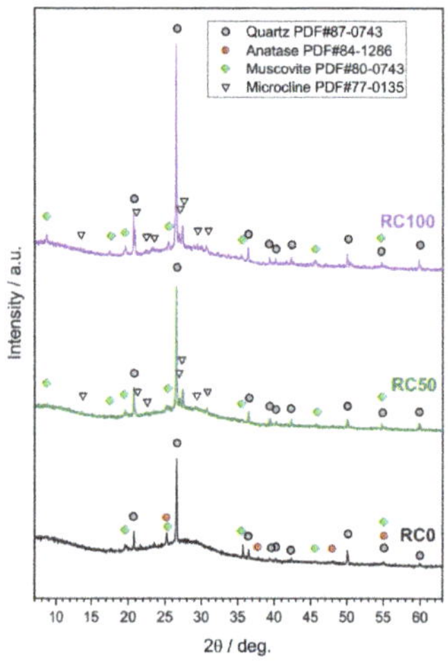

Figure 10. A comparison between the XRD patterns of the binders synthesized with progressive replacement (0, 50, and 100%) of metakaolin by local calcined clay after 28 days of curing.

3.3. Alkali Activated Mortars

Mortar samples were produced and characterized to preliminarily assess the practical interest and industrial feasibility of the developed CRC binders. Fundamental properties evaluated in a fresh state included fresh density and workability. The workability of the mortars was investigated by the flow table test. The spread values were slightly reduced with the increment in the CRC content in the mix design (Table 6). However, all of the results are in line with the spread values used in regular hydraulic mortars (non-self-leveling/self-compacting) that typically lie in the range of 150–180 mm. Therefore, all mortars demonstrated adequate workability levels without the use of any additional additives and could meet the standard flowability requirements. The fresh density values were nearly constant, albeit slightly higher values were observed in the MR100 mortars. This effect was even more pronounced in the AD values at 28 days. The polymerization reaction seems to dictate the unit weight with progressive increases in fresh density and AD as the CRC dose rises (MR100, $AD_{28dmax} = 1.99 \pm 0.01$ MPa).

Table 6. The fresh and hardened properties of mortars produced with different dosages of local calcined clays at 28 days.

Code	SV (mm) *	FD (g/cm^3) *	AD (g/cm^3) *	EM (GPa) *	WA (%) *	CI *
MR0	180 ± 5	2.01	1.78 ± 0.00	15.4 ± 0.0	6.7 ± 0.0	0.118 ± 0.002
MR25	175 ± 5	1.99	1.85 ± 0.02	16.8 ± 0.2	7.0 ± 0.3	0.138 ± 0.002
MR50	175 ± 10	2.01	1.88 ± 0.07	16.3 ± 0.5	7.5 ± 0.3	0.120 ± 0.007
MR75	175 ± 5	2.00	1.91 ± 0.01	16.8 ± 0.1	4.1 ± 0.7	0.128 ± 0.006
MR100	170 ± 5	2.05	1.99 ± 0.01	14.7 ± 0.1	3.0 ± 0.2	0.101 ± 0.000

* Properties abbreviate as: SV—spread value; FD—fresh density; EM—elastic modulus; WA—water absorption; CP—capillarity index.

As with the binders, CS and OP were the utmost important properties that must be investigated in AAM mortars. Figure 11 describes the relationship between the CS and OP levels in mortars produced with progressive dosages of CRC. The CS values decreased almost linearly as the CRC increased. Interestingly, a similar trend was not observed in the CRC binders (Figure 8), suggesting aggregate interference in the polymerization reactions/products and/or structural weaknesses located in the interfacial transition zones. The OP increased slightly with an increasing CRC up to 50 wt.%, but beyond this threshold, considerable decreases were observed. The values for the WA and elastic modulus were greatly affected by pores and voids, following, in general, a pattern similar to that observed for OP (Table 6). The capillarity index values were found to be inconsistent and without any visible correlation to other properties. Therefore, the relationships between the density, porosity, CRC content, and strength development in the CRC mortars seem much more intricate with respect to the binders only, and effectively rely on a larger number of complex reactions and confounding variables. In Figure 11, the existence of three distinct zones is enunciated based on the MK replacement level. In zone 1, the increases in porosity seemed to be partially mitigated by the formation of slightly more resistant binders, while in zone 2, the exact opposite seemed to happen, but to a lesser extent. In zone 3, the motifs that led to mortars with denser structures and produced from binders more resistant to develop poorer mechanical performances are still unclear at this point and will be the subject of research in future works. However, all mortars produced presented low OP and WA (below 15% and 8%, respectively) and CS values above 28 MPa, which provide good performance indicators regarding their mechanical behavior and durability and well illustrates their potential for applications in construction.

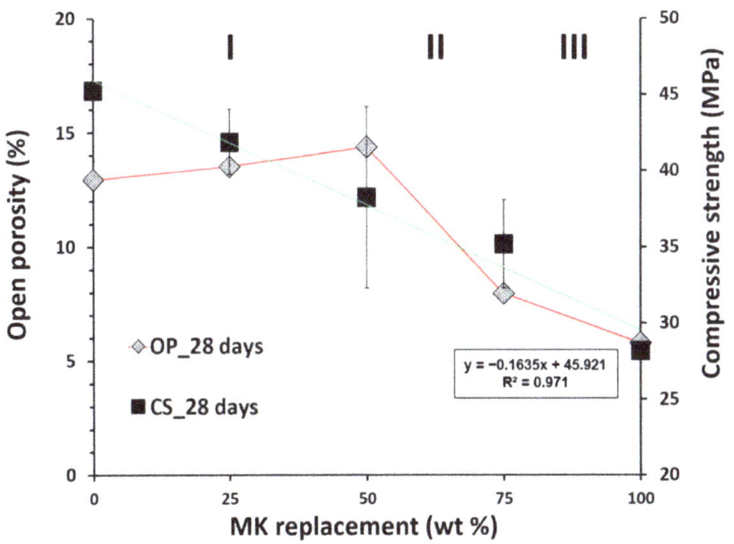

Figure 11. A comparison of the open porosity and compressive strength values of mortars produced with the progressive replacement of metakaolin by local calcined clays after 28 days of curing. I, II and III indicate ranges of MK replacement levels with different relationships between open porosity and compressive strength.

4. Conclusions

In this work, the factors affecting the properties of the alkali activated binders and mortars produced from different thermally modified clays were investigated. Commercial metakaolin was taken as the benchmark clay precursor and an endogenous common clay was explored as a representative example of potential local alternatives for greater environmental and economic benefits. The resulting MK binders showed compressive strengths above 10 MPa, low density (AD > 1.40 g/cm^3), and low OP and WA values (30% and 20%, respectively) after 28 days of curing. Replacing MK with locally available clays was found to be feasible as the key fundamental properties were maintained at levels comparable to those of the reference system, thus making the mortars produced thereof (spread >170 mm; CS >28 MPa; 1.78 g/cm^3 > AD > 1.99 g/cm^3; OP >15%; WA >8%) potentially viable candidates for various applications in construction, rehabilitation, and architectural works. The results obtained in this study also showed that clay-based AAMs exhibit adaptable properties and that resilient manufacturing processes can be designed by proper control of the mix design, for example, MK and CRC can be alternated as the main aluminosilicate source, whereas additional Si can be delivered either by the addition of Na_2SiO_3 or silica fume without major deleterious effects. A prolonged curing period at slightly elevated temperatures is recommended for yielding the maximum strength results.

Overall, the results shown here demonstrate that novel building materials can be produced from an as yet unexplored myriad of common clay resources, thus making an initial contribution to the development of more sustainable manufacturing practices. This work also aimed to shed light on emergent clay binders to stimulate further new scientific developments and industrial advances in clay activation technology with the meaningful co-creation of economic and environmental added value in the region.

Author Contributions: Conceptualization, G.A.; Methodology, G.A.; Validation, G.A.; Formal analysis, G.A. and E.B.; Investigation, G.A.; Resources V.M.F.; Data curation, G.A.; Writing—original draft preparation, G.A.; Writing—review and editing, G.A., E.B. and V.M.F.; Visualization, G.A.; Funding acquisition, V.M.F. All authors have read and agreed to the published version of the manuscript.

Funding: This research work was also supported by the Foundation for Science and Technology (FCT)—Aveiro Research Center for Risks and Sustainability in Construction (RISCO), Universidade de Aveiro, Portugal [FCT/UIDB/ECI/04450/2020]. This publication reflects only the authors' views, exempting the funding agency from any liability.

Institutional Review Board Statement: Not applicable.

Informed Consent Statement: Not applicable.

Data Availability Statement: The data presented in this study are available on request from the corresponding author.

Acknowledgments: The authors wish to thank Rui M. Novais and João Carvalheiras for their assistance during the clay calcination protocols and analytical testing.

Conflicts of Interest: The authors declare no conflict of interest. The funders had no role in the design of the study; in the collection, analyses, or interpretation of data; in the writing of the manuscript, or in the decision to publish the results.

References

1. EPA USA. Inventory of U.S. *Greenhouse Gas Emissions and Sinks. 8 February 2017*. Available online: https://www.epa.gov/ghgemissions/inventory-us-greenhouse-gas-emissions-and-sinks (accessed on 8 June 2022).
2. Habert, G.; Miller, S.A.; John, V.M.; Provis, J.L.; Favier, A.; Horvath, A.; Scrivener, K.L. Environmental impacts and decarbonization strategies in the cement and concrete industries. *Nat. Rev. Earth Environ.* **2020**, *1*, 559–573. [CrossRef]
3. Scrivener, K.L.; John, V.M.; Gartner, E.M. Eco-efficient cements: Potential economically viable solutions for a low-CO2 cement-based materials industry. *Cem. Concr. Res.* **2018**, *114*, 2–26. [CrossRef]
4. Abergel, T.; Dean, B.; Dulac, J.; Hamilton, I.; Wheeler, T. Global Status Report: Towards a Zero-Emission, Efficient and Resilient Buildings and Construction Sector. 2018. Available online: https://www.worldgbc.org/sites/default/files/2018%20GlobalABC%20Global%20Status%20Report.pdf (accessed on 10 June 2022).
5. Favier, A.; De Wolf, C.; Scrivener, K.; Habert, G. *A Sustainable Future for the European Cement and Concrete Industry: Technology Assessment for Full Decarbonisation of the Industry by 2050*; ETH Zurich: Zurich, Switzerland, 2018.
6. Amran, M.; Debbarma, S.; Ozbakkaloglu, T. Fly ash-based eco-friendly geopolymer concrete: A critical review of the long-term durability properties. *Constr. Build. Mater.* **2021**, *270*, 121857. [CrossRef]
7. Gartner, E.; Sui, T. Alternative cement clinkers. *Cem. Concr. Res.* **2018**, *114*, 27–39. [CrossRef]

8. Hertel, T.; Van den Bulck, A.; Onisei, S.; Sivakumar, P.P.; Pontikes, Y. Boosting the use of bauxite residue (red mud) in cement—Production of an Fe-rich calciumsulfoaluminate-ferrite clinker and characterisation of the hydration. *Cem. Concr. Res.* **2021**, *145*, 106463. [CrossRef]
9. Ribeiro, F.R.C.; Modolo, R.C.E.; Kulakowski, M.P.; Brehm, F.A.; Moraes, C.A.M.; Ferreira, V.M.; Mesquita, E.F.T.; de Azevedo, A.R.G.; Monteiro, S.N. Production of Belite Based Clinker from Ornamental Stone Processing Sludge and Calcium Carbonate Sludge with Lower CO_2 Emissions. *Materials* **2022**, *15*, 2352. [CrossRef]
10. Sabbah, A.; Zhutovsky, S. Effect of sulfate content and synthesis conditions on phase composition of belite-ye'elimite-ferrite (BYF) clinker. *Cem. Concr. Res.* **2022**, *155*, 106745. [CrossRef]
11. Tan, B.; Okoronkwo, M.U.; Kumar, A.; Ma, H. Durability of calcium sulfoaluminate cement concrete. *J. Zhejiang Univ. Sci. A* **2020**, *21*, 118–128. [CrossRef]
12. Meyer, V.; de Cristofaro, N.; Bryant, J.; Sahu, S. Solidia Cement an Example of Carbon Capture and Utilization. *Key Eng. Mater.* **2018**, *761*, 197–203. [CrossRef]
13. Bishnoi, S. *Calcined Clays for Sustainable Concrete*; Springer: Berlin/Heidelberg, Germany, 2019.
14. Dhandapani, Y.; Joseph, S.; Bishnoi, S.; Kunther, W.; Kanavaris, F.; Kim, T.; Irassar, E.; Castel, A.; Zunino, F.; Machner, A.; et al. Durability performance of binary and ternary blended cementitious systems with calcined clay: A RILEM TC 282 CCL review. *Mater. Struct.* **2022**, *55*, 145. [CrossRef]
15. Ez-zaki, H.; Marangu, J.M.; Bellotto, M.; Dalconi, M.C.; Artioli, G.; Valentini, L. A Fresh View on Limestone Calcined Clay Cement (LC3) Pastes. *Materials* **2021**, *14*, 3037. [CrossRef] [PubMed]
16. Sharma, M.; Bishnoi, S.; Martirena, F.; Scrivener, K. Limestone calcined clay cement and concrete: A state-of-the-art review. *Cem. Concr. Res.* **2021**, *149*, 106564. [CrossRef]
17. Ascensão, G.; Marchi, M.; Segata, M.; Faleschini, F.; Pontikes, Y. Reaction kinetics and structural analysis of alkali activated Fe–Si–Ca rich materials. *J. Clean. Prod.* **2020**, *246*, 119065. [CrossRef]
18. Novais, R.M.; Pullar, R.C.; Labrincha, J.A. Geopolymer foams: An overview of recent advancements. *Prog. Mater. Sci.* **2020**, *109*, 100621. [CrossRef]
19. Palomo, A.; Maltseva, O.; Garcia-Lodeiro, I.; Fernández-Jiménez, A. Portland Versus Alkaline Cement: Continuity or Clean Break: "A Key Decision for Global Sustainability". *Front. Chem.* **2021**, *9*, 705475. [CrossRef] [PubMed]
20. Abdulkareem, M.; Havukainen, J.; Nuortila-Jokinen, J.; Horttanainen, M. Environmental and economic perspective of waste-derived activators on alkali-activated mortars. *J. Clean. Prod.* **2021**, *280*, 124651. [CrossRef]
21. Abdulkareem, M.; Havukainen, J.; Horttanainen, M. How environmentally sustainable are fibre reinforced alkali-activated concretes? *J. Clean. Prod.* **2019**, *236*, 117601. [CrossRef]
22. Fořt, J.; Mildner, M.; Keppert, M.; Černý, R. Waste solidified alkalis as activators of aluminosilicate precursors: Functional and environmental evaluation. *J. Build. Eng.* **2022**, *54*, 104598. [CrossRef]
23. Habert, G.; Ouellet-Plamondon, C. Recent update on the environmental impact of geopolymers. *RILEM Tech. Lett.* **2016**, *1*, 17. [CrossRef]
24. Mascarin, L.; Ez-zaki, H.; Garbin, E.; Bediako, M.; Valentini, L. Mitigating the ecological footprint of alkali-activated calcined clays by waste marble addition. *Cem. Concr. Compos.* **2022**, *127*, 104382. [CrossRef]
25. Djobo, J.N.Y.; Elimbi, A.; Tchakouté, H.K.; Kumar, S. Volcanic ash-based geopolymer cements/concretes: The current state of the art and perspectives. *Environ. Sci. Pollut. Res.* **2017**, *24*, 4433–4446. [CrossRef] [PubMed]
26. Hamid, M.A.; Yaltay, N.; Türkmenoğlu, M. Properties of pumice-fly ash based geopolymer paste. *Constr. Build. Mater.* **2022**, *316*, 125665. [CrossRef]
27. Tan, J.; Dan, H.; Ma, Z. Metakaolin based geopolymer mortar as concrete repairs: Bond strength and degradation when subjected to aggressive environments. *Ceram. Int.* **2022**, *48*, 23559–23570. [CrossRef]
28. Novais, R.M.; Ascensão, G.; Seabra, M.P.; Labrincha, J.A. Waste glass from end-of-life fluorescent lamps as raw material in geopolymers. *Waste Manag.* **2016**, *52*, 245–255. [CrossRef] [PubMed]
29. Ascensão, G.; Seabra, M.P.; Aguiar, J.B.; Labrincha, J.A. Red mud-based geopolymers with tailored alkali diffusion properties and pH buffering ability. *J. Clean. Prod.* **2017**, *148*, 23–30. [CrossRef]
30. Giels, M.; Hertel, T.; Gijbels, K.; Schroeyers, W.; Pontikes, Y. High performance mortars from vitrified bauxite residue; the quest for the optimal chemistry and processing conditions. *Cem. Concr. Res.* **2022**, *155*, 106739. [CrossRef]
31. Arnout, L.; Beersaerts, G.; Liard, M.; Lootens, D.; Pontikes, Y. Valorising Slags from Non-ferrous Metallurgy into Hybrid Cementitious Binders: Mix Design and Performance. *Waste Biomass Valorization* **2021**, *12*, 4679–4694. [CrossRef]
32. Ascensão, G.; Beersaerts, G.; Marchi, M.; Segata, M.; Faleschini, F.; Pontikes, Y. Shrinkage and Mitigation Strategies to Improve the Dimensional Stability of $CaO\text{-}FeO_x\text{-}Al2O_3\text{-}SiO_2$ Inorganic Polymers. *Materials* **2019**, *12*, 3679. [CrossRef]
33. Klima, K.M.; Schollbach, K.; Brouwers, H.J.H.; Yu, Q. Thermal and fire resistance of Class F fly ash based geopolymers—A review. *Constr. Build. Mater.* **2022**, *323*, 126529. [CrossRef]
34. Li, X.; Bai, C.; Qiao, Y.; Wang, X.; Yang, K.; Colombo, P. Preparation, properties and applications of fly ash-based porous geopolymers: A review. *J. Clean. Prod.* **2022**, *359*, 132043. [CrossRef]
35. Novais, R.M.; Ascensão, G.; Tobaldi, D.M.; Seabra, M.P.; Labrincha, J.A. Biomass fly ash geopolymer monoliths for effective methylene blue removal from wastewaters. *J. Clean. Prod.* **2018**, *171*, 783–794. [CrossRef]

36. Novais, R.M.; Buruberri, L.H.; Ascensão, G.; Seabra, M.P.; Labrincha, J.A. Porous biomass fly ash-based geopolymers with tailored thermal conductivity. *J. Clean. Prod.* **2016**, *119*, 99–107. [CrossRef]
37. Rakhimova, N.R. A review of calcined clays and ceramic wastes as sources for alkali-activated materials. *Geosystem Eng.* **2020**, *23*, 287–298. [CrossRef]
38. Beersaerts, G.; Ascensão, G.; Pontikes, Y. Modifying the pore size distribution in Fe-rich inorganic polymer mortars: An effective shrinkage mitigation strategy. *Cem. Concr. Res.* **2021**, *141*, 106330. [CrossRef]
39. Ferone, C.; Liguori, B.; Capasso, I.; Colangelo, F.; Cioffi, R.; Cappelletto, E.; Di Maggio, R. Thermally treated clay sediments as geopolymer source material. *Appl. Clay Sci.* **2015**, *107*, 195–204. [CrossRef]
40. Khalifa, A.Z.; Pontikes, Y.; Elsen, J.; Cizer, Ö. Comparing the reactivity of different natural clays under thermal and alkali activation. *RILEM Tech. Lett.* **2019**, *4*, 74–80. [CrossRef]
41. Valentini, L.; Mascarin, L.; Dalconi, M.C.; Garbin, E.; Ferrari, G.; Artioli, G. Performance and Properties of Alkali-Activated Blend of Calcined Laterite and Waste Marble Powder. In *Calcined Clays for Sustainable Concrete*; Bishnoi, S., Ed.; RILEM Bookseries; Springer: Singapore, 2020; Volume 25, pp. 375–380, ISBN 9789811528057.
42. Valentini, L.; Mascarin, L. Assessing the dimensional stability of alkali-activated calcined clays in the fresh state: A time-lapse X-ray imaging approach. *Mater. Struct.* **2021**, *54*, 35. [CrossRef]
43. Khalifa, A.Z.; Cizer, Ö.; Pontikes, Y.; Heath, A.; Patureau, P.; Bernal, S.A.; Marsh, A.T.M. Advances in alkali-activation of clay minerals. *Cem. Concr. Res.* **2020**, *132*, 106050. [CrossRef]
44. Lopes, C.; Lisboa, V.; Carvalho, J.; Mateus, A.; Martins, L. Challenges to access and safeguard mineral resources for society: A case study of kaolin in Portugal. *Land Use Policy* **2018**, *79*, 263–284. [CrossRef]
45. Reichl, C.; Schatz, M.; Zsak, G. World Mining Data 2018 Iron and Ferro Alloy Metals Non-Ferrous Metals Precious Metals Industrial Minerals Mineral Fuels. 2018. Available online: https://www.world-mining-data.info/wmd/downloads/PDF/WMD2018.pdf (accessed on 28 June 2022).
46. Peys, A.; White, C.E.; Rahier, H.; Blanpain, B.; Pontikes, Y. Alkali-activation of CaO-FeOx-SiO2 slag: Formation mechanism from in-situ X-ray total scattering. *Cem. Concr. Res.* **2019**, *122*, 179–188. [CrossRef]
47. Simon, S.; Gluth, G.J.G.; Peys, A.; Onisei, S.; Banerjee, D.; Pontikes, Y. The fate of iron during the alkali-activation of synthetic (CaO-)FeOx-SiO2 slags: An Fe K -edge XANES study. *J. Am. Ceram. Soc.* **2018**, *101*, 2107–2118. [CrossRef]
48. Ascensão, G.; Faleschini, F.; Marchi, M.; Segata, M.; Van De Sande, J.; Rahier, H.; Bernardo, E.; Pontikes, Y. High-Temperature Behavior of CaO-FeOx-Al2O3-SiO2-Rich Alkali Activated Materials. *Appl. Sci.* **2022**, *12*, 2572. [CrossRef]
49. *EN 1015-11:2019*; Methods of Test for Mortar for Masonry—Part 11: Determination of Flexural and Compressive Strength of Hardened Mortar. European Committee for Standardization: Brussels, Belgium, 2019.
50. ASTM International. Standard Test Method for Obtaining and Testing Drilled Cores and Sawed Beams of Concrete. Available online: https://www.astm.org/c0042_c0042m-12.html (accessed on 15 July 2022).
51. *EN 1015-3*; Methods of Test for Mortar for Masonry—Part 3: Determination of Consistence of Fresh Mortar (by Flow Table). European Committee for Standardization: Brussels, Belgium, 2007. Available online: https://www.en-standard.eu/din-en-1015-3-methods-of-test-for-mortar-for-masonry-part-3-determination-of-consistence-of-fresh-mortar-by-flow-table-includes-amendments-a1-2004-a2-2006/ (accessed on 15 July 2022).
52. *EN 1015-6*; Methods of Test for Mortar for Masonry Determination of Bulk Density of Fresh Mortar. European Committee for Standardization. European Committee for Standardization: Brussels, Belgium, 2007. Available online: https://www.en-standard.eu/search/?q=EN+1015-6 (accessed on 15 July 2022).
53. *EN 12504-4*; Testing Concrete in Structures—Part 4: Determination of Ultrasonic Pulse Velocity. European Committee for Standardization. European Committee for Standardization: Brussels, Belgium, 2021. Available online: https://www.en-standard.eu/csn-en-12504-4-testing-concrete-in-structures-part-4-determination-of-ultrasonic-pulse-velocity/ (accessed on 29 July 2022).
54. *EN 12504-4*; EN 1015-18:2002 Methods of Test for Mortar for Masonry Determination of Water Absorption Coefficient Due to Capillary Action of Hardened Mortar. European Committee for Standardization: Brussels, Belgium, 2002. Available online: https://www.en-standard.eu/bs-en-1015-18-2002-methods-of-test-for-mortar-for-masonry-determination-of-water-absorption-coefficient-due-to-capillary-action-of-hardened-mortar/ (accessed on 15 July 2022).
55. Merabtene, M.; Kacimi, L.; Clastres, P. Elaboration of geopolymer binders from poor kaolin and dam sludge waste. *Heliyon* **2019**, *5*, e01938. [CrossRef]
56. Ferraz, E.; Andrejkovičová, S.; Hajjaji, W.; Velosa, A.L.; Silva, A.S.; Rocha, F. Pozzolanic activity of metakaolins by the French standard of the modified Chapelle test: A direct methology. *Acta Geodyn. Geomater.* **2015**, *12*, 289–298. [CrossRef]
57. Oshani, F.; Allahverdi, A.; Kargari, A.; Norouzbeigi, R.; Mahmoodi, N.M. Effect of preparation parameters on properties of metakaolin-based geopolymer activated by silica fume-sodium hydroxide alkaline blend. *J. Build. Eng.* **2022**, 104984. [CrossRef]
58. Sukontasukkul, P.; Chindaprasirt, P.; Pongsopha, P.; Phoo-Ngernkham, T.; Tangchirapat, W.; Banthia, N. Effect of fly ash/silica fume ratio and curing condition on mechanical properties of fiber-reinforced geopolymer. *J. Sustain. Cem. Based Mater.* **2020**, *9*, 218–232. [CrossRef]
59. Park, S.; Yu, J.; Oh, J.E.; Pyo, S. Effect of Silica Fume on the Volume Expansion of Metakaolin-Based Geopolymer Considering the Silicon-to-Aluminum Molar Ratio. *Int. J. Concr. Struct. Mater.* **2022**, *16*, 20. [CrossRef]
60. Albidah, A.; Alghannam, M.; Abbas, H.; Almusallam, T.; Al-Salloum, Y. Characteristics of metakaolin-based geopolymer concrete for different mix design parameters. *J. Mater. Res. Technol.* **2021**, *10*, 84–98. [CrossRef]

61. Supit, S.W.M.; Olivia, M. Compressive strength and sulfate resistance of metakaolin-based geopolymer mortar with different ratio of alkaline activator. *Mater. Today Proc.* **2022**, S2214785322045333. [CrossRef]
62. Zhao, Y.; Wnag, H.; He, Y.; Yang, L.; Wu, H. Effect of Na+ on hydration degree of alkali activated metakaolin polymer. *Mag. Civ. Eng.* **2022**, *112*, 11208. [CrossRef]
63. Dimas, D.; Giannopoulou, I.; Panias, D. Polymerization in sodium silicate solutions: A fundamental process in geopolymerization technology. *J. Mater. Sci.* **2009**, *44*, 3719–3730. [CrossRef]

applied sciences

MDPI

Article

Physical, Compressive Strength, and Microstructural Characteristics of Alkali-Activated Engineered Composites Incorporating MgO, MWCNTs, and rGO

Mohammad Ali Hossain and Khandaker M. A. Hossain *

Department of Civil Engineering, Toronto Metropolitan University, Toronto, ON M5B 2K3, Canada; mohammadali.hossain@torontomu.ca
* Correspondence: ahossain@torontomu.ca

Featured Application: The developed different classes of AAECs with smart functional properties can be used as self-healing, self-sensing, piezoresistive, conductive, and 3D-prining materials, especially in bridge and building infrastructures.

Abstract: Thirty-two ambient cured alkali-activated engineered composites (AAECs) were developed by incorporating MgO, multi-walled carbon nanotubes (MWCNTs), reduced graphene oxide (rGO), and polyvinyl alcohol (PVA) fiber with a one-part dry mix technique using powder-based activators/reagents. The effects of material variables, namely binary or ternary combination source materials (fly ash C or F and ground granulated blast furnace slag 'GGBFS'), two types of reagents with varying chemical ratios and dosages of additives (from 0 to 5% MgO and from 0 to 6% MWCNT/rGO), on the physical (slump flow, flow time, flow velocity, and density), hardness (compressive strength from 0 to 180 days and 28-day ultrasonic pulse velocity 'UPV'), and micro-structural (SEM/EDS, XRD and FTIR) properties were evaluated. All these variables, individually or combined, influenced the properties and microstructural aspects of AAECs. Problems associated with the dispersion and agglomeration of nanomaterials, which could disrupt the microstructure and weaken its mechanical/physical properties, were avoided through the use of defined ultra-sonication with a high-shear mixing protocol. All AAECs achieved a 28-day compressive strength ranging from 26.0 MPa to 48.5 MPa and a slump flow > 800 mm, satisfying the criteria for flowable structural concrete. The addition of 5% MgO and up to 0.3% MWCNT/rGO increased the compressive strength/UPV of AAECs with MgO-MWCNT or rGO combination provided an improved strength at a higher dosage of 0.6%. A linear correlation between compressive strength and UPV was derived. As per SEM/EDS and XRD analyses, besides common C-A-S-H/N-C-A-S-H or C-A-S-H/C-S-H gels, the addition of MgO led to the formation of magnesium-aluminum hydrotalcite (Ht) and M-S-H (demonstrating self-healing potential), while the incorporation of rGO produced zeolites which densified the matrix and increased the compressive strength/UPV of the AAECs. Fourier transform infrared spectrometer (FTIR) analysis also suggested the formation of an aluminosilicate network in the AAECs, indicating a more stable structure. The increased UPV of MWCNT/rGO-incorporated AAECs indicated their better conductivity and ability of self-sensing. The developed AAECs, incorporating carbon-nano materials and MgO additive, have satisfactory properties with self-healing/-sensing potentials.

Keywords: alkali-activated engineered composites; MgO; MWCNT; rGO; slump flow; workability; compressive strength; ultrasonic pulse velocity; microstructure

1. Introduction

The high population growth and new global standards of modern construction require more demanding and sustainable infrastructures. Statistical trends estimate an increase in global cement production from 4.3 in 2015 to nearly 6.1 billion metric tons in 2050 [1,2], which would involve demolishing reserves of limestone, increasing global CO_2 emissions by around 7%, and conducting an energy-consuming process with an energy intensity of about 4.8 GJ [3]. Geopolymers, also named "alkali-activated (AA) binders", have been introduced as a promising alternative to Portland cement with a lower environmental impact, producing 60% less energy and approximately 80–90% less greenhouse gas [4–6]. The mechanism of producing geopolymers is a polymerization process that involves the chemical reaction of alumina-silicate materials (such as fly ash (FA), volcanic materials, ground granulated blast furnace slag (GGBFS), metakaolin (MK), etc.) in the presence of alkaline activators (such as sodium sulphate, calcium hydroxide, sodium silicate, etc.), resulting in three-dimensional Si-O-Al chains/bonds: poly sialate (-Si-O-Al-O-), poly sialate-siloxo (Si-O-Al-O-Si-O), and poly sialate-disiloxo (Si-O-Al-O-Si-OSi-O) [1,2,7–10]. Geopolymers/AA binders provide high mechanical strength, better resistance to corrosion/fire, and lower creep/shrinkage [11]. However, one major constraint associated with the use of geopolymer mortars on a commercial scale is their low-strength development at ambient temperature curing and the need for heat curing [12]. To overcome this problem, additives such as slag, lime, and ultrafine fly ash are being added during their geo-polymerization [13,14].

Research has been conducted on the fresh state and hardened properties of AA/geopolymer materials and, to improve such properties, nanomaterials and other additives have been added [15–18]. The shrinkage deformation and cracking are more significant in geopolymer paste than in cement due to its lower porosity/pore diameter and quick curing/hardening. This elucidates the importance of improving the volume stability and durability of this material by adding shrinkage compensation compounds such as calcium oxide, calcium sulphoaluminate, and MgO [19–22]. Calcium oxide and calcium sulphoaluminate react violently in a high alkali environment and cannot achieve an excellent compensating shrinkage effect in geopolymers, suggesting the use of MgO, which can induce a self-healing ability [22–24].

More recently, cement-free alkali-activated engineered composites (AAECs) using polyvinyl alcohol (PVA) and other fibers that exhibit strain hardening and micro-cracking characteristics with satisfactory fresh state, strength, durability, and shrinkage characteristics have been developed as more green, sustainable alternatives to traditional engineered cementitious composites (ECCs) [1,4,25–30]. The total substitution of cement with industrial by-products in AAECs decreases the energy investment, pollution, and use of virgin materials, which facilitates infrastructure sustainability through the simultaneous enhancement of durability and material greenness [1,31]. One-part ambient-cured AAECs produced using binary combinations of fly ash class C (FA-C) and ground-granulated blast furnace slag (GGBFS) with powder-form alkaline reagents and PVA fiber exhibited higher compressive strengths and ultrasonic pulse velocities, due to the formation of a combination of reaction products (C-S-H/C-A-S-H), compared to the mixture of amorphous (N-C-A-S-H/N-A-S-H) and crystalline (C-A-S-H/C-S-H) binding phases in their ternary (FA-C + FA class F 'FA-F' + GGBFS) counterparts [1]. An investigation on engineered geopolymer composites (EGCs) made of fly ash, GGBFS, fine silica sand, and a 2% volume fraction of two fiber types showed a significant increase in the tensile strength with the use of steel compared to polyethylene fibers [32].

In recent years, there has been a significant increase in the use of nanomaterials in concrete to achieve enhanced properties. Nanomaterials are specifically incorporated into AA mortars and composites to improve their mechanical and durability characteristics [33–35].

Most of the prior studies have shown that increasing the proportion of nano-additives improves mechanical properties such as the compressive and flexural strength [18,36,37]. The addition of multi-wall 0.1 wt% carbon nanotubes (CNTs), pre-sonicated in water with a polycarboxylate-based superplasticizer, increased the compressive strength of AA slag mortars, and X-ray diffraction (XRD) and Fourier transform infrared spectroscopy (FTIS) showed that CNTs enhanced the amorphous geopolymer structure [38]. However, a higher CNT ratio resulted in its agglomeration and limited the propagation of the geopolymerization reaction, causing a negative effect on the material's physical and mechanical properties. An increase of approximately 32% for the 28-day compressive test was observed for 0.5% multi-walled CNTs (MWCNTs) incorporated into metakaolin (MK) that was based on AA composites that had NaOH and Na_2SiO_3 as activators [35]. Li et al. [39] investigated the effect of adding 0.05–0.15 wt% of functionalized MWCNTs grafted by carboxyl groups, 0.5–0.25% PVA fibers, and 10–30 wt% of GGBFS on the mechanical properties of FA-based AAC. Among the three variable precursors, changes in the GGBFS followed by the MWCNTs and PVA fibers have the greatest effect on changes in the compressive strength. They also stated that the optimal value of MWCNTs was 0.1%, which resulted in a denser microstructure; however, adding more than that reduced the compressive strength significantly. Rovnaník et al. [40] declared that the addition of from 0.05 to 0.5% MWCNTs increased the compressive strength while 1% showed about a 13% decrease. XRD analysis demonstrated an increase in the amorphous phase and geopolymerization after the addition of MWCNTs to the FA-incorporated geopolymer paste, showing a dense matrix and a lower content of unreacted fly ash [41]. No noticeable influence of MWCNTs on the reaction products has been observed via XRD or Fourier transform infrared spectrometer (FTIR) analysis, even though that MWCNTs can act as nucleation sites and accelerate polymerization [35,38,42]. The effective dispersion of CNTs is crucial for maximizing their benefits in AA concrete composites, as strong attractive forces between CNT particles cause rapid clumping (agglomeration), which disrupts the microstructure and weakens the mechanical and physical properties [43]. Ultra-sonication with high-shear mixing (mechanical) [44] or the functionalization approach (chemical) of modifying the CNT surface with compatible chemicals [45] are proposed to reduce attraction and promote better dispersion.

Revathi and Jeyalakshmi [46] observed an increase in the compressive strength of two types (one hardened with phosphoric acid and the other with sodium silicate) of MK-based geopolymers with 2% reduced graphene oxide (rGO). Zhou et al. [47] added 0.7% large and small GO to MK-based geopolymers and indicated that large GO has a smaller increasing effect on the compressive and tensile strength of geopolymer due to blocking chemical reactions between geopolymer particles. Ranjbar et al. [48] reported that, when increasing the graphene nanoplatelets (GNPs) content from 0.1 to 1 wt% in FA-based geopolymer, the compressive strength steadily increased, showing the highest enhancement of 144% at 1% GNPs. On the other hand, Dong et al. [49] revealed that the compressive strength decreased steadily when the GO content was increased from 0 to 0.12% in alkali-activated slag with an alkaline activator consisting of 6 wt% NaOH. Matalkah and Soroushian [50] used from 0 to 0.3 vol% GNPs in AA concrete based on a ternary blend of FA, GGBFS, and albite, and found that the compressive strength decreased with the addition of GNPs at 0.1 and 0.3 vol and slightly increased at 0.2%. The 28-day compressive strength of fly ash geopolymer was improved by 23% with the addition of GO admixed at 0.02% by mass of fly ash, where a nuclear magnetic resonance (*NMR*) study showed that the GO improved the polymerization degree by increasing the total Q3 and Q4 Si-tetrahedrons, suggesting potential for improving the immobilization of heavy metals in fly ash [51]. Saafi et al. [17] investigated the mechanical properties, morphological changes, and chemical functional group changes of fly ash geopolymer composites by incorporating rGO and GO in fly ash geopolymer composites and found an increase of 134% in flexural strength, 376% in the Young's modulus, and 56% in toughness due to the restacking of rGO as a nano filler.

The use of MgO as a shrinkage-reducing mineral additive dates back to the mid-1970s [52]. The chemical, autogenous, and drying shrinkage decrease with the addition of reactive MgO in the presence of water, as it reacts to form fine $Mg(OH)_2$ crystals in geopolymer paste to produce uniform expansion, refine the pore size, and increase the compressive strength [53]. An increase in compressive strength of 26% at an early age was observed for AA slag (AAS) pastes with 5% MgO [54]. The effect of MgO in AAS systems has been investigated, both in terms of its varying natural content in different slag compositions [55] and its use as an additive [56]. Studies on this topic revealed that, although the main hydration product is still C–S–H gel, MgO reacts with the slag to form hydrotalcite 'Ht' ($Mg_6Al_2(OH)16CO_3{\cdot}4H_2O$)-like phases whose content increases with increases in the MgO content. The formation of low-density Ht/M-S-H phase (compared to CSH) was also confirmed by X-ray diffraction (XRD), thermogravimetric analysis (TGA), and scanning electron microscopy (SEM) analyses, and was shown to cause healing for cracks resulting from the shrinkage of AAS [55,57].

The literature contains previous research on the fresh state and mechanical properties of AA/geopolymer binders but limited research on AAECs incorporating CNTs/MWCNTs, with no significant studies using rGO, MgO, or a combination of MgO-MWCNTs/rGO. Moreover, the current literature does not provide specific directives to assess changes in the complicated fresh state, mechanical, or micro-structural properties of AAECs containing nanomaterials (such as MWCNTs or rGO), self-healing agents like MgO, or their combinations. In addition, most of the currently available literature is concentrated on the performance of heat-cured AAECs using the two-part (wet mix) rather than the one-part (dry mix) technique with solution-based reagents (which are corrosive and difficult to handle) and low calcium-based systems.

This paper focuses on the development and evolution of the fresh state and mechanical properties of PVA-fiber-reinforced ambient-cured AAECs containing healing agents (such as MgO) and nanomaterials (MWCNT and rGO). The novel aspects of this study addressing the current research gaps are the use of binary and ternary combinations of FA and GGBFS as precursors, two combinations of reagents (type 1 and type 2), high-calcium-based activators/regents rather than traditional Na-based ones, and a one-part dry mix technique with ambient curing as well as a comparative performance study considering variable mix design parameters and materials. The influences of the mix ingredients, such as the combinations of reagents, binary/ternary blends, fundamental chemical ratios present in the precursors/reagents, and the MWCNT/rGO/MgO dosages, on the physical properties in terms of workability (slump flow, flow time, flow velocity, and setting times) and mechanical properties (compressive strength and ultrasonic pulse velocity 'UPV') of AAECs have been described with reference to micro-structural characterization using SEM/EDS, XRD, and FTIR analyses. Moreover, correlations among the fresh state and mechanical properties (compressive strength-UPV) of the studied materials are developed. The results of this study will benefit engineers and scientists in developing and characterizing green low-carbon AAECs with additional smart functional properties (such as self-healing, self-sensing, electrical conductivity, piezoresistivity, and usability in 3D-printing) compared to traditional cementitious ECC for structural applications in bridge and building infrastructures [58]. The use of MgO, MWCNTs, and rGO at small doses will increase the cost of AAECs but the induced multifunctional properties will provide cost-compensation with aided benefits during their entire life cycle.

2. Experimental Program, Materials, Mix Designs, Mix Preparation, and Test Methods

This research was conducted to develop ambient-cured AAECs with multi-functional properties based on four basic mixes developed in previous research studies [1,59] by incorporating binary/ternary blends of source materials activated with two different types (type 1 and type 2)—high-calcium powdered reagent combinations and various combinations of MgO expensive agent (for self-healing) and carbon nanomaterials (MWCNTs and rGO)—for inducing conductive and self-sensing properties. The rationale for choosing these base alkali-activated mixes is to maintain continuity with the previous research study conducted at Toronto Metropolitan University that aimed to develop alkali-activated mixes using powder reagent and ambient curing, where the chosen combinations performed better than controls. A one-part dry mix technique using powder-based reagent under ambient conditions was chosen considering its benefits of easy handling, cost-effectiveness, and environment-friendliness. A total of 32 AAECs were produced and their workability or fresh state (slump flow, slump flow with time, fresh density), hardened state (compressive strength, UPV and dry density), and microstructural characteristics were evaluated using scanning electron microscopy (SEM) coupled with energy-dispersive Spectrometer EDS, X-ray diffraction (XRD) and Fourier transform Infrared Spectroscopy (FTIR) analysis.

2.1. Materials

The AAEC mixes were prepared using two combinations of FA and GGBFS as source materials. The binary (designated as B) AAEC mixes were prepared by mixing high-calcium FA type C (FA-C) and GGBFS, whereas the ternary ones (designated as T) were developed by mixing high-calcium FA type C (FA-C), low-calcium FA type F (FA-F), and GGBFS. The rationales for selecting these industrial waste-based source materials are a reduction in CO_2 emissions and an improvement in the properties of AAEC (particularly strength) due to the resulting high calcium and considerable amount of silicate contents. A high-range water reducer admixture (HRWRA) was required to ensure flowable AAM mixes. A polycarboxylate ether-based HRWRA with a solid content of 40%, pH of 6, and specific gravity of 1.06 g/cm^3 was used. Silica sand with a maximum particle size of 600μ was used as the fine aggregate. The MgO used in this study was prepared by burning $MgCO_3$ for two hours at 900 °C, and was classified as lightly burnt [22]. The AAEC mixes were prepared by reinforcement with 1.2% oil-coated polyvinyl alcohol (PVA) fibers. The PVA fibers were approximately 8 mm in length and 39 μm in diameter and could impart strain-hardening characteristics with multiple micro crack formations. The physical properties and chemical compositions (as per the X-ray fluorescence spectrometer 'XRF' analysis) of the FAs, GGBFS, silica sand, and MgO are presented in Table 1, with gradations shown in Figure 1.

As per the dry mix technique [60,61], two combinations of different alkaline reagents (obtained from Westlab Canada) were used in powder form to prepare two types of high-calcium-based activators to activate the source materials. Activator type 1 was prepared by mixing calcium hydroxide ($CaOH_2$ with specific gravity = 2.24 and pH = 12.4–12.6) with sodium meta silicate ($Na_2SiO_3{\cdot}5H_2O$ with modular ratio, SiO_2/Na_2O = 1, specific gravity = 0.7, pH = 11.3) in a ratio of 1:2.5 ($Ca(OH)_2$:$Na_2SiO_3{\cdot}5H_2O$ = 1:2.5), and the modular ratio (SiO_2/Na_2O) of this mixture was 3.22. Activator type 2 was prepared by combining calcium hydroxide ($Ca(OH)_2$) and sodium sulfate (Na_2SO_4 with specific gravity = 2.70 and pH = 7) beads/powder in a ratio of 2.5:1 ($Ca(OH)_2$:Na_2SO_4). The physical and chemical properties of both activator combinations are listed in Table 1 [59–61].

Table 1. Chemical compositions and physical characteristics of materials.

Chemical Composition (%)	Fly Ash Type C (FA-C)	Fly Ash Type F (FA-F)	Ground Granulated Blast Furnace Slag (GGBFS)	Silica Sand	HRWRA	Magnesium Oxide (MgO)
SiO_2	36.53	55.66	35.97	99.70		2.02
Al_2O_3	18.26	22.09	9.18	0.14		6.124
Fe_2O_3	5.66	4.26	0.50	0.016		0.94
CaO	20.97	7.97	38.61	0.01		2.40
MgO	5.08	1.16	10.99	0.01		92.26
K_2O	0.68	1.49	0.36	0.04		-
Na_2O	4.04	4.10	0.28	0.01		-
MnO	0.03	0.03	0.25	0.00		-
TiO_2	1.26	0.61	0.39	0.00		-
P_2O_5	0.96	0.43	0.01	0.00		-
L.O.I.	2.18	1.05	0.74	0.00		1.14
pH					6.00	
Density (g/cm^3)	2.61	2.02	2.87	2.65	1.06	3.58
Retained on 45μ, %	-	18.00	-	3.00		
Blaine fineness (m^2/kg)	315.00	306.00	489.30	-		

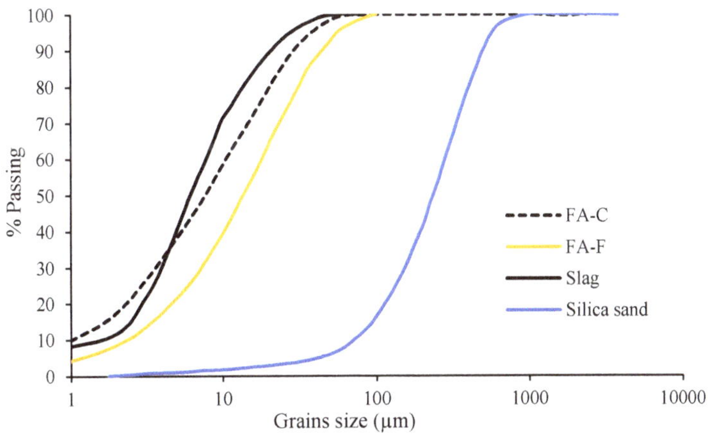

Figure 1. Grain size distributions of SCMs and silica sand.

Two different carbon-based nanomaterials (MWCNTs and rGO) were incorporated at different dosages within the AAEC mixes as conductive and self-sensing agents. MWCNTs are hollow and cylindrically shaped carbon allotropes with high aspect ratios. They usually consist of layers of rolled concentric nanotubes and walls of multiple one-atom-thick sheets of carbon. The MWCNTs had a diameter of 20–30 nm, length of 10–30 μm, Blaine fineness of 110 m^2/g, density of 1.2 g/cm^3, electrical conductivity greater than 10^{-2} s/cm, and purity of 95% or more, containing less than 1.5% ash by weight per the supplier Cheaptubes Inc. (Grafton, VT, USA) [62]. rGO usually undergoes a reduction process, following the microwave technique, after its production. This process aids in removing functional groups, which promotes dispersion and restores the carbon structure. rGO consists of randomly aggregated sheets, corrugations, and scrolling because its 2D membrane structure becomes thermodynamically stable via blending [63,64]. rGO (black powder form with a purity of 99% or more) with 8–10 layers, a length of 1–10 μm, a thickness of 3–6 μm, a diameter of 0.5 μm, a density of 0.03 g/cm^3, an electrical conductivity around 560 s/cm, and a Blaine fineness of 130 m^2/g was used. The elements contained in the rGO are 91% C, <8% O_2, <1% S, 2% H, and 0.4% N. Images of all aforementioned raw materials used for this study are presented in Figure 2.

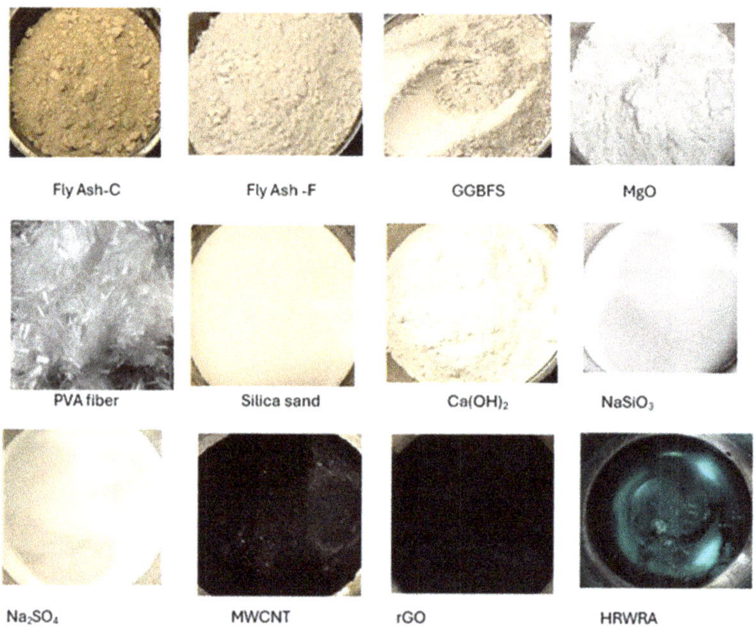

Figure 2. Source, sand, activator, additive, fiber, and admixture materials.

2.2. Mix Proportions

The experimental study consisted of a total of 32 AAEC mixes with two types of muti-component activators (Type 1 and Type 2) and three different fillers/additive materials (MgO, MWCNT, and rGO). All AAEC mixes were prepared following either the "one part" or "dry mix" technique. The mix proportions for all 32 one-part AAEC mixes are presented in Table 2, along with their mix designations. Four control mixes (B1, B2, T1, T2) without additions of MgO, MWCNT, or rGO were produced based on previous studies [1,65–68], in addition to the other 24 AAEC mixes that contained MgO, MWCNT, and rGO. The binary (B) mixes contained the combinations of FA-C and GGBFS, whereas the ternary (T) mixes were prepared with FA-C, FA-F, and GGBFS. The FA and GGBFS contents were varied from 52% to 60% and from 38% to 45%, respectively, and the silica sand content was kept constant at 30% by mass of the total binder content. The water-to-binder ratio was varied from 0.35 to 0.4 to achieve a minimum slump flow diameter of 500 mm. The dosage of HRWRA was kept constant (0.02 wt% of binder) due to its acidic nature and to prevent the effect of its alkalinity on the varying mix proportions. The PVA fiber was used at 2% by volume for all mixes to induce strain hardening and multiple-microcracking characteristics.

Table 2. Mix proportions of 32 AAEC mixes.

AAECs Mix ID.	Total SCMs (Binder *)	MgO/ MWCNT/ rGO	SCMs			Reagent Component Ratio	R./B	Chemical Ratios (SCMs + Reagent)			
			FA-C	FA-F	GGBFS			SiO_2/Al_2O_3	Na_2O/SiO_2	CaO/SiO_2	Na_2O/Al_2O_3
Four basic AAEC mixes (with 0% MgO/MWCNT/rGO)											
B1	1	0	0.55	0	0.45	1:2.5	0.09	2.62	0.09	0.84	0.23
B2	1	0	0.55	0	0.45	2.5:1	0.12	2.56	0.14	1.02	0.35
T1	1	0	0.25	0.35	0.40	1:2.5	0.09	2.75	0.08	0.59	0.22
T2	1	0	0.25	0.35	0.40	2.5:1	0.12	2.69	0.12	0.73	0.32
Four AAEC mixes with 5% MgO											
B1M5	1	0.05	0.52	0	0.43	1:2.5	0.09	2.58	0.09	0.85	0.23
B2M5	1	0.05	0.52	0	0.43	2.5:1	0.12	2.51	0.14	1.03	0.35
T1M5	1	0.05	0.24	0.33	0.38	1:2.5	0.09	2.97	0.05	0.54	0.14
T2M5	1	0.05	0.24	0.33	0.38	2.5:1	0.12	2.97	0.05	0.54	0.14
Eight AAEC mixes with 0.3% and 0.6% MWCNT											
B1C3	1	0.003	0.55	0	0.45	1:2.5	0.09	2.62	0.09	0.84	0.23
B2C3	1	0.003	0.55	0	0.45	2.5:1	0.12	2.56	0.14	1.02	0.35
T1C3	1	0.003	0.25	0.35	0.40	1:2.5	0.09	2.75	0.08	0.59	0.22
T2C3	1	0.003	0.25	0.35	0.40	2.5:1	0.12	2.69	0.12	0.73	0.32
B1C6	1	0.006	0.55	0	0.45	1:2.5	0.09	2.62	0.09	0.84	0.23
B2C6	1	0.006	0.55	0	0.45	2.5:1	0.12	2.56	0.14	1.02	0.35
T1C6	1	0.006	0.25	0.35	0.40	1:2.5	0.09	2.75	0.08	0.59	0.22
T2C6	1	0.006	0.25	0.35	0.40	2.5:1	0.12	2.69	0.12	0.73	0.32
Eight AAEC mixes with 0.3% and 0.6% rGO											
B1R3	1	0.003	0.55	0	0.45	1:2.5	0.09	2.62	0.09	0.84	0.23
B2R3	1	0.003	0.55	0	0.45	2.5:1	0.12	2.56	0.14	1.02	0.35
T1R3	1	0.003	0.25	0.35	0.40	1:2.5	0.09	2.75	0.08	0.59	0.22
T2R3	1	0.003	0.25	0.35	0.40	2.5:1	0.12	2.69	0.12	0.73	0.32
B1R6	1	0.006	0.55	0	0.45	1:2.5	0.09	2.62	0.09	0.84	0.23
B2R6	1	0.006	0.55	0	0.45	2.5:1	0.12	2.56	0.14	1.02	0.35
T1R6	1	0.006	0.25	0.35	0.40	1:2.5	0.09	2.75	0.08	0.59	0.22
T2R6	1	0.006	0.25	0.35	0.40	2.5:1	0.12	2.69	0.12	0.73	0.32
Eight AAEC with 5%MgO and MWCNT or rGO (0.3% or 0.6%))											
B2M5C3	1	0.05/0.003	0.52	0	0.43	2.5:1	0.12	2.56	0.14	1.02	0.35
T2M5C3	1	0.05/0.003	0.24	0.33	0.38	2.5:1	0.12	2.97	0.05	0.54	0.14
B2M5C6	1	0.05/0.003	0.52	0	0.43	2.5:1	0.12	2.56	0.14	1.02	0.35
T2M5C6	1	0.05/0.003	0.24	0.33	0.38	2.5:1	0.12	2.97	0.05	0.54	0.14
B2M5R3	1	0.05/0.006	0.52	0	0.43	2.5:1	0.12	2.56	0.14	1.02	0.35
T2M5R3	1	0.05/0.006	0.24	0.33	0.38	2.5:1	0.12	2.97	0.05	0.54	0.14
B2M5R6	1	0.05/0.006	0.52	0	0.43	2.5:1	0.12	2.56	0.14	1.02	0.35
T2M5R6	1	0.05/0.006	0.24	0.33	0.38	2.5:1	0.12	2.97	0.05	0.54	0.14

* All numbers are mass ratios of binder; binder denotes source materials (SMs): supplementary cementitious materials (SCMs) such as FA-C, FA-F, GGBFS, and activator; MgO: magnesium oxide; MWCNT: multi-wall carbon nano tube; rGO: reduced graphene oxide; B1, B2: binary AAEC with activator type 1 and type 2, respectively; T1, T2: ternary AAEC with activator type 1 and type 2, respectively; M5: AAEC mixes with 5% MgO by wt. of binder, C3/C6: AAEC mixes with MWCNT content of 0.3% and 0.6% wt. of binder, R3/R6: AAEC mixes with rGO of 0.3% and 0.6% wt. of the binder.

The AAEC mixes incorporating MgO were prepared using 5% MgO, while those with MWCNT and rGO were prepared by incorporating 0.3% and 0.6% MWCNT and rGO by wt of binder contents. The activator/reagent components and the initial chemical ratios in the mix compositions of all AAECs are also presented in Table 2. Activator type 1 has a reagent component ratio (calcium hydroxide to sodium metasilicate) of 1:2.5, while activator type 2 has a reagent component ratio (calcium hydroxide to sodium sulfate) of 2.5:1. These ratios were chosen based on previous research studies on AABs [60,61,65]. The fundamental chemical ratios in terms of silicon oxide/aluminum oxide, sodium oxide/silicon oxide, calcium oxide/silicon oxide, and sodium oxide/aluminum oxide were evaluated based on their chemical compositions of reagents and source materials. These chemical ratios were all found to be within the acceptable range as per research studies on fly ash- and slag-based mortars [60,61,65].

2.3. Mixing, Dispersion of MWCNT/rGO, Casting, and Curing of Specimens

The powdered reagent components were first mixed thoroughly and then added to the rigorously blended source materials (and MgO, if needed). The complete binder system was then dry mixed for about 5 min in a shear mixer before two-thirds of the required water was gradually added to the mix while mixing continued for 3–4 min. Then, HRWRA mixed with the remaining amount of water was gradually added for 4–5 min to make a flowable paste. After that, silica sand was added slowly for 3–4 min as per the proportions given in Table 2, then PVA fibers were added slowly for 3–4 min and mixed for an additional one minute to make control and MgO AAECs.

For AAEC mixes with MWCNT and rGO, two-thirds water and 50% HRWRA were mixed for 1–2 min by hand stirring in the beaker, after which MWCNT/rGO was added and hand stirred for 2–3 min. The beaker was then placed inside the sonicator, a probe was inserted into the beaker, and sonication was performed for 30–40 min. The amount of sonication energy was 50 J/mL~75 J/mL to effectively disperse MWCNT/rGO to the water [35,38,69]. The reason behind using this range of sonication energy is the shortening effect in MWCNTs for energy amounts over 75 J/mL [65]. Suave et al. [70] reported that a lower sonication energy with a longer sonication time provides better conditions than a higher sonication energy with a shorter sonication time in terms of preparing MWCNTs functionalized with carboxylic groups. However, the optimal level of sonication energy for dispersion is related to the CNT concentration. Still, when the concentration reaches a certain level, it cannot be well dispersed, requiring different technologies with different degrees of dispersion and a different optimal sonication energy. The mixture of sonicated MWCNT/rGO with water and HRWRA was then slowly added into the already-made dry mix (as explained previously in control mix preparation) for 4–5 min while mixing continued in shear mixer, after which silica sand was gradually added to prepare MWCNT/rGO-incorporated AAECs. Figure 3 shows MWCNTs before, during, and after sonication and the sonication device. The use of the dry-mixing method allowed us to avoid the handling of solution-based corrosive alkaline reagents and used less powder-based reagent (0.35 of total binder content) than the solution-based one [4,60,61]. Thus, it demonstrated its enhanced sustainability, user-friendliness, and commercial viability in developing AAECs on a large scale compared to the conventional two-part method.

The total mixing procedure lasted about 20–25 min. At least 15 cube specimens with dimensions of 50 mm × 50 mm × 50 mm were prepared for each AAEC mix. The molds/specimens were kept in the curing room at 23 ± 3 °C and 95 ± 5% relative humidity (RH). The molds were de-molded after 24 h of casting and were kept in the curing chamber until the day of testing.

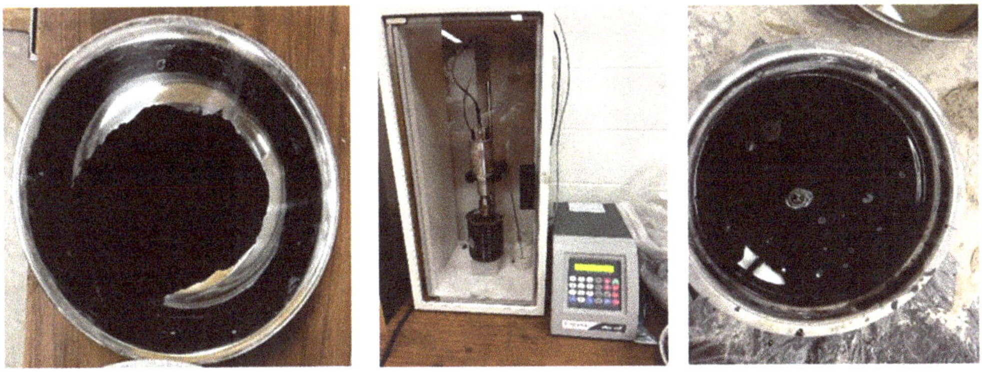

(a) (b) (c)

Figure 3. MWCNT at three stages: (**a**) before sonication (dry), (**b**) during sonication, and (**c**) after sonication.

2.4. Test Methods

The physical workability properties in terms of the slump flow, flow time (T50), and fresh density of AAECs with MgO and MWCNTs/rGO and MgO-MWCNTs/rGO were evaluated immediately after mixing. The slump flow and fresh density of AAECs were measured as per the ASTM C1611 [71] and ASTM C138 [72] standards, respectively. The hardened density of the AAECs was evaluated following the conventional weight and volume measurement methods. The slump flow spread characteristics of the mixes, of which a typical one is presented in Figure 4a, were assessed through a slump cone test in compliance with ASTM C1437 [73]. Additionally, the slump flow time (T50) to reach 500 mm slump flow was recorded to calculate the slump flow velocity. A compressive strength test using cube specimens (Figure 4b) at 7, 14, 28, 56, and 180 days was conducted according to ASTM C109/C109M [74] by applying a loading rate of 1 kN/s. An ultrasonic pulse velocity (UPV) test was conducted on the same cube specimens using a Portable Ultrasonic Non-Destructive Digital Indicating Tester as per ASTM C597 [75]. At least three specimens were tested at each age for the compressive and UPV tests.

(a) (b)

Figure 4. (**a**) Slump flow test, (**b**) compressive strength test.

SEM and EDS analyses were performed on the AAECs to determine the reaction products by taking chip specimens (approximately 10 mm × 5 mm × 5 mm) from the core of the failed compression test cubes at 180 days followed by coating them gold to make the surface conductive. The fracture surface was studied using secondary electrons (SEs) and backscattered electron (BSE) microscopy at 20 kV. The specimens' morphologies were studied at 100× (100 μm), and the assessment of the reaction products was performed at 2000× (10 μm).

The specimen preparation for the XRD analysis consisted of grinding the specimen taken from the core of the 180-day failed compression cubes. The ground specimen was passed through a 200-mesh sieve. A Bruker D8 Endeavor diffractometer equipped with a Cu X-ray source and operating at 40 kV and 40 mA; a range of 5–70 deg θ; a step size of 0.02 deg 2°; a time per step of 0.5 s; a fixed divergence slit, an angle of 0.30; a sample rotation of one rev/s was used to identify the mineral phases using the PDF4/Minerals ICDD database.

Like the XRD analysis, the powder from the AAEC specimens was subjected to Fourier transform infrared spectroscopy (FTIR) analysis. The FTIR spectra were recorded with the Perkin Elmer Spectrum 400 spectrometer in mid-IR mode. Before beginning the analysis, rubbing alcohol was used to clean the sample platform. Approximately 5 mg of the powder specimen was decanted onto the platform for further analysis. The spectrum was obtained over a range from 4000 to 400 cm^{-1}. FTIR measurements were performed on a Fourier transform infrared spectrophotometer (Jasco 4200 Type A, Jasco, Easton, MD, USA). The FTIR spectra were obtained within a wavenumber range from 400 to 4000 cm^{-1} at a resolution of 4 cm^{-1}. The KBr pellet was used to carry out the FTIR measurements and the pellets were prepared by pressing a mixture of the sample in a die set. In FTIR analysis, the sample was irradiated with infrared light over a range of frequencies. The interaction between the infrared light and the material caused the absorption or transmission of specific wavelengths, which was detected and analyzed to identify the functional groups and chemical bonds within the material.

3. Results and Discussion

Table 3 summarizes the workability (slump flow/flow time (T50)/flow velocity/fresh density) and hardness (28-day compressive strength/density/UPV) properties of all 32 AAECs. The following subsections present the influence of the MgO, nanomaterials (MWCNT/rGO), and reagent types on the workability, compressive strength, UPV, and microstructural characteristics of the AAECs.

Table 3. Physical workability and hardened properties of AAECs.

AAECMix ID	Fresh Density (kg/m^3)	Slump Flow (mm)	Slump Flow Time (s)	Slump Flow Velocity (mm/s)	* Hard/Dry Density kg/m^3	* 28-Day UPV (m/s)	* 28-Day Compressive Strength (MPa)
B1	1921	700	4	175.0	2000	3432	38.5
B2	1924	705	5	141.0	2032	3067	40.8
T1	1930	710	3	236.7	2120	3231	30.5
T2	1914	750	3	250.0	2064	3717	33.4
B1M5	1924	690	5	138.0	2016	3845	46.2
B2M5	1923	685	5	137.0	2072	3896	48.5
T1M5	1927	700	4	175.0	1984	3762	38.2
T2M5	1933	705	4	176.3	2056	3954	41.1
B1C3	1930	720	4	180.0	2008	3566	41.5
B2C3	1930	660	4	165.0	2080	3436	44.6

Table 3. *Cont.*

AAECMix ID	Fresh Density (kg/m^3)	Slump Flow (mm)	Slump Flow Time (s)	Slump Flow Velocity (mm/s)	* Hard/Dry Density kg/m^3	* 28-Day UPV (m/s)	* 28-Day Compressive Strength (MPa)
T1C3	1903	710	3	236.7	2016	3605	33.2
T2C3	1928	725	3	241.7	2096	4068	36.4
B1C6	1950	700	5	140.0	2044	3582	35.5
B2C6	1950	670	5	134.0	2036	3659	36.4
T1C6	1935	630	4	157.5	1960	3634	26.0
T2C6	1938	710	4	177.5	2168	3785	28.6
B1R3	1939	650	4	162.5	2032	3738	43.0
B2R3	1978	620	4	155.0	2152	3582	45.3
T1R3	1882	625	3	200.0	2060	3450	35.0
T2R3	1911	720	3	240.0	2096	3943	37.9
B1R6	1945	620	4	155.0	2060	3282	40.5
B2R6	1954	660	4	165.0	2060	3122	42.8
T1R6	1953	735	3	245.0	2068	3278	32.5
T2R6	1953	780	2	260.0	2176	3823	35.4

* mean value of at least three specimens; deviation from mean: density ~± 5 kg/m^3, UPV ~± 10 m/s; compressive strength ~± 0.5 MPa.

3.1. Physical Characteristics: Slump Flow, Flow Time, Flow Velocity and Density

The intent of this investigation was to find stable AAEC mixes with good workability considering the raw materials, curing conditions, and activator types used, as well as the optimum mix proportions [76]. Visual inspections were also conducted to observe segregation characteristics in terms of the separation between solid and liquid components of the mix. All the mixes showed stable performance with no segregation, as can be seen from the slump flow shown in Figure 4a. The fresh density of the AAECs ranged from 1911 kg/m^3 to 1954 kg/m^3 while the dry (hardened) density ranged from 1984 kg/m^3 to 2176 kg/m^3 (Table 3).

Figure 5 shows the influence of the MWCNT, rGO, and MgO content on the slump flow of the AAEC mixes. The slump flow value of the AAEC mixes decreased with an increase in MWCNT content (Figure 5a) from 0 to 0.6%, except for the binary mixes with reagent 2 (B2). The slump flow (Figure 5b) decreased with the increase in rGO content from 0 to 0.3% and then showed an increasing trend up to 0.6%, except for the binary mixes with reagent type 1 (B1). The binary and ternary mixes with 0.3% rGO content generated the minimum slump flow as compared to their control mixes (with 0% rGO or MWCNTs) and, mostly, those with 0.6% rGO content. However, for the binary and ternary mixes, the MgO content had an insignificant influence on the slump flow of the AAEC mixes, although a slight decrease in the slump flow with the addition of 5% MgO was observed (Figure 5c). Among all the mixes (Table 3), a ternary mix with 0.6% rGO (T2R6) showed the highest 800 mm slump flow while the lowest slump flow of 620 mm was recorded for the binary rGO mix (B2R6).

As per the EFNARC (2002) [75] guidelines, concrete and composites possess good filling and self-consolidating ability if their slump flow ranges between 650 mm and 800 mm. Almost all the AAEC mixes satisfied the EFNARC [77] criteria by showing a slump flow value of ≥600 mm.

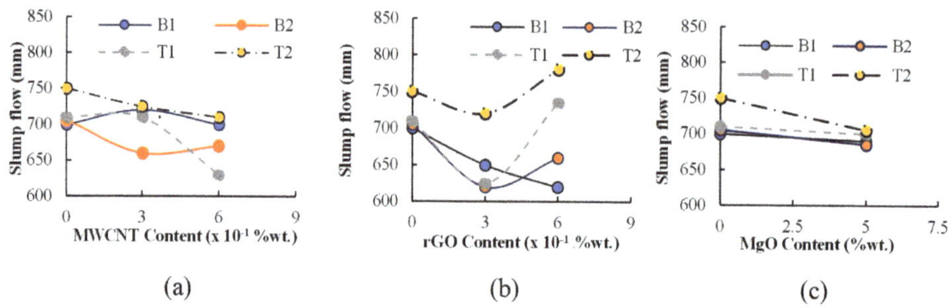

Figure 5. Influence of (**a**) MWCNT, (**b**) rGO, and (**c**) MgO contents on slump flow characteristics.

In the case of control AAECs, the ternary mixes (T1 or T2) exhibited higher slump flows (Table 3) than their binary counterparts (B1 or B2) due to the replacement of FA-C by FA-F with smooth round shape particles (associated with a decrease in FA-C and GGBFS contents as per Table 2) that enhance flowability. As discussed earlier, activator type 2 results in less viscous mixes owing to a lower silica ratio and low desolation rate, and, thus, can be theorized to yield a higher slump diameter. This trend can be observed for the control mixes made with activator type 2 (B2 and T2). However, the addition of MgO/MWCNTs/rGO, in almost all cases (specifically binary mixes), showed less significant but lower or almost similar slump values for the mixes with activator type 2 compared to the mixes with activator type 1 (Table 3). The increase in the solid load, finer surface of fine particles, water affinity, and agglomeration tendency of MgO, MWCNTs, and rGO can be theorized to contribute to such variations and the reduced workability. In the case of the ternary mixes, however, the mixes with activator type 2 have a slightly higher slump flow value due to the presence of round fly ash combined with the lower colloidal formation of the low-silica-based activator type 2. An increase in the MWCNT content induces a higher solid ratio. Moreover, the agglomeration and high water affinity of MWCNT particles are also essential contributing factors. There was about a 4–5% decrease in the slump flow (or workability) for an increase in MWCNT content from 0.3% to 0.6% for the AAECs. Such a decrease in workability with the proportional increase in the MWCNT additive was reported in earlier studies [78].

During the slump flow diameter measurement, the flow time (T_{50}) was recorded for the AAECs to spread up to a 500 mm diameter—a lesser T_{50} time suggests good flowability, as per EFNARC [77]. As per Table 3, the AAEC mixes achieved T_{50} values within a time range from 2 s to 5 s, which is considered to satisfy the criteria for good-flowability concrete [77]. A minimum slump flow time of 2 s was observed for the AAEC mix T2R6 (with activator type 2 and 0.6% rGO), whereas a maximum flow time of 5 s was observed for the binary mixes with MWCNT and MgO (Table 3).

As for the control mixes, the binary variants took more time to reach a 500 mm diameter than the ternary ones. This corresponded to their lower slump velocity compared to the ternary ones (about 26% and 46% less for binary mixes with activator type 1 and 2, respectively), due to the absence of smooth round fly ash F particles that can reduce interparticle friction, which is described as being connected to the slump flow. The reduction in slump velocity is more pronounced with the use of activator type 2 due to the presence of more colloidal particles compared to type 1. The addition of 5% MgO increased the amount of finer particles with a high water affinity and thus reduced the slump velocity by about 19~20% compared to the control mixes (Table 4). Similarly, the addition of MWCNTs/rGO and increase in their content caused a reduction in the slump velocity due to inducing a higher solid ratio in the mixes. Moreover, the agglomeration and high water affinity of the MWCNT particles were also important factors.

Table 4. Compressive strength changes over time of different mixes.

AAEC Mix ID	Compressive Strength (MPa) *					Strength Change with Respect to Control (%) (+ve: Increase, −ve: Decrease)				
	7 Days	14 Days	28 Days	56 Days	180 Days	7 Days	14 Days	28 Days	56 Days	180 Days
Control AAECs										
B1	14.3	28.5	38.5	45.8	50.3	0	0	0	0	0
B2	15.6	30.8	40.8	48.1	52.6	0	0	0	0	0
T1	14.9	22.5	30.5	37.8	42.3	0	0	0	0	0
T2	15	25.4	33.4	40.7	45.2	0	0	0	0	0
5% MgO Incorporated AAECs										
B1M5	20.1	35.5	46.2	52.6	56.8	40.4	24.6	20	14.8	12.9
B2M5	21.4	37.8	48.5	54.9	59.1	37.1	22.7	18.9	14.1	12.4
T1M5	20.7	29.5	38.2	44.6	48.8	38.9	31.1	25.2	18	15.4
T2M5	20.7	32.4	41.1	47.5	51.7	38.6	27.6	23.1	16.7	14.4
MWCNT (0.3% and 0.6%) incorporated AAEC										
B1C3	17.9	30.5	41.5	49.4	53.9	25.1	7	7.8	7.8	7.1
B2C3	19.4	34.6	44.6	51.9	56.4	24	12.2	9.2	7.8	7.1
T1C3	18.4	26	33.2	41.3	45.8	23.5	15.6	8.9	9.3	8.3
T2C3	18.5	28.9	36.4	44.2	48.7	23.4	13.8	9	8.6	7.7
B1C6	15.3	29.5	35.5	44.2	48.4	7	3.5	−7.8	−3.5	−3.8
B2C6	16.6	31.8	36.4	45.3	50.3	6.4	3.2	−10.8	−5.8	−4.4
T1C6	15.9	20.2	26	36.2	40.4	6.7	−10.2	−14.8	−4.2	−4.5
T2C6	16	21.1	28.6	39.1	42.7	6.7	−16.9	−14.4	−3.9	−5.5
rGO (0.3% and 0.6%) incorporated AAEC										
B1R3	18.8	33	43	50.3	54.8	31.5	15.8	11.7	9.8	8.9
B2R3	20.1	35.3	45.3	52.6	57.1	28.8	14.6	11	9.4	8.6
T1R3	19.4	27	35	42.3	46.8	30.3	20	14.8	11.9	10.6
T2R3	19.5	29.9	37.9	45.2	49.7	30.1	17.7	13.5	11.1	10
B1R6	16.3	30.5	40.5	47.8	52.3	14	7	5.2	4.4	4
B2R6	17.6	32.8	42.8	50.1	54.6	12.8	6.5	4.9	4.2	3.8
T1R6	16.9	24.5	32.5	39.8	44.3	13.4	8.9	6.6	5.3	4.7
T2R6	17	27.4	35.4	42.7	47.2	13.4	7.9	6	4.9	4.4
5% MgO and MWCNT or rGO (0.3% and 0.6%) incorporated AAECs										
B2M5C3	23.8	40.2	50.9	57.3	61.5	52.4	30.5	24.8	19.1	16.9
T2M5C3	23.1	34.8	43.5	49.9	54.1	54.3	36.8	30.1	22.5	19.6
B2M5C6	21.6	38	48.9	55.1	59.3	38.6	23.5	20	14.6	12.8
T2M5C6	21	32.7	41.4	47.8	52	40.4	28.6	23.8	17.3	15
B2M5R3	25.18	41.6	52.3	58.7	62.9	61.4	35.1	28.2	22	19.6
T2M5R3	24.61	36.27	44.97	51.37	55.57	64.5	42.8	34.6	26.2	22.9
B2M5R6	22.18	38.6	49.3	55.7	59.9	42.2	25.3	20.8	15.8	13.9
T2M5R6	21.7	33.4	42.1	48.5	52.7	45.3	31.5	26	19.2	16.6

* mean value of at least three specimens (deviation from mean: ± 0.5 MPa).

As can be seen, the flowability of all geopolymer binders increases with the increase in the Na_2O/SiO_2 ratio [60,61]. This also means that a higher alkalinity provides a higher flowability of the mixtures. The setting times of the representative mortar portions of all the AAECs without PVA fiber were also studied, where an increase in the Na_2O/SiO_2 ratio led to quicker setting times, meaning that the higher the alkalinity, the faster the agglomeration and poly-condensation. The same trend was observed by Gado et al. [79] and Allahverdi and Kani [80]. The addition of MgO and MWCNTs yielded comparatively lower initial and final setting times compared to the controls. The initial setting times ranged from 190 min to 311 min while the final setting times ranged from 230 min to 369 min, depending on the reagent type (type 1 and type 2) and the use of MgO/MWCNTs/rGO. Both MgO and functionalized MWCNTs have a high water affinity due to an increased surface area, which absorbs a significant quantity of water within the surface, resulting in a more viscous matrix and producing lower setting times. The lower water affinity of the rGO particles ensured a higher workability with a low viscosity, causing a lower variation in setting times.

3.2. *Hardened Characteristics: Compressive Strength and UPV*

This section describes the evolution of the compressive strength over the time periods of 7, 14, 28, 56, and 180 days with strength enhancement at various ages with respect to the control (summarized in Table 4) and the UPV characteristics, as well as to the co-relation between the UPV and compressive strength at 28 days. The influences of the MgO/MWCNTs/rGO and other various mix design parameters (reflecting binary and ternary combinations of source materials, reagent types, and chemical ratios) on the compressive strength and UPV are analyzed. These parameters are described to illustrate the influence of various mix design parameters, reflecting the binary and ternary combinations of SCMs as well as the types and proportions, of powder-based reagents and chemical ratios.

3.2.1. Influence of MgO, MWCNT and rGO on Compressive Strength Development

All AAECs achieved a 28-day compressive strength ranging from 26.0 MPa to 48.5 MPa > 18 MPa (as per Tables 3 and 4), which is the range specified for structural concrete as per ACI 318 [81], and which is also comparable to other conventional two-part FA and GGBFS-based geopolymers [4,61]. The binary AAECs (FA-C + GGBFS designated as B) that had higher CaO/SiO_2 ratios (as presented in Table 2) obtained higher compressive strengths than their ternary (FA-C + FA-F + GGBFS designated as T) counterparts (Table 3) due to the formation of additional CSH binding phases/gels with C-A-S-H/N-C-A-S-H, as is evident from the microstructural analysis (discussed later). All the mixes exhibited a ratio of Na_2O/Al_2O_3 of less than 1 (Table 2), which prevented efflorescence, as observed in previous studies where $Na_2O/Al_2O_3 < 1$ resulted in sodium consumption in the reaction process and, in turn, prevented efflorescence.

The evolution of the compressive strength with the age (0, 7,14, 28, 56, and 180 days) of control AAEC mixes (B1, B2, T1, and T2) without any MWCNTs/rGO or MgO are presented in Figure 6 and Table 4. The compressive strength increased with age, and the rate of increase was higher than that of controls up to 56 days (due to a higher rate of polymerization) and then slowed down, showing a steady increase up to 180 days (Figure 6). The compressive strength was higher for the binary mixes (B1, B2) than their ternary counterparts (T1 and T2) and, also, the reagent 1 mixes developed a lower strength than their reagent 2 counterparts throughout the aging process. Binary mix B2 with reagent 2 exhibited the highest compressive strength (with 40.8 MPa, 48.1 MPa, and 52.6 MPa at 28 days, 56 days, and 180 days, respectively) throughout the aging followed by B1, T2, and T1. Reagent 2 ($Ca(OH)_2$:Na_2SO_4 = 2.5:1) performed better than reagent

1 ($Ca(OH)_2$:$Na_2SiO_3 \cdot 5H_2O$ = 1:2.5) in terms of its compressive strength development. The reagent 2 component ratio of $Ca(OH)_2$:Na_2SO_4 = 2.5:1 was judged to be the optimal composition based on the strength results (52.6 MPa at 180 days) of binder system B2. The high alkalinity of sodium metasilicate (pH = 14) enhanced the dissolution of Si and Al ions from the FA-F, which otherwise showed low reactivity. The high-calcium precursors (GGBFS and FA-C) also formed C-A-S-H acute phases, which compacted the amorphous N-A-S-H, resulting in a dense microstructure and high compressive strengths. However, for reagent 2, a $Ca(OH)_2/Na_2SO_4$ = 2.5:1 component ratio was determined to be superior, and also outperformed reagent 1, as depicted in Figure 6. The additional C-S-H gel formation from the composition with reagent 2 led to higher compressive strengths due to the high calcium levels in the system. B2 exhibited a higher strength (28% higher at 28-days) than its counterpart T2 due to its 30% higher fly ash C content. This can also be attributed to the high Ca^{2+} content in the matrix, coming from the FA-C, the GGBFS, and the activator $Ca(OH)_2$ component, which formed a denser microstructure.

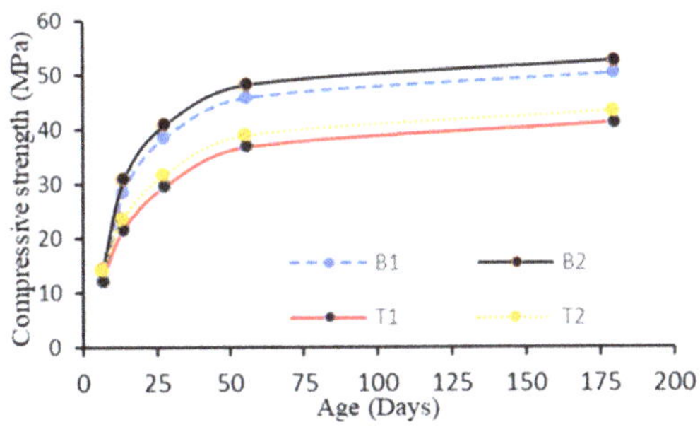

Figure 6. Evolution of compressive strength of AAECs (without MgO/MWCNTs/rGO) with age.

The MgO-incorporated AAEC mixes (B1M5, B2M5, T1M5, T2M5) exhibited higher strength development with age (Figure 7) compared to their control counterparts (B1, B2, T1, and T2). Adding 5% MgO in the AAEC mixes increased the compressive strength (for example, 19~20% and 23.1~25.2% at 28 days for the binary and ternary mixes, respectively, Table 4). Overall, compared to the control mixes, strength increases of 37.1~40.4%, 22.7~31.1%, 18.9~25.25%, 14.15~19.05%, and 12.45~15.4% were observed at 7, 14, 28, 56, and 180 days, respectively (Table 4), showing strength increases with increases in the age and that the % of strength increase decreased with increases in the age. The geopolymers solidify quickly and generate C-S-H and C-A-S-H gels, leading to considerable chemical, autogenous, and drying shrinkage at the early stage. The highly reactive MgO accelerates the early age hydration of slag as a result of fast heat release during the dissolution process of MgO, which accelerates the hydration reaction, leading to the formation of more hydration products [57]. The addition of fine reactive MgO generated a large number of well-dispersed, fine, worm-like $Mg(OH)_2$ crystals in a high-alkalinity liquid phase environment, which did not overgrow, resulting in uniform-volume micro expansion, effectively compensating for volume shrinkage in the hardening process, matching the shrinkage process, refining the matrix pore size, and increasing the compressive strength, as observed in this study [53].

Figures 8–10 show the evolution of the compressive strength of the MWCNT-, rGO-, MgO-MWCNT-, and MgO-rGO-incorporated AAECs with ages up to 180 days compared to their control counterpart. The strength increased with the increase in age and the trend of the rate of strength development was similar to those of the control and MgO-incorporated AAECs.

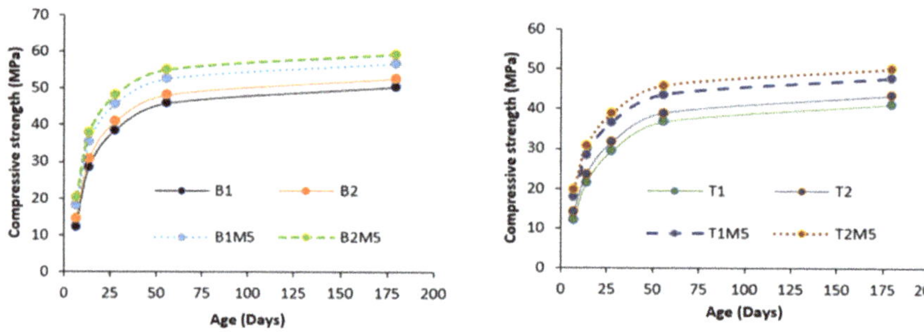

(a) (b)

Figure 7. Evolution of compressive strength with age for MgO-incorporated (**a**) binary and (**b**) ternary AAEC mixes with reagents 1 and 2.

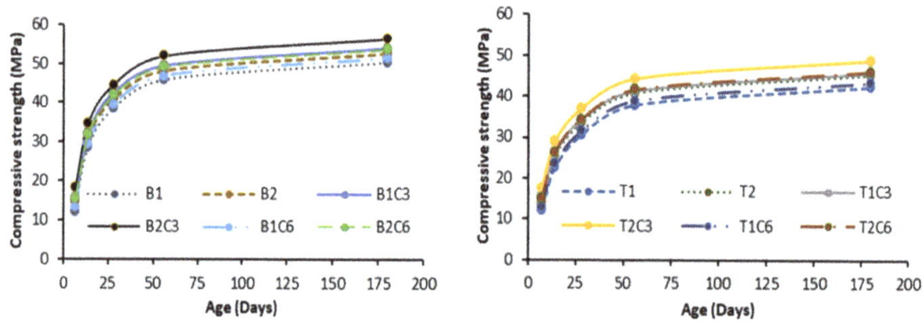

(a) (b)

Figure 8. Evolution of compressive strength with age for MWCNT-incorporated (**a**) binary and (**b**) ternary AAEC mixes with reagents 1 and 2.

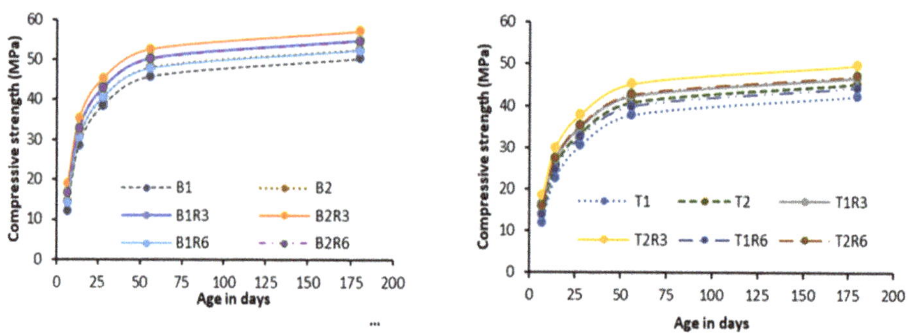

(a) (b)

Figure 9. Evolution of compressive strength with age for rGO-incorporated (**a**) binary and (**b**) ternary AAEC mixes with reagents 1 and 2.

For the MWCNT-incorporated mixes, the compressive strength was increased (+ve value)/decreased (−ve value), respectively, with the use of 0.3% and 0.6% MWCNTs by 15.1~23.4% and 6.4~7.0% at 7 days, 7.0~15.6% and 3.2~16.9% at 14 days, 7.8~9.2% and 7.8 ~−14.8% at 28 days, 7.8~9.3% and −3.5~−5.8% at 56 days, and 7.1~8.3% and −3.8~−5.5% at 180 days (Table 4). The compressive strength decreased at a higher 0.6% MWCNT content at 28 days and beyond. For the rGO-incorporated mixes, the compressive strength was increased, respectively, with the use of 0.3% and 0.6% rGO by 28.8~31.5%

and 12.8~14.0% at 7 days, 14.6~20% and 6.5~8.9% at 14 days, 1~14.8% and 4.9~6.6% at 28 days, 9.4~11.9% and 4.2~5.3% at 56 days, and 8.6~10.6% and 3.8~4.7% at 180 days. The rGO incorporated mixes exhibited higher compressive development than their MWCNT counterparts at all ages, showing no decrease.

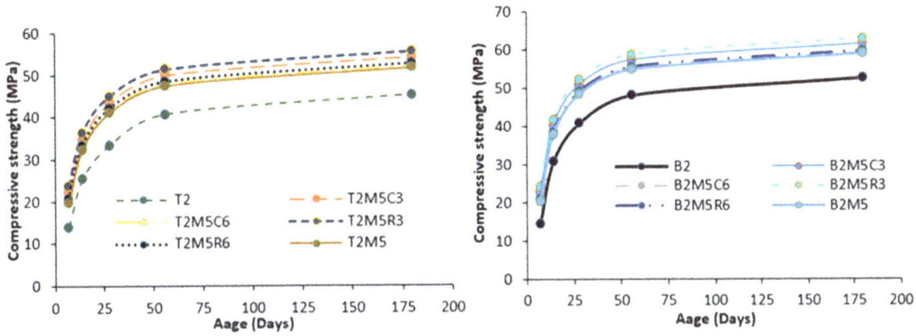

(a) (b)

Figure 10. Evolution of compressive strength with age for MgO-MWCNT/rGO-incorporated (**a**) binary and (**b**) ternary AAEC mixes with reagents 1 and 2.

For the MgO-MWCNT-incorporated mixes, the compressive strength was increased with the use of 0.3% and 0.6% MWCNTs, respectively, by 52.4~54.3% and 38.6~40.4% at 7 days; 30.5~36.8% and 23.5~28.6% at 14 days; 24.8~30.1% and 20~23.8% at 28 days; 19.1~22.5% and 14.6~17.3% at 56 days; 16.9~19.6% and 12.8~15.0% at 180 days (Table 4). For the MgO-rGO-incorporated mixes, the compressive strength was increased with the use of 0.3% and 0.6% rGO, respectively, by 24.2~25.2% and 21.7~22.2% at 7 days; 35.1~42.8% and 25.3~31.5% at 14 days; 28.2~34.6% and 20.8~26% at 28 days; 22~26.2% and 15.8~19.2% at 56 days; 19.6~22.9% and 13.9~16.6% at 180 days. The MgO-MWCNT/rGO combination significantly increased the compressive strength development, showing a higher % increase in strength at all ages for the AAECs. Moreover, the negative influence of higher dosages (>0.3%) of MWCNTs was overcome by using a 5% MgO-0.3%/0.6% MWCNT combination, changing the decrease in compressive strength to an increase.

Figure 11a–c shows the influence of the MWCNT, rGO, and MgO content on the 28-day compressive strength of AAECs. The increase in MgO content from 0 to 5% increased the 28-day compressive strength of the AAECs (Figure 11a). On the other hand, the 28-day compressive strength increased with the increase in MWCNT or rGO content to 0.3% then showed a decreasing trend up to 0.6% (Figure 11b,c). Therefore, it can be concluded that 0.3% is an optimum dose for MWCNTs/rGO to achieve better compressive strength.

These results indicate that the compressive strength was decreased by adding a higher amount (>0.3% wt) of MWCNTs/rGO, although this effect can be compensated/reversed by using MgO-MWCNT/rGO combinations. Previous studies by Rovnaník et al. [40] also confirmed the compressive strength enhancement of fly ash-based geopolymer composites with the addition of low dosages (0.2 wt%) of MWCNTs. An increase in the flexural strength, modulus of elasticity, and toughness was also observed with the use of 0.35% of rGO as a nano-filler in geopolymer composites [17]. This was attributed to the strong dispersion of rGO sheets between fly ash particles due to their malleable property causing them to fill voids and produce a denser matrix. A lower dosage of MWCNTs/rGO can be uniformly dispersed in the matrix, which in turn will improve the compressive strength, as observed in [17]. Although an appropriate amount of rGO can be evenly dispersed in the matrix, overlap and agglomeration may still occur in excess, limiting the development of compressive strength. Furthermore, the rGO with a wrinkled texture cross-linked around geopolymer particles altered the morphology and reduced the matrix's porosity, which positively impacted its fracture toughness [17]. Also, a smaller-sized MWCNT resulted in a higher mortar compressive strength, as smaller MWCNTs were distributed at a much finer scale, filling nanopore space within the matrix more efficiently [40].

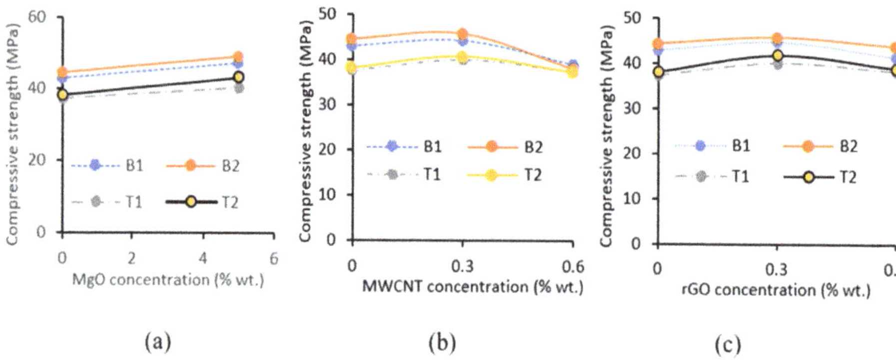

Figure 11. (**a**–**c**) Influence of (**a**) MgO, (**b**) MWCNT, and (**c**) rGO on 28-day compressive strength of AAEGC mixes with reagents 1 and 2.

3.2.2. Influence of MgO, MWCNT, and rGO on 28-Day UPV

A UPV test was conducted to determine the quality of concrete, including detecting the presence of cavities, joints, and cracks in addition to other properties over time. A high UPV indicates a lower number of pores, that the particles are closer to each other, a higher matrix homogeneity, and a lower wave travel time. Table 3 summarizes the 28-day UPV values, which ranged between 3067 m/s and 3717 m/s, between 3762 m/s and 3954 m/s, between 3436 m/s and 4068 m/s, and between 3122 m/s and 3943 m/s for control and the MgO-, MWCNT-, and rGO-incorporated AAECs, respectively. Higher UPV values were obtained for the AAEC mixes containing MgO, MWCNTs, and rGO than their control counterparts (Table 3 and Figure 12a–c).

Figure 12a shows that, by adding 5% MgO in the AAECs, the UPV values were increased from 6.4% (T2M5) to 27% (B2M5). The reason for this improvement was due to additional reactions and extra C–A–S–H formation, which densified the matrix and reduced the time for wave transmission. In addition, the formation of $Mg(OH)_2$ crystals produced uniform expansion and refined the pore size and pore structure, which also contributed to the higher UPV and compressive strength, as discussed earlier. These results supported the findings of previous studies by other researchers [22,82].

Figure 12b,c shows that adding 0.3% MWCNT/rGO increased the maximum UPV values up to 12% (for B2C3). This indicated CNTs'/rGO's filling capability, whichreduces the porosity to produce a denser matrix (as discussed earlier), which in turn led to stronger UPV and conductive properties. However, the addition of 0.6% MWCNTs/rGO did not produce a linear increase in UPV; rather, it tended to produce lower UPV values, indicating that 0.3% MWCNT/rGO is the optimum concentration (as observed in the case of compressive strength), with the exception of B2C6 (showing an increase of 19% in UPV). This can happen due to the improper dispersion of MWCNTs/rGO at higher concentrations beyond the optimum doses in the AAECs. This finding supports the results of a previous study by Asil and Ranjbar [83].

Figure 13 shows a linear correlation (dotted line) between the UPV and compressive strength of AAEC mixes at 28 days. The observed trend reveals an increase in compressive strength with an increase in UPV, as expected. The increase in compressive strength often corresponds to a denser and more homogeneous concrete matrix. A denser matrix transmits ultrasonic waves more efficiently, resulting in higher UPV values [84,85].

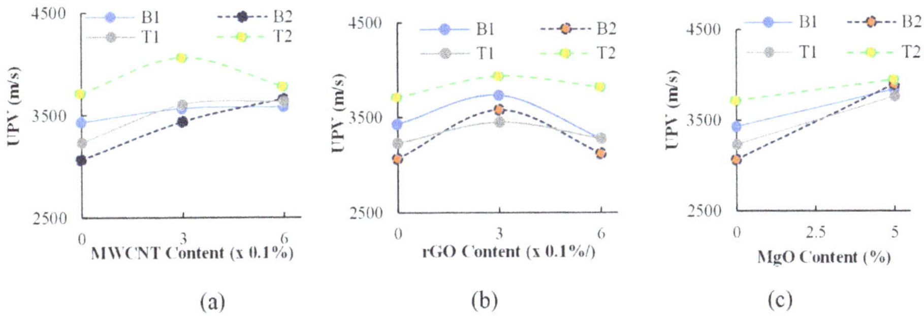

Figure 12. Influence of (**a**) MWCNTs, (**b**) rGO, and (**c**) MgO on 28-day UPV of AAEC mixes with reagents 1 and 2.

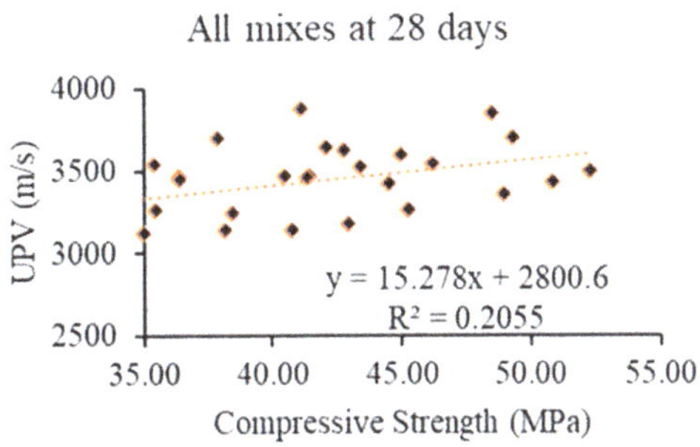

Figure 13. Correlation between (UPV) and compressive strength at 28 days.

It is important to note that, while Figure 13 provides a positive correlation, additional analysis considering influencing factors related to the mix design, MWCNT/rGO dispersion, and curing conditions needs to be conducted for a comprehensive understanding of the relationship between the UPV and compressive strength of AAECs. Nevertheless, the derived correlation suggests the potential use of UPV as a reliable non-destructive testing method for assessing the compressive strength of developed AAECs.

3.3. *Microstructural Characteristics*

The morphology and microstructural characteristics of the AAECs were studied by SEM analysis, while their reaction products and elemental compositions were determined and discussed using SEM/EDS, FTIR, and XRD analyses, which are detailed in the following subsections.

3.3.1. SEM/EDS Analyses

Figure 14a–d presents the morphology of the control AAECs (B1, B2, T1, and T2). As expected, the PVA fibers embedded in the AAEC matrix remained intact without any rupture or breaking in the failed cube specimens tested under compression. The morphology of the binary mixes appears to be denser, with a smaller number of unhydrated/partially hydrated fly ash particles than their ternary counterparts, as shown in Figure 14a,c. The fly ash particles are embedded in the matrix of ternary mixes, as indicated in Figure 14b,d. This can be attributed to the relatively lower reactivity of fly ash class F particles.

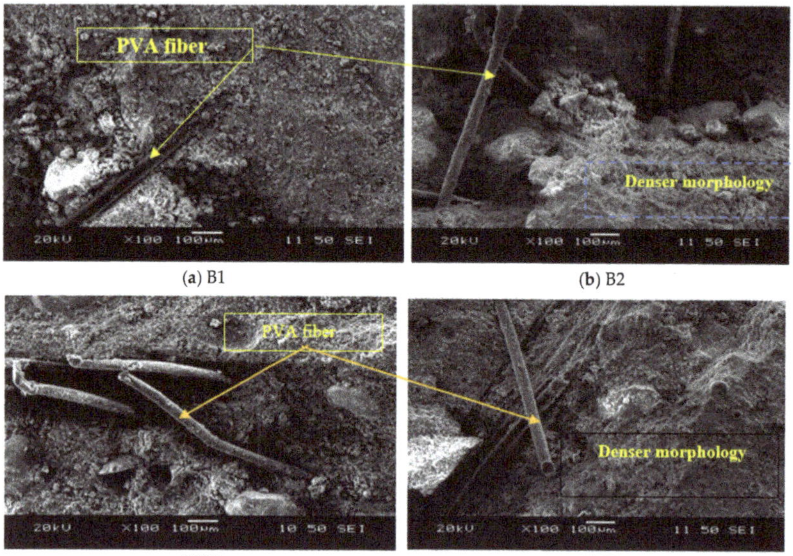

(**a**) B1 (**b**) B2 (**c**) T1 (**d**) T2

Figure 14. Morphology of the control mixes: (**a**) B1, (**b**) B2, (**c**) T1, (**d**) T2.

The MgO-added mix exhibited the highest compressive strength (as described in the previous section, Figure 11a) and showed a denser morphology than the control mixes. The MgO-added mixes appeared to be the most compact among all the mixes, with some of the particles acting as inert material, filling up the void spaces, as noted in Figure 15a–d.

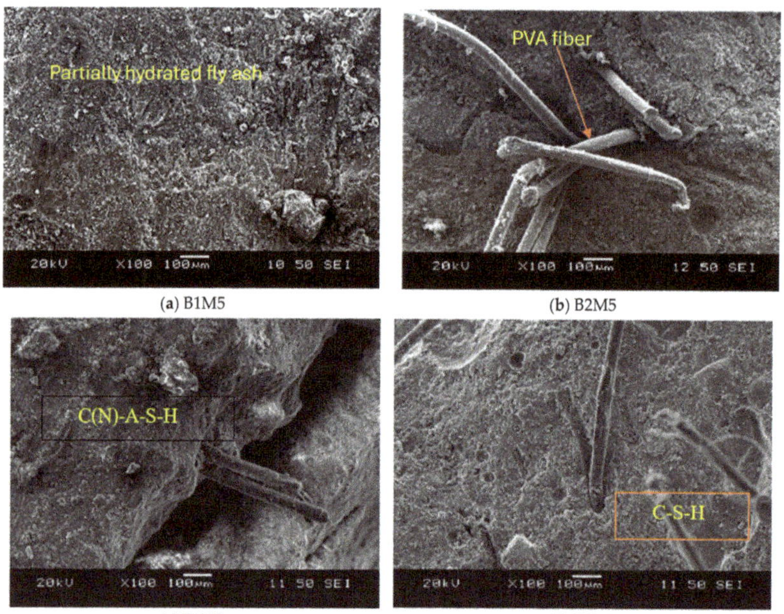

(**a**) B1M5 (**b**) B2M5 (**c**) T1M5 (**d**) T2M5

Figure 15. *Cont.*

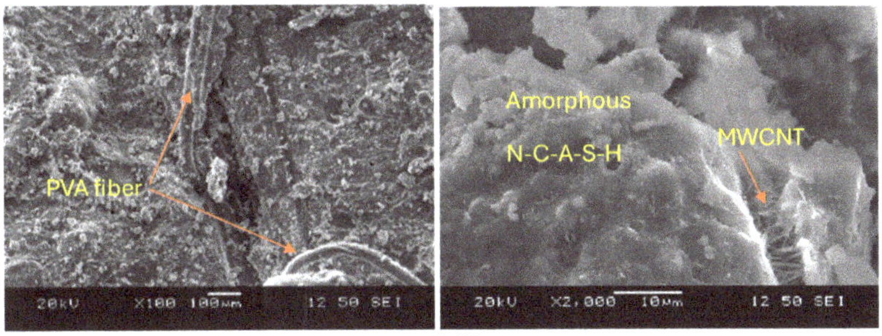

(**e**) B2C3

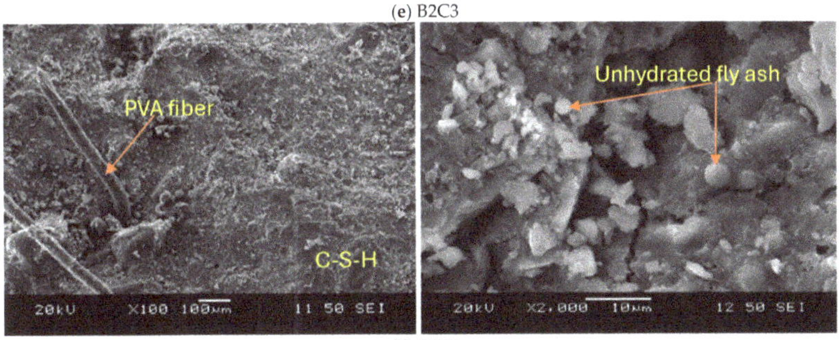

(**f**) B2C6

(**g**) B2R3

(**h**) B2R6

Figure 15. Morphology of (**a**) B1M5, (**b**) B2M5, (**c**) T1M5, (**d**) T2M5, (**e**) B2C3, (**f**) B2C6, (**g**) B2R3, and (**h**) B2R6.

Good dispersion can be seen in Figure 15e,f, and individual MWCNTs are distributed throughout the matrix uniformly. In addition to filling pores and voids, the MWCNTs bridge hollow spaces and micro cracks. MWCNT bridging was observed, as shown in Figure 15e, in a 2000k zoom picture from the SEM test. The crack-bridging effect of MWCNTs contributes to the deacceleration of the initiation and extension of cracks [35]. The mechanical properties of MWCNT-incorporated AAEC mixes are better (especially the compressive strength and UPV for 0.3% wt. MWCNT dosage, as discussed in the previous section, Figures 11b and 12a) than those of their control counterparts. The propagation of micro cracks and branching must pass the nanotubes or occur in areas with a weak matrix–MWCNT interaction bond state, which is essential to the macro properties of AAECs. The absence of a prominent interfacial layer between the MWCNTs and the geopolymer matrix, as can be inferred from the microstructure, demonstrates a healthy bond, indicating the chemical stability of MWCNTs under geopolymerization. Figure 15g,h shows the good morphology of the rGO-incorporated AAEC mixes (B2R3, B2R6) with uniform dispersion.

The main reaction product for binary mix B1 consisted of C-A-S-H and calcium-rich N-C-A-S-H acute phases with traces of MgO, as indicated in the SEM/EDS graphs presented in Figure 16a Their presence can be confirmed from the elements (Ca = 10.8%, Si = 8.3%, Al = 5.6%, C = 14.7%, Mg = 5.5%, Na = 1.5%, and O = 53.6%) noted in the EDS graph. The partially hydrated round fly ash class C and angular GGBFS particles can be observed in the SEM micrograph. The dominant reaction product for binary mix B2 consisted of C-A-S-H, as shown by the elements (Ca = 17.9%, Si = 7.7%, Al = 4.1%, C = 12.8%, O = 48.8%) with higher percentages in the EDS analysis presented in Figure 16b. An additional binding phase composed of C-S-H can be observed in the SEM micrograph because of the higher calcium content in the system (for B2 with reagent 2) than that in mix B1. The higher calcium content is attributed to the higher calcium content in reagent 2 ($Ca(OH)_2$:Na_2SO_4 = 2.5:1) compared to that in reagent 1 ($Ca(OH)_2$:$Na_2SiO_3 \cdot 5H_2O$ = 1:2.5). The higher calcium content in reagent 2 resulted mainly in the formation of C-A-S-H/C-S-H gels and comparatively less dissolution of silicate and aluminum ions, which are responsible for polymerization/alkali activation. As a result, the reagent 1 mixes developed lower strengths than their reagent 2 counterparts. The binary and ternary mixes developed C-A-S-H, N-C-A-SH, and C-S-H reaction product combinations depending on whether they incorporated reagent type 1 or 2. However, the compressive strength was higher for the binary mixes (B1, B2) than their ternary counterparts (T1 and T2) due to the higher fly ash C content, as described earlier.

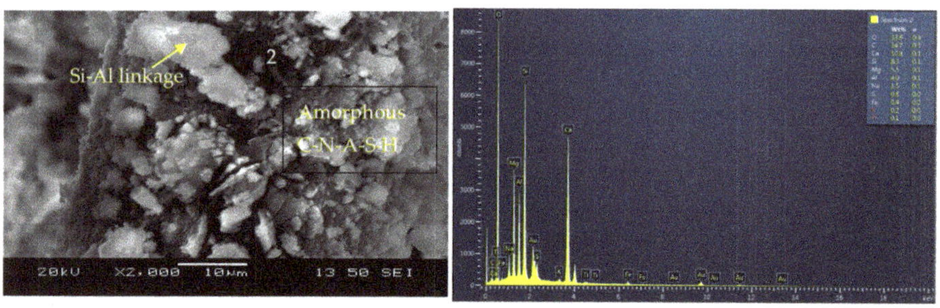

(**a**) B1

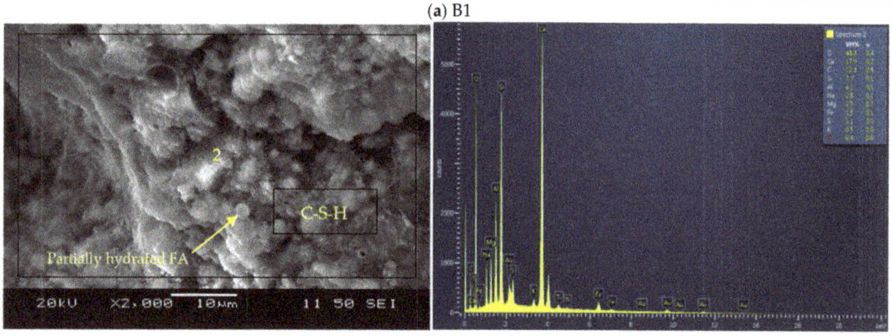

(**b**) B2

Figure 16. *Cont.*

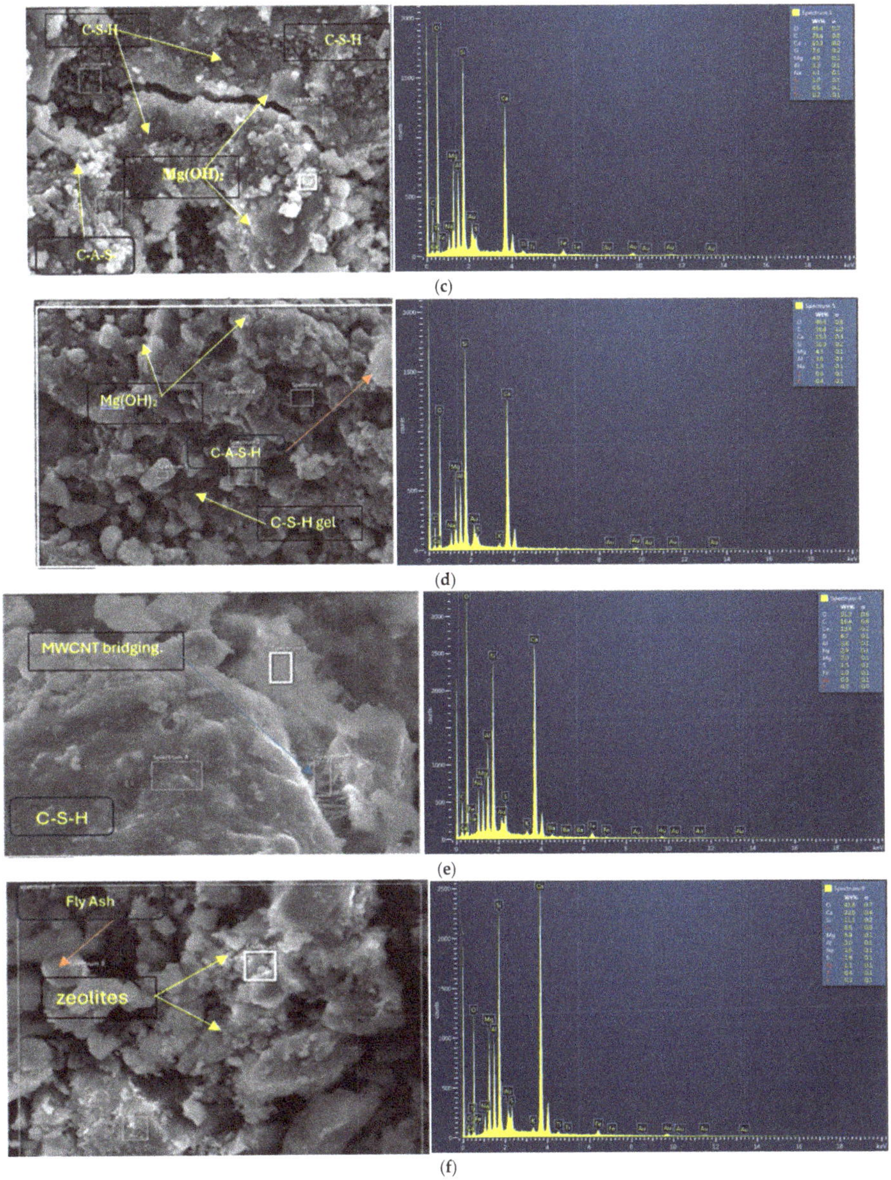

Figure 16. (**a**,**b**) SEM micrographs and EDS analysis: (**a**) B1 and (**b**) B2 [61]. (**c**) SEM micrographs and EDS analysis of the mix B1M5. (**d**) SEM micrographs and EDS analysis of the mix B2M5. (**e**) SEM micrographs and EDS analysis of the mix B2C3. (**f**) SEM micrographs and EDS analysis of the mix B2R3.

The representative SEM and EDS images of AAECs with varying levels of MgO, MWCNTs, rGO, and reagents, as shown in Figure 16c–f, exhibited a denser microstructure and almost fully reacted matrices with some unreacted chemosphere and pedosphere FA particles. It was reported that the unreacted particles do not act as filler in the mixture but lead to an increase in the strength of the matrix with age [86].

Similar reaction products and combinations (C-A-S-H, N-C-A-S-H, and C-S-H) are developed in GO-incorporated binary/ternary mixes depending on the reagent types, with the additional formation of $Mg(OH)_2$ with traces of MgO, which can be identified from the EDS and XRD analyses shown in Figure 16c,d. The presence of these dominant reaction products/compounds can be confirmed from the elements in the binary B1M5 (Ca = 10.3%, Si = 7.6%, Al = 3.3%, Mg = 4.0%, C = 23.4%, and O = 48.4%; Figure 16c) and in binary B2M5 (Ca = 15.3%, Si = 10.9%, Al = 3.6%, Mg = 4.5%, C = 16.8%, and O = 46.6%; Figure 16d) mixes, as noted in the EDS graphs. Partially hydrated round fly ash class C and angular GGBFS particles can also be observed in the SEM micrograph. Compared with the control mixes, AAEC mixes with 5% (wt%) MgO contain almost identical CaO/SiO_2 but a bit higher silica and alumina contents. Moreover, the addition of fine MgO results in a higher surface area, which eventually leads to a higher water absorption on the particle surface, leaving a less porous and denser matrix. These primary factors contributed towards an increase in compressive strength up to 25.2% MPa due to the introduction of MgO as an additive to the matrix. Moreover, the resulting denser matrix with lesser porosity was also validated by the higher UPV values of the MgO-incorporated mixes, as discussed earlier. The SEM results indicated that the hydration of MgO particles, when they are combined with water, forms $Mg(OH)_2$ crystals, which can produce self-expansive stresses on the walls of the concrete pores. Excessive nano-MgO additives will result in higher expansive stresses that push on the pore walls. Once the pores cannot locally restrain the stresses, macro cracks will occur in the matrix, limiting the dosage of MgO [87].

The presence of reaction compounds, as mentioned earlier, in the MWCNT-incorporated mixes (such as B2C3) above can be confirmed from the elements (Ca = 13.6%, Si = 7.66%, Al = 3.8%, Mg = 2.0%, C = 16.4%, and O = 51.3%) noted in the EDS graph. Figure 16e also shows some embedded MWCNT bridging in the crack area, which means that they did not react with the source materials or act in the polymerization process. The presence of MWCNTs in some areas of crack openings of the fractured surface bridging micro cracks, with the existence of good bonding between the 0.3% MWCNTs and the surrounding matrix, can be noted from the SEM images of the fractured surface of B1C3 mixes with pull-out of the MWCNTs. Such well-dispersed MWCNT bridging with good MWCNT-matrix bonding can also be attributed to the compressive strength increase (in 0.3% MWCNT mix), as well as the reaction products. This argument supports previous research [40]. As the geopolymerization proceeds, fly ash grains are dissolved in an alkaline environment, and the products diffuse into a pore solution with dispersed MWCNTs. The dissolved species then undergo polycondensation reactions to form amorphous geopolymer gel in which the nanotubes remain embedded. Some of the nanotubes are partially pulled out of the matrix during the breaking/fracture process, even though they are well bonded to the aluminosilicate matrix via hydrogen bonds of carboxymethyl cellulose [40]. MWCNTs can be distinguished in some areas in crack openings of the fractured surface in Figure 16e,f. A good bonding between the 0.3% MWCNTs and the surrounding AAEC is shown obviously in the SEM images of the fractured surface of B1C3 mixes. The images indicate the pull-out of MWCNTs and show that many MWCNTs are bridging the micro crack of the AAECs. The compressive strength increased with an increasing MWCNT concentration up to 0.3 wt%, and then subsequently dropped to a strength close to that of the cube specimens at 0.6 wt%. This indicates that a percolation threshold was achieved at 0.3 wt%, where there was no significant increase in the strength at MWCNT contents above 0.5 wt% due to agglomeration of nanotubes. The increase in strength was achieved by the cube specimens containing 0.3 wt% and 0.6% MWCNTs, with a 9.2% increase and 10.8% decrease for the B2C3 and B2C6 mixes, respectively.

The presence of the reaction compounds, as mentioned earlier, can be confirmed from the elements (Ca = 22.6%, Si = 11.1%, Al = 5.0%, Mg = 2.0%, C = 8.6%, and O = 41.8%) noted in the EDS graph (Figure 16f) for the rGO-incorporated AAEC mixes (such as B2R3). rGO reduces the porosity of the matrix by accelerating the growth of geopolymer gel and the filling of nanoscale pores with self-adsorbed zeolites ($Na_2Al_2Si_3O_8{\cdot}2H_2O$) and other materials, as shown in Figure 16f. The process of forming zeolites within the geopolymer involves the absorption of energy [88], and factors including the activation energy and mass transfer rate influence its rate. Graphene oxide (GO) is a material with high electron activity that attracts free ions in the gel and speeds up the mass transfer rate during crystal growth [16]. When GO comes in contact with the crystal nucleus, it is likely to aid in the growth of the zeolite-like phase and promote the formation of a monocrystalline layer [89]. This is also a contributing factor for the compressive strength increase of rGO-incorporated mixes, as observed in this study. Overall, as evident from the EDS analyses, the carbon content varied from 1% to 23%, showing especially higher values in the MWCNT- and rGO-incorporated AAECs, as expected.

In general, the mixes with reagent 2 exhibited a denser microstructure, as evident from the formation of crystalline C-A-S-H with additional C-S-H gels, than their counterparts with reagent 1 (developing C-A-S-H and comparatively amorphous N-C-A-S-H)d due to the higher calcium content in the mix compositions. This resulted in enhanced mechanical and durability characteristics in terms of the higher compressive strength, lower shrinkage/expansion, and better resistance to freeze–thaw cycles of mixes with reagent 2 compared to their counterparts with reagent 1 [1]. The formation of reaction products on the PVA fibers can be observed in Figure 15a–d. The primary reaction products/acute phases developed on the fibers consisted of C-A-S-H with si-al solid linkages. Similar observations were made in previous studies where the adhesion or chemical bonding of the reaction products with the fibers was improved by incorporating fly ash and slag in geopolymer composites [4].

3.3.2. XRD Analysis

The XRD patterns of the representative control binary/ternary mixes incorporating reagents 1 and 2 (B1, T1 and T2) are illustrated in Figure 17a,b. A mix of peaks and broad humps can be identified in the diffractograms for the binary and ternary mixes, signifying a combination of reaction products composed of crystalline and amorphous phases. For instance, the peak at 36.7° 2θ for all compositions consisted of multiple minerals (quartz, wadalite, gehlenite, hydrotalcite, and gypsum). The dominant crystalline phases of quartz (Q or E′) were seen for all the mix compositions, with the highest peak being seen at about 26.7° 2θ. The sharpest and highest intensity peak for the quartz phase was observed in the T2 diffractogram (indicating a dominant crystalline binding phase), which had more peaks than its counterparts (T1) and other binary (B1) mixes with reagent 1. Also, portlandite (calcium hydroxide) can be seen around 34.2° 2θ in the ternary mix T2 diffractogram because of the composition of reagent 2 and is responsible for forming additional binding phases. This peak characterization is consistent with the previous investigations on fly ash/slag binder [90]. The primary binding phases consisted of C-A-S-H and calcium-rich N-C-A-S-H for the binary mixes. In the ternary mixes, C-A-S-H and N-C-A-S-H were observed for the mixes with reagent 1, and a blend of N-A-S-H and N-C-A-S-H was identified with traces of C-S-H for the mixes with the reagent 2. The presence of these acute phases in the XRD analysis validated the SEM/EDS analysis of the binary and ternary mixes. Small reflections of ettringite were seen for ternary mix T2, with its maximum peak at 15.5° 2θ, as presented in Figure 18b. Traces of gypsum and periclase (MgO) were also determined in all compositions. The presence of gypsum prevented the flash setting of these mixes.

The MgO content in the mixes is known to reduce shrinkage and facilitate the self-healing behavior of cementitious composites due to its inherent expansive characteristics.

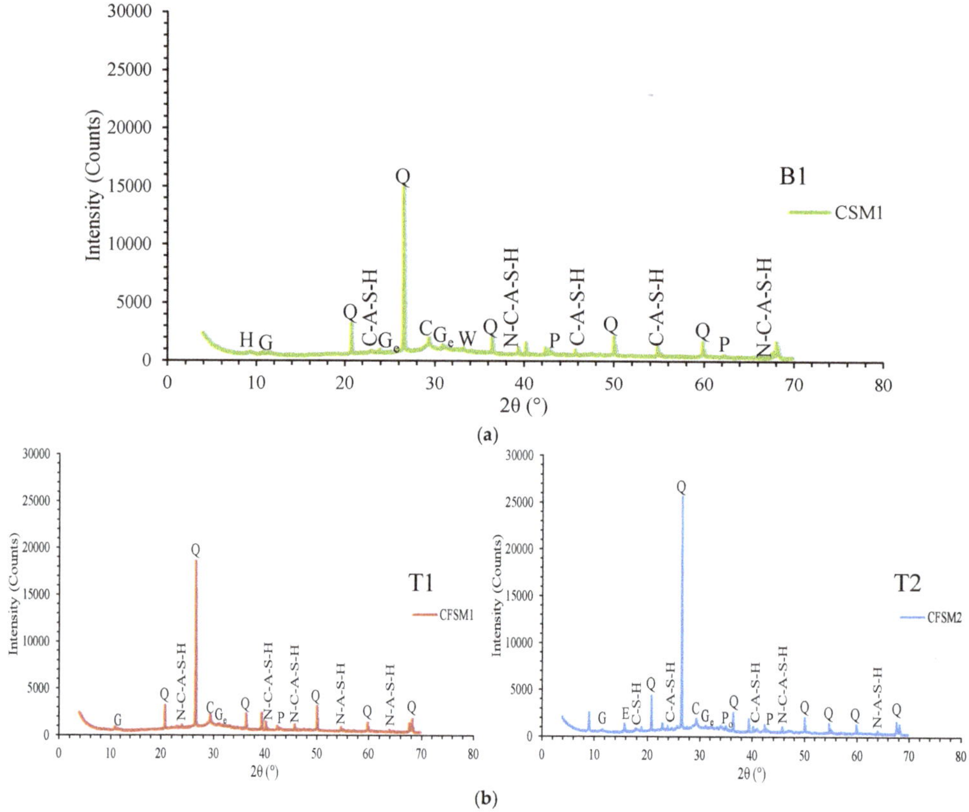

Figure 17. XRD diffractograms: (**a**) B1, (**b**) T1 and T2; (Q—quartz, G—gypsum, E—ettringite, G_e—gehlenite, P_o—portlandite, P—periclase, H—hydrotalcite, C—calcite, W—wadalite) [61].

Figure 18 shows the material compositions and phase compositions of the B2M5 and T2M5 mixes. The dominant crystalline phases of quartz (E′) were seen for all compositions. The highest peak, representing calcite, quartz, and periclase (p′), for B2M5 was identified at 29.54° 2θ, (Figure 18a). In contrast, quartz and periclase (MgO or P′) showed the highest peak for T2M5 at 27.11° and 29.49° 2θ (Figure 18b). Both mixes exhibited a dominant element phase consisting of oxygen, carbon, and silicon. They exhibited domination crestline products of calcium carbonate and silicon oxide, which enhanced the compressive strength of the MgO-incorporated AAECs (Figure 11a). The XRD analysis of T2M5 (Figure 18b) shows a 45% crestline product of tris boroxine and 25.5% silicon oxide, which help to enhance the compressive strength, and the peak count was identified at 27.11° and 29.49° 2θ. This peak characterization is consistent with the previous investigations on fly ash/slag mixes [91]. MgO impacted the XRD pattern of the AAECs, and the presence of MgO can lead to the formation of magnesium silicate phases such as forsterite (Mg_2SiO_4) and enstatite ($MgSiO_3$), which can be identified by their characteristic peaks in the XRD pattern. The traces of gypsum and periclase were also found in all compositions. The XRD pattern also confirms the previous SEM/EDS result shown in Figure 16c,d. It is expected that the presence of gypsum will prevent the flash setting and MgO is expected to reduce shrinkage and facilitate the self-healing behavior of the AAECs due to its inherent expansive characteristics, as observed in previous research studies on cementitious composites [22,92]. The XRD analysis (Figure 17) also revealed the formation of magnesium-aluminum hydrotalcite ($Mg_6Al_2(OH)_{16}CO_3{\cdot}4H_2O$)-like phases due to the reaction of MgO with the slag in addition to C-S-H, as confirmed from previous studies by Ben Haha et al. (2011) [55] on the effect of natural MgO on alkali-activated slag. The content of these hydrotalcite (Ht or L′)-like phases increases with the dosage of MgO and is more voluminous than that of C–S–H, which results in a higher strength (as observed in the current study) and the long-term stabilization of polymers, leading to excellent performance. In general, MgO hydrolyzed on the surface either reacts with the broken Si–O or Al–O to form magnesium silicate hydrate (M–S–H) or Ht, hindering the precipitation of brucite. These findings are also confirmed by early studies that found that Mg is quickly consumed to form Ht or M–S–H in combination with silica fume or slag [93,94], although M–S–H is hard to detect with XRD [95].

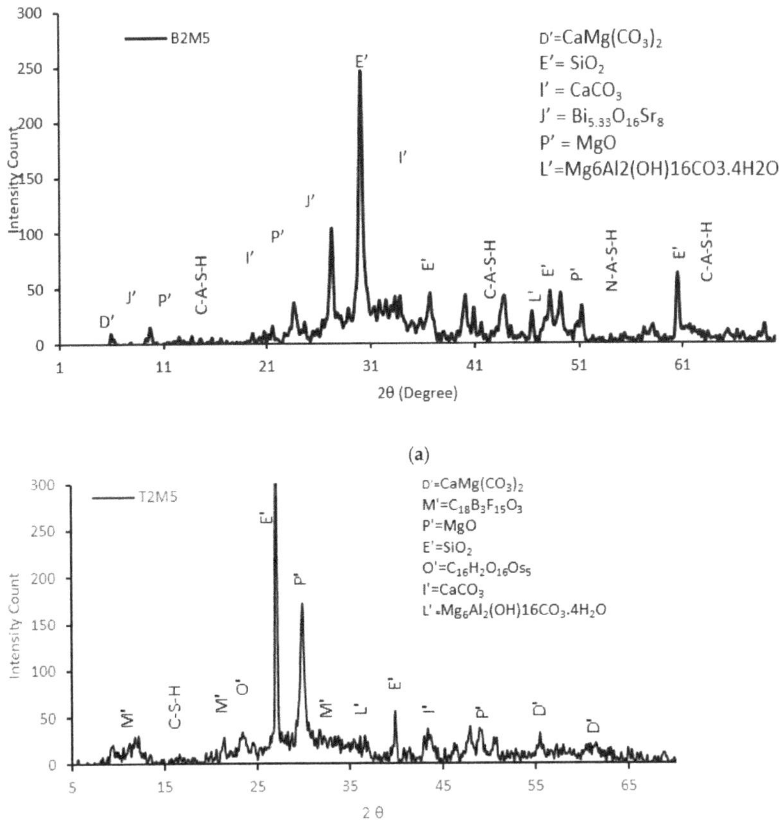

Figure 18. XRD diffractograms of (**a**) B2M5 and (**b**) T2M5.

For the MWCNT-incorporated AAEC mix B2C6, the highest peak representing quartz (E′) is identified at 26.43° 2θ (Figure 19), and the highest peak for quartz (E′) and penta barium tetra niobium oxide (V′) is identified at 26.43° and 29.43° 2θ, respectively, for the rGO-incorporated mix B2R6 (Figure 20). The dominant element phase consisting of oxygen, and silicon was observed for both mixes. In previous studies, XRD analysis demonstrated an increase in the amorphous phase, C-S-H, and geopolymerization with the addition of MWCNTs, exhibiting a dense matrix with a lower content of unreacted fly ash particles and an increase in the compressive strength (low dosages of 0.3% as observed) [41]. However, no influence of MWCNTs on the reaction products of AAECs has been observed via XRD even though MWCNTs can act as nucleation sites and accelerate polymerization, similar to the effects observed in previous studies [35,38,42]. This can be attributed to the minimal dosages of MWCNTs which cannot effectively be detected and the hindrance of deep and quantitative analyses due to the highly amorphous characteristics of geopolymer with coexisting terpolymeric gel (three-component monomer gel) and partially reacted or unreacted precursors. More advanced technologies are then required for further investigation.

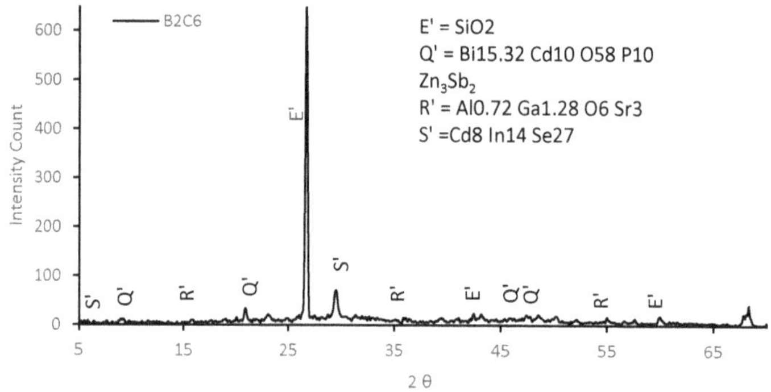

Figure 19. XRD analysis of MWCNT incorporated AAEC (B2C6).

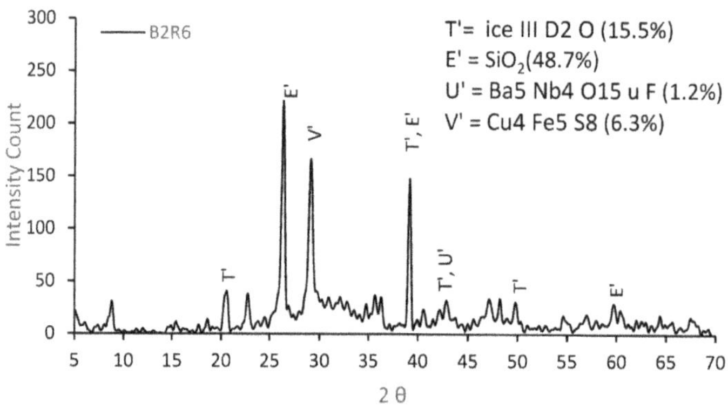

Figure 20. XRD analysis of rGO-incorporated AAEC (B2R6).

The introduction of rGO leads to an increase in the peak intensity of quartz (E′) and zeolites within the geopolymer, representing the development of crystalline phases. FA/GGBFS compositions predominantly comprise amorphous glassy microspheres and crystalline quartz, mullite, and hatrurite. Geopolymerization is the process of dissolving the glassy phase of source materials (FA or GGBFS) in an alkaline solution to form geopolymer gels. rGO acts as a catalyst to accelerate electron/mass transfer during alkali-activation reaction and promotes the dissolution of amorphous phases of FA/GGBFS microspheres by transmitting free electrons. rGO promotes the transformation of $[SiO_2(OH)_2]_2$ and $Al(OH)_4$ to the $AlSiO_4$ or $AlSi3O_8$ structure and accelerates the growth of geopolymer gels [96] to produce zeolite crystals with a higher diffraction intensity (Figure 20).

3.3.3. Fourier Transform Infrared Spectroscopy (FTIR) Analysis

In FTIR spectroscopy, a sample is exposed to a beam of infrared radiation that interacts with the chemical bonds within the material. FTIR analysis provided information on the formation of geopolymeric gels and the presence of specific functional groups such as Si-O and Al-O, as well as changes in the molecular structure of the material during curing and over time. The resulting spectrum shows the absorption or transmission of radiation at different frequencies, which corresponds to specific functional groups and chemical bonds within the material. The transmittance (%) in FTIR spectroscopy is influenced by the analyzed material's presence and the strength of different functional groups and chemical bonds within it. Different chemical bonds and functional groups absorb radiation at different frequencies, which can be identified by their characteristic absorption peaks in the FTIR spectrum. Overall, the relationship between the % of transmittance and the material composition in FTIR spectroscopy is based on the absorption or transmission of infrared radiation by specific chemical bonds and functional groups within the material, which is directly related to the molecular structure and chemical composition of the material. The % of transmittance is an important parameter in FTIR spectroscopy because it provides information about the sample's absorption properties. A sample that absorbs a large amount of radiation will have a low % of transmittance, while a sample that absorbs very little radiation will have a high % of transmittance.

Figure 21 shows the FTIR spectrum (% transmittance vs. wavelength) of the AAEC samples at the age of 180 days, which shows major bands at 3380 cm^{-1}, 2072 cm^{-1}, 1423 cm^{-1}, 957 cm^{-1}, 875 cm^{-1}, and 670 cm^{-1}. The broadband of the AAEC specimen at 3297 cm^{-1} and 2072 cm^{-1} characterizes the O-H stretching and H-O-H bending vibrations, respectively. Overall, AAECs with MWCNTs showed the lowest % of transmittance (highest radiation absorbing capacity) followed by those with rGO and MgO. B2C6 had a lower % of transmittance at 3380 cm^{-1}, which indicated a weaker bond headed by less compressive strength. These vibration modes are due to the weakly bound water molecules that were adsorbed on the surface or trapped in large cavities [88]. Water plays an important role in the process of polymerization as it is associated with the destruction of solid particles and the hydrolysis of dissolved Al^{3+} and Si^{4+} ions [97]. The absorption band at 1429 cm^{-1} is attributed to the presence of sodium carbonate (Na_2CO_3), which corresponds to the stretching vibration of the O-C-O bond [98] (Swanepoel and Strydom, 2002). Álvarez-Ayuso et al. [99] suggested that the presence of Na_2CO_3 is due to the atmospheric carbonation of alkaline media. The strongest band of the AAEC specimens at 958 cm^{-1}, representing Si-O-Si and Si-O-Al, corresponds to the asymmetric stretching vibrations of Si-O-T (T = Si or Al). Moreover, the stretching modes are sensitive to the Si-Al composition of the framework and a lower frequency is obtained with an increasing number of tetrahedral aluminum atoms [100]. Muzˇek et al. [97] suggested that the amorphous aluminosilicate gel formed at this phase is due to the depolymerization and structural reorganization of the amorphous phases in geopolymer materials. Mix B2C6 exhibited a lower % of transmission at 905 wavelengths. The bands in the region between 800 cm^{-1} and 550 cm^{-1} correlate with the tetrahedral vibrations of the secondary building unit (SBU) and fragments of the aluminosilicate [9]. The bands at 957 cm^{-1} observed in Figure 21 are assigned to quartz as the crystalline phase in the original FA [9], whereas the band located at 566 cm^{-1} indicates the presence of mullite. Referring to the finding by Mollah et al. [101], the spectral band at 566 cm^{-1} is attributed to the symmetric stretching of Al-O-Si. The band below 550 cm^{-1} is an indication of the degree of amorphization of the material, since its intensity does not depend on the degree of crystallization [102,103]. The symmetrical stretching vibrations of Si-O located between 687 cm^{-1} and 749, as well as between 1140 cm^{-1} and 1192 cm^{-1}, are linked to the existence of unreacted and/or partially reacted quartz [104]. The main band from 940 cm^{-1} to 970 cm^{-1}, ascribed to the asymmetric stretching vibration of Si-O-Si and Si-O-Al (Wang et al., 2020), confirms more intensive peaks and a higher formation of the reaction products of Si-O-Si and the Si-O-Al composition [105]. This peak shifted to a higher wavenumber, revealing the substitution of Al^{3+} with Si^{4+} species and the growth of more stable Si-O-Si bonds. According to Wan et al. [106], geopolymer strengthening is enhanced when Si tetrahedrals replace Al species in Si-rich gels.

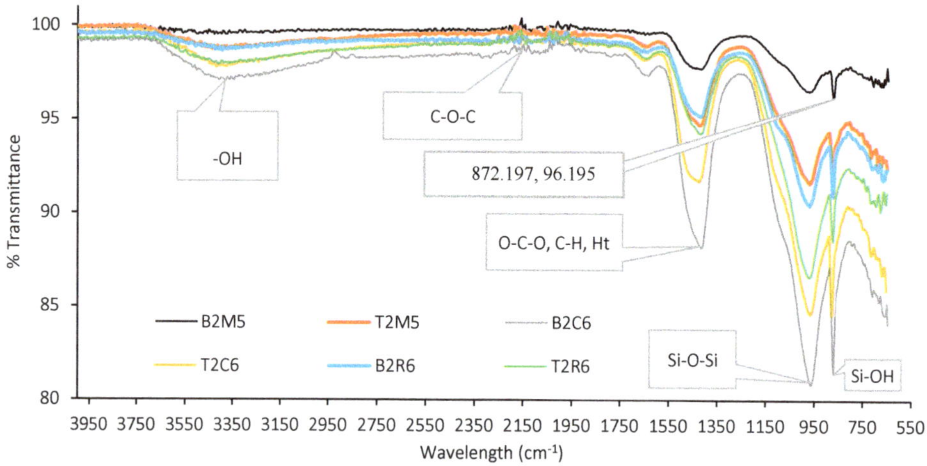

Figure 21. FTIR infrared spectrum of 180-day powder specimens.

4. Conclusions

This study evaluates the effects of MgO, MWCNT, and rGO on the physical characteristics related to the workability (slump flow, flow time, and flow velocity), compressive strength, UPV, and microstructural characteristics of developed AAECs that incorporate binary and ternary combinations of source materials and two types of reagents. Based on the findings, the following conclusions are drawn:

- The AAEC mixes satisfied the EFNARC [77] criteria for self-consolidating ability. The initial and final setting times ranged from 190 min to 311 min and from 230 min to 369 min, respectively;
- The ternary (fly ash C 'FA-C' + FA-F + ground granulated blast furnace slag 'GGBFS') mixes exhibited higher slump flows than their binary (FA-C + GGBFS) counterparts due to having a lower FA-C/GGBFS content, while the activator/reagent 2 (calcium hydroxide: sodium sulfate = 2.5:1) mixes resulted in higher slump flows due to having lower silica ratio than the mixes that used reagent 1 (calcium hydroxide: sodium metasilicate = 1:2.5);
- The slump flow of the AAECs decreased with an increase in the MWCNT/rGO content, with a few exceptions depending on mix/reagent types, while the addition of MgO showed a less significant decrease. The higher surface area or water affinity of MgO/MWCNTs and higher agglomeration tendency of MWCNTs/rGO were considered contributing factors;
- All AAECs achieved a 28-day compressive strength ranging from 26.0 MPa to 48.5 MPa > 18 MPa and satisfied the criteria for structural concrete. Control binary AAECs obtained higher 28-day compressive strengths than their ternary counterparts, due to the dominant formation of C-A-S-H/C-S-H compared to the dominant formation of amorphous N-C-A-S-H/N-A-S-H in their ternary counterparts. Binary mix B2

with reagent 2 exhibited a 22% higher 28-day compressive strength than its ternary counterpart T2 due to having higher FA-C and GGBFS contents;
- The AAECs with reagent 2 generally produced a higher compressive strength compared to their reagent 1 counterparts due to the formation of additional C-S-H gel (in addition to C-A-S-H/N-C-A-S-H or C-A-S-H/C-S-H) and more compact microstructures, as per XRD-SEM analysis, which showed the presence of sharper crystalline peaks of quartz and calcite;
- The addition of 5% MgO increased the 28-day compressive strength (by 19~25%), and UPV exhibited potential self-healing capability due to the formation of well-dispersed fine $Mg(OH)_2$, hydrotalcite, and M-S-H, which compensated for volume shrinkage and refined the matrix pore size, as confirmed from XRD-SEM/EDS;
- The 28-day compressive strength increased with the increase in MWCNT/rGO content up to 0.3%, indicating it as an optimum dosage. The decrease in strength at higher dosages (>0.3% wt) of MWCNTs/rGO can be compensated for/reversed by using MgO-MWCNT/rGO combinations that show beneficial effects of MgO addition;
- The addition of 0.3% MWCNTs and rGO increased the compressive strength by 9% and 14.8%, respectively, showing a higher effectiveness of rGO in improving strength. The increase in compressive strength and UPV was attributed to the uniform dispersion and good MWCNT-matrix bond bridging cracks (with no newly formed MWCNT-induced reaction products), while the formation of additional zeolites caused matrix densification with the addition of rGO, as confirmed from the SEM/EDS analysis. Proper dispersion of the MWCNTs/rGO is important to avoid agglomeration and negative effects on the workability, mechanical, and micro-structural properties;
- Fourier transform infrared spectrometer (FTIR) analysis suggested the formation of an aluminosilicate network in AAECs with a higher concentration of Si-O-Si bonds, indicating a more stable structure. A lower % of transmittance at 3380 cm^{-1} in binary AAECs with reagent 2 and 0.6% MWCNTs (B2C6), compared to their counterpart with 5% MgO (B2M5), confirmed a weaker bond and lower compressive strength even though MWCNTs can act as nucleation sites and accelerate polymerization;
- The UPV values ranged from 3067 m/s to 4068 m/s, showing an increase with the addition MWCNTs/rGO/MgO, with the highest values (19% higher) being observed for MWCNT-incorporated AAECs. This is an indication of a higher capacity to transmit ultrasonic waves, better conductivity, and a better self-sensing ability. A linear correlation revealed an increase in compressive strength with an increase in the UPV content of AAEC mixes at 28 days;
- This research demonstrated the viability of producing ambient-cured powder-based AAECs, as more green and sustainable alternatives to conventional ECCs, that incorporate carbon-nano materials and MgO additives, as they have satisfactory workability, mechanical, and microstructural characteristics. The developed multi-functional AAECs have the potential to be used for structural applications and to produce high-performance resilient durable bridge-building infrastructures with self-healing/sensing and 3-D printing potentials.

Author Contributions: Conceptualization, K.M.A.H. and M.A.H.; Methodology, K.M.A.H. and M.A.H.; Formal analysis, K.M.A.H. and M.A.H.; Investigation, K.M.A.H. and M.A.H.; Resources, K.M.A.H.; Writing—original draft, K.M.A.H. and M.A.H.; Writing—review & editing, K.M.A.H.; Supervision, K.M.A.H.; Project administration, K.M.A.H.; Funding acquisition, K.M.A.H. All authors have read and agreed to the published version of the manuscript.

Funding: This research was funded by Natural Sciences and Engineering Research Council (NSERC) Canada, grant number RGPIN-5613-2019.

Institutional Review Board Statement: Not applicable.

Informed Consent Statement: Not applicable.

Data Availability Statement: The data presented in this study are available on request from the corresponding author.

Acknowledgments: The authors gratefully acknowledge the financial support provided by NSERC, Canada. Supports provided by the technical staff of Concrete and Advanced Concrete Material laboratories of Toronto Metropolitan University are also acknowledged.

Conflicts of Interest: The authors declare no conflicts of interest.

References

1. Hossain, K.M.A.; Sood, D. The Strength and Fracture Characteristics of One-Part Strain-Hardening Green Alkali-Activated Engineered Composites. *Materials* **2023**, *16*, 5077. [CrossRef]
2. Huseien, G.F.; Shah, K.W. Durability and life cycle evaluation of self-compacting concrete containing fly ash as GBFS replacement with alkali activation. *Constr. Build. Mater.* **2020**, *235*, 117458. [CrossRef]
3. Naqi, A.; Jang, J.G. Recent Progress in Green Cement Technology Utilizing Low-Carbon Emission Fuels and Raw Materials: A Review. *Sustainability* **2019**, *11*, 537. [CrossRef]
4. Nematollahi, B.; Sanjayan, J.; Shaikh, F.U.A. Tensile Strain Hardening Behavior of PVA Fiber-Reinforced Engineered Geopolymer Composite. *J. Mater. Civ. Eng.* **2015**, *27*, 0001242. [CrossRef]
5. Samantasinghar, S.; Singh, S. Effects of curing environment on strength and microstructure of alkali-activated fly ash-slag binder. *Constr. Build. Mater.* **2020**, *235*, 117481. [CrossRef]
6. Shah, S.F.A.; Chen, B.; Oderji, S.Y.; Haque, M.A.; Ahmad, M.R. Improvement of early strength of fly ash-slag based one-part alkali activated mortar. *Constr. Build. Mater.* **2020**, *246*, 118533. [CrossRef]
7. Davidovits, P.J. 30 Years of Successes and Failures in Geopolymer Applications. In Proceedings of the Geopolymer 2002 Conference, Melbourne, VIC, Australia, 28–29 October 2002; pp. 1–16.
8. Huseien, G.F.; Shah, K.W. Performance evaluation of alkali-activated mortars containing industrial wastes as surface repair materials. *J. Build. Eng.* **2020**, *30*, 101234. [CrossRef]
9. Fernández-Jiménez, A.; Palomo, A. Mid-infrared spectroscopic studies of alkali-activated fly ash structure. *Microporous Mesoporous Mater.* **2005**, *86*, 207–214. [CrossRef]
10. Xu, J.; Wang, X. Self-healing of concrete cracks by use of bacteria-containing low alkali cementitious material. *Constr. Build. Mater.* **2018**, *167*, 1–14. [CrossRef]
11. Mehta, A.; Siddique, R.; Ozbakkaloglu, T.; Uddin Ahmed Shaikh, F.; Belarbi, R. Fly ash and ground granulated blast furnace slag-based alkali-activated concrete: Mechanical, transport and microstructural properties. *Constr. Build. Mater.* **2020**, *257*, 119548. [CrossRef]
12. Khan, M. Evidence on the functional form of inflation and output growth variability relationship in European economies. *Int. Econ.* **2016**, *146*, 119548. [CrossRef]
13. Khan, M.Z.N.; Shaikh, F.; Uddin, A.; Hao, Y.; Hao, H. Synthesis of high strength ambient cured geopolymer composite by using low calcium fly ash. *Constr. Build. Mater.* **2016**, *125*, 809–820. [CrossRef]
14. Herbert, S.J.; Sakthieswaran, N.; Shiny, B.G. Review on geopolymer concrete with different additives. *Int. J. Eng. Res.* **2015**, *1*, 21–31.
15. Shah, S.P.; Konsta-Gdoutos, M.S.; Metaxa, Z.S.; Mondal, P. Nanoscale Modification of Cementitious Materials. In *Nanotechnology in Construction 3: Proceedings of the NICOM3*; Springer: Berlin, Germany, 2009; pp. 125–130. [CrossRef]
16. Guo, X.; Shi, H.; Dick, W.A. Compressive strength and microstructural characteristics of class C fly ash geopolymer. *Cem. Concr. Compos.* **2010**, *32*, 142–147. [CrossRef]
17. Saafi, M.; Tang, L.; Fung, J.; Rahman, M.; Liggat, J. Enhanced properties of graphene/fly ash geopolymeric composite cement. *Cem. Concr. Res.* **2015**, *67*, 292–299. [CrossRef]
18. Zhang, C.; Khorshidi, H.; Najafi, E.; Ghasemi, M. Fresh, mechanical and microstructural properties of alkali-activated composites incorporating nanomaterials: A comprehensive review. *J. Clean. Prod.* **2023**, *384*, 135390. [CrossRef]
19. Abdollahnejad, Z.; Dalvand, A.; Mastali, M.; Luukkonen, T.; Illikainen, M. Effects of waste ground glass and lime on the crystallinity and strength of geopolymers. *Mag. Concr. Res.* **2019**, *71*, 1218–1231. [CrossRef]
20. Mastali, M.; Kinnunen, P.; Dalvand, A.; Mohammadi Firouz, R.; Illikainen, M. Drying shrinkage in alkali-activated binders–A critical review. *Constr. Build. Mater.* **2018**, *190*, 533–550. [CrossRef]
21. Kheradmand, M.; Abdollahnejad, Z.; Pacheco-Torgal, F. Shrinkage Performance of Fly Ash Alkali-activated Cement Based Binder Mortars. *KSCE J. Civ. Eng.* **2018**, *22*, 1854–1864. [CrossRef]

22. Sherir, M.A.A.; Hossain, K.M.A.; Lachemi, M. Self-healing and expansion characteristics of cementitious composites with high volume fly ash and MgO-type expansive agent. *Constr. Build. Mater.* **2016**, *127*, 80–92. [CrossRef]
23. MacKenzie, K.J.D.; Bolton, M.J. Electrical and mechanical properties of aluminosilicate inorganic polymer composites with carbon nanotubes. *J. Mater. Sci.* **2009**, *44*, 2851–2857. [CrossRef]
24. Xu, J.; Zhang, D. Multifunctional structural supercapacitor based on graphene and geopolymer. *Electrochim. Acta* **2017**, *224*, 105–112. [CrossRef]
25. Wang, S.; Li, V. Engineered Cementitious Composites with High-volume Fly Ash. *ACI Mat. J.* **2007**, *104*, 233–241.
26. Bhutta, A.; Borges, P.H.R.; Zanotti, C.; Farooq, M.; Banthia, N. Flexural behavior of geopolymer composites reinforced with steel and polypropylene macro fibers. *Cem. Concr. Compos.* **2017**, *80*, 31–40. [CrossRef]
27. Nematollahi, B.; Sanjayan, J.; Shaikh, F.U.A. Matrix design of strain hardening fiber reinforced engineered geopolymer composite. *Compos. Part B Eng.* **2016**, *89*, 253–265. [CrossRef]
28. Nguyen, H.; Carvelli, V.; Adesanya, E.; Kinnunen, P.; Illikainen, M. High performance cementitious composite from alkali-activated ladle slag reinforced with polypropylene fibers. *Cem. Concr. Compos.* **2018**, *90*, 150–160. [CrossRef]
29. Ranjbar, N.; Mehrali, M.; Mehrali, M.; Alengaram, U.J.; Jumaat, M.Z. High tensile strength fly ash based geopolymer composite using copper coated micro steel fiber. *Constr. Build. Mater.* **2016**, *112*, 629–638. [CrossRef]
30. Shaikh, F.U.A. Deflection hardening behaviour of short fibre reinforced fly ash based geopolymer composites. *Mater. Des.* **2013**, *50*, 674–682. [CrossRef]
31. Yu, J.; Chen, Y.; Leung, C.K.Y. Mechanical performance of Strain-Hardening Cementitious Composites (SHCC) with hybrid polyvinyl alcohol and steel fibers. *Compos. Struct.* **2019**, *226*, 111198. [CrossRef]
32. Alrefaei, Y.; Dai, J.-G. Tensile behavior and microstructure of hybrid fiber ambient cured one-part engineered geopolymer composites. *Constr. Build. Mater.* **2018**, *184*, 419–431. [CrossRef]
33. Bahari, A.; Sadeghi-Nik, A.; Roodbari, M.; Sadeghi-Nik, A.; Mirshafiei, E. Experimental and theoretical studies of ordinary Portland cement composites contains nano LSCO perovskite with Fokker-Planck and chemical reaction equations. *Constr. Build. Mater.* **2018**, *163*, 247–255. [CrossRef]
34. Sadeghi-Nik, A.; Berenjian, J.; Bahari, A.; Safaei, A.S.; Dehestani, M. Modification of microstructure and mechanical properties of cement by nanoparticles through a sustainable development approach. *Constr. Build. Mater.* **2017**, *155*, 880–891. [CrossRef]
35. Abbasi, S.M.; Ahmadi, H.; Khalaj, G.; Ghasemi, B. Microstructure and mechanical properties of a metakaolinite-based geopolymer nanocomposite reinforced with carbon nanotubes. *Ceram. Int.* **2016**, *42*, 15171–15176. [CrossRef]
36. Nazar, S.; Yang, J.; Khan, M.S.; Quazi, M.F.; Li, H.; Rahim, A. Investigating the effects of MWCNTs on sustainable alkali-activated mortar: Physical, rheological, and mechanical analysis. *Mater. Today Sustain.* **2023**, *24*, 100519. [CrossRef]
37. Rovnaník, P.; Šimonová, H.; Topolář, L.; Bayer, P.; Schmid, P.; Keršner, Z. Carbon nanotube reinforced alkali-activated slag mortars. *Constr. Build. Mater.* **2016**, *119*, 223–229. [CrossRef]
38. Khater, H.M.; Abd El Gawaad, H.A. Characterization of alkali activated geopolymer mortar doped with MWCNT. *Constr. Build. Mater.* **2016**, *102*, 329–337. [CrossRef]
39. Li, F.; Yang, Z.; Zheng, A.; Li, S. Properties of modified engineered geopolymer composites incorporating multi-walled carbon Nanotubes(MWCNTs) and granulated blast furnace Slag(GBFS). *Ceram. Int.* **2021**, *47*, 14244–14259. [CrossRef]
40. Rovnaník, P.; Šimonová, H.; Topolář, L.; Schmid, P.; Keršner, Z. Effect of Carbon Nanotubes on the Mechanical Fracture Properties of Fly Ash Geopolymer. *Procedia Eng.* **2016**, *151*, 321–328. [CrossRef]
41. Jittabut, P.; Horpibulsuk, S. Physical and Microstructure Properties of Geopolymer Nanocomposite Reinforced with Carbon Nanotubes. *Mater. Today: Proc.* **2019**, *17*, 1682–1692. [CrossRef]
42. Yuan, J.; He, P.; Jia, D.; Fu, S.; Zhang, Y.; Liu, X.; Cai, D.; Yang, Z.; Duan, X.; Wang, S.; et al. In situ processing of MWCNTs/leucite composites through geopolymer precursor. *J. Eur. Ceram. Soc.* **2017**, *37*, 2219–2226. [CrossRef]
43. Hilding, J.; Grulke, E.A.; George Zhang, Z.; Lockwood, F. Dispersion of Carbon Nanotubes in Liquids. *J. Dispers. Sci. Technol.* **2003**, *24*, 1–41. [CrossRef]
44. Schaefer, D.W.; Brown, J.M.; Anderson, D.P.; Zhao, J.; Chokalingam, K.; Tomlin, D.; Ilavsky, J. Structure and dispersion of carbon nanotubes. *J. Appl. Crystallogr.* **2003**, *36*, 553–557. [CrossRef]
45. Vaisman, L.; Wagner, H.D.; Marom, G. The role of surfactants in dispersion of carbon nanotubes. *Adv. Colloid Interface Sci.* **2006**, *128–130*, 37–46. [CrossRef] [PubMed]
46. Revathi, T.; Jeyalakshmi, R. Fly ash–GGBS geopolymer in boron environment: A study on rheology and microstructure by ATR FT-IR and MAS NMR. *Constr. Build. Mater.* **2021**, *267*, 120965. [CrossRef]
47. Zhou, G.-X.; Li, C.; Zhao, Z.; Qi, Y.-Z.; Yang, Z.-H.; Jia, D.-C.; Zhong, J.; Zhou, Y. 3D printing geopolymer nanocomposites: Graphene oxide size effects on a reactive matrix. *Carbon* **2020**, *164*, 215–223. [CrossRef]
48. Ranjbar, N.; Mehrali, M.; Mehrali, M.; Alengaram, U.J.; Jumaat, M.Z. Graphene nanoplatelet-fly ash based geopolymer composites. *Cem. Concr. Res.* **2015**, *76*, 222–231. [CrossRef]

49. Dong, Q.; Wan, L.; Luan, C.; Wang, J.; Du, P. Effect of Graphene Oxide on Properties of Alkali-Activated Slag. *Materials* **2021**, *14*, 6107. [CrossRef] [PubMed]
50. Matalkah, F.; Soroushian, P. Graphene nanoplatelet for enhancement the mechanical properties and durability characteristics of alkali activated binder. *Constr. Build. Mater.* **2020**, *249*, 118773. [CrossRef]
51. Xu, G.; Zhong, J.; Shi, X. Influence of graphene oxide in a chemically activated fly ash. *Fuel* **2018**, *226*, 644–657. [CrossRef]
52. Gao, P.; Lu, X.; Tang, M. Shrinkage and expansive strain of concrete with fly ash and expansive agent. *J. Wuhan Univ. Technol. Mater. Sci. Ed.* **2009**, *24*, 150–153. [CrossRef]
53. Li, Z.; Zhang, W.; Wang, R.; Chen, F.; Jia, X.; Cong, P. Effects of reactive MgO on the reaction process of geopolymer. *Materials* **2019**, *12*, 526. [CrossRef]
54. Abdel-Gawwad, H.A.; Abd El-Aleem, S. Effect of reactive magnesium oxide on properties of alkali-activated slag geopolymer cement pastes. *Res. Gate* **2015**, *59*, 37–47.
55. Haha, M.B.; Lothenbach, B.; Le Saout, G.; Winnefeld, F. Influence of slag chemistry on the hydration of alkali-activated blast-furnace slag—Part I: Effect of MgO. *Cem. Concr. Res.* **2011**, *41*, 955–963. [CrossRef]
56. Shen, W.; Wang, Y.; Zhang, T.; Zhou, M.; Li, J.; Cui, X. Magnesia modification of alkali-activated slag fly ash cement. *J. Wuhan Univ. Technol. Mater. Sci. Ed.* **2011**, *26*, 121–125. [CrossRef]
57. Jin, F.; Gu, K.; Al-Tabbaa, A. Strength and drying shrinkage of reactive MgO modified alkali-activated slag paste. *Constr. Build. Mater.* **2014**, *51*, 395–404. [CrossRef]
58. Ružek, V.; Dostayeva, A.M.; Walter, J.; Grab, T.; Korniejenko, K. Carbon Fiber-Reinforced Geopolymer Composites: A Review. *Fibers* **2023**, *11*, 17. [CrossRef]
59. Hossain, M.A.; Hossain, K.M.A. Rheological, Fresh State, and Strength Characteristics of Alkali-Activated Mortars Incorporating MgO and Carbon Nanoparticles. *Materials* **2024**, *17*, 5931. [CrossRef] [PubMed]
60. Sood, D.; Hossain, K.M.A. Fresh State, Rheological and Microstructural Characteristics of Alkali-Activated Mortars Developed Using Novel Dry Mixing Technique under Ambient Conditions. *Appl. Sci.* **2021**, *11*, 8920. [CrossRef]
61. Sood, D.; Hossain, K.M.A. Strength, Fracture and Durability Characteristics of Ambient Cured Alkali—Activated Mortars Incorporating High Calcium Industrial Wastes and Powdered Reagents. *Crystals* **2021**, *11*, 1167. [CrossRef]
62. Cheaptubes. April 2024. Available online: https://www.cheaptubes.com/product-category/multi-walled-carbon-nanotubes/ (accessed on 22 October 2024).
63. Yang, S.; Yue, W.; Huang, D.; Chen, C.; Lin, H.; Yang, X. A facile green strategy for rapid reduction of graphene oxide by metallic zinc. *RSC Adv.* **2012**, *2*, 8827. [CrossRef]
64. Fan, Z.-J.; Kai, W.; Yan, J.; Wei, T.; Zhi, L.-J.; Feng, J.; Ren, Y.; Song, L.-P.; Wei, F. Facile Synthesis of Graphene Nanosheets via Fe Reduction of Exfoliated Graphite Oxide. *ACS Nano* **2011**, *5*, 191–198. [CrossRef]
65. Chen, J.; Akono, A.-T. Influence of multi-walled carbon nanotubes on the fracture response and phase distribution of metakaolin-based potassium geopolymers. *J. Mater. Sci.* **2021**, *56*, 19403–19424. [CrossRef]
66. Luz, G.; Gleize, P.J.P.; Batiston, E.R.; Pelisser, F. Effect of pristine and functionalized carbon nanotubes on microstructural, rheological, and mechanical behaviors of metakaolin-based geopolymer. *Cem. Concr. Compos.* **2019**, *104*, 103332. [CrossRef]
67. Jindal, B.B.; Sharma, R. The effect of nanomaterials on properties of geopolymers derived from industrial by-products: A state-of-the-art review. *Constr. Build. Mater.* **2020**, *252*, 119028. [CrossRef]
68. Wang, W.; Fan, C.; Wang, B.; Zhang, X.; Liu, Z. Workability, rheology, and geopolymerization of fly ash geopolymer: Role of alkali content, modulus, and water–binder ratio. *Constr. Build. Mater.* **2023**, *367*, 130357. [CrossRef]
69. Davoodabadi, M.; Vareli, I.; Liebscher, M.; Tzounis, L.; Sgarzi, M.; Paipetis, A.; Yang, J.; Cuniberti, G.; Mechtcherine, V. Thermoelectric Energy Harvesting from Single-Walled Carbon Nanotube Alkali-Activated Nanocomposites Produced from Industrial Waste Materials. *Nanomaterials* **2021**, *11*, 1095. [CrossRef]
70. Suave, J.; Coelho, L.A.F.; Amico, S.C.; Pezzin, S.H. Effect of sonication on thermo-mechanical properties of epoxy nanocomposites with carboxylated-SWNT. *Mater. Sci. Eng. A* **2009**, *509*, 57–62. [CrossRef]
71. *ASTM C1611*; Standard Test Method for Slump Flow of Self-Consolidating Concrete. ASTM International: West Conshohocken, PA, USA, 2014; pp. 1–6.
72. *ASTM C138/C138M*; Standard Test Method for Density (Unit Weight), Yield, and Air Content (Gravimetric) of Concrete. ASTM International: West Conshohocken, PA, USA, 2017.
73. *ASTM C1437*; Standard Test Method for Flow of Hydraulic Cement Mortar. ASTM International: West Conshohocken, PA, USA, 2015.
74. *ASTM C109/C109M*; Standard Test Method for Compressive Strength of Hydraulic Cement Mortars (Using 2-in. or [50-mm] Cube Specimens). ASTM International: West Conshohocken, PA, USA, 2016.
75. *ASTM C597-16*; Standard Test Method for Pulse Velocity Through Concrete. ASTM: West Conshohocken, PA, USA, 2016.
76. Kastiukas, G.; Ruan, S.; Liang, S.; Zhou, X. Development of precast geopolymer concrete via oven and microwave radiation curing with an environmental assessment. *J. Clean. Prod.* **2020**, *255*, 120290. [CrossRef]

77. EFNARC Specification. *Guidelines for Self-Compacting Concrete*; Report from EFNARC; Association House: London, UK, 2002; p. 32.
78. Abbas, S.; Qissab, M. Effect of Multi-Walled Carbon Nanotubes on Mechanical Properties of Concrete. *Coll. Eng. J. (NUCEJ)* **2016**, *91*, 6192–6194.
79. Gado, R.A.; Hebda, M.; Łach, M.; Mikuła, J. Alkali Activation of Waste Clay Bricks: Influence of The Silica Modulus, SiO_2/Na_2O, H_2O/Na_2O Molar Ratio, and Liquid/Solid Ratio. *Materials* **2020**, *13*, 383. [CrossRef] [PubMed]
80. Allahverdi, A.; Kani, E.N. Use of construction and demolition waste (CDW) for alkali-activated or geopolymer cements. In *Handbook of Recycled Concrete and Demolition Waste*; Elsevier: Amsterdam, The Netherlands, 2013; pp. 439–475. [CrossRef]
81. ACI 318. *Building Code Requirements for Structural Concrete*; American Concrete Institute: Farmington Hills, MI, USA, 2019.
82. Kaya, M.; Koksal, F.; Gencel, O.; Munir, M.J.; Kazmi, S.M.S. Influence of micro Fe_2O_3 and MgO on the physical and mechanical properties of the zeolite and kaolin based geopolymer mortar. *J. Build. Eng.* **2022**, *52*, 104443. [CrossRef]
83. Asil, M.B.; Ranjbar, M.M. Hybrid effect of carbon nanotubes and basalt fibers on mechanical, durability, and microstructure properties of lightweight geopolymer concretes. *Constr. Build. Mater.* **2022**, *357*, 129352. [CrossRef]
84. Ofuyatan, O.; Olowofoyeku, A.; Oluwafemi, J.; Ighalo, J. Predicting the Compressive Strength of Concrete By Ultrasonic Pulse Velocity. *IOP Conf. Ser. Mater. Sci. Eng.* **2021**, *1036*, 012053. [CrossRef]
85. Wei, J.; Yang, Q.; Jiang, Q.; Li, X.; Liu, S.; Li, K.; Wang, Q. Mechanical properties of basalt fiber reinforced ambient-cured lightweight expanded polystyrene geopolymer concrete. *J. Build. Eng.* **2023**, *80*, 108072. [CrossRef]
86. Ryu, G.S.; Lee, Y.B.; Koh, K.T.; Chung, Y.S. The mechanical properties of fly ash-based geopolymer concrete with alkaline activators. *Constr. Build. Mater.* **2013**, *47*, 409–418. [CrossRef]
87. Mo, L.; Deng, M.; Wang, A. Effects of MgO-based expansive additive on compensating the shrinkage of cement paste under non-wet curing conditions. *Cem. Concr. Compos.* **2012**, *34*, 377–383. [CrossRef]
88. Bakharev, T. Geopolymeric materials prepared using Class F fly ash and elevated temperature curing. *Cem. Concr. Res.* **2005**, *35*, 1224–1232. [CrossRef]
89. Li, H.; Liu, X.; Qi, S.; Xu, L.; Shi, G.; Ding, Y.; Yan, X.; Huang, Y.; Geng, J. Graphene Oxide Facilitates Solvent-Free Synthesis of Well-Dispersed, Faceted Zeolite Crystals. *Angew. Chem.* **2017**, *129*, 14278–14283. [CrossRef]
90. Hanjitsuwan, S.; Hunpratub, S.; Thongbai, P.; Maensiri, S.; Sata, V.; Chindaprasirt, P. Effects of NaOH concentrations on physical and electrical properties of high calcium fly ash geopolymer paste. *Cem. Concr. Compos.* **2014**, *45*, 9–14. [CrossRef]
91. Pan, Z.; Tao, Z.; Cao, Y.F.; Wuhrer, R.; Murphy, T. Compressive strength and microstructure of alkali-activated fly ash/slag binders at high temperature. *Cem. Concr. Compos.* **2018**, *86*, 9–18. [CrossRef]
92. Sherir, M.A.A.; Hossain, K.M.A.; Lachemi, M. Permeation and Transport Properties of Self-Healed Cementitious Composite Produced with MgO Expansive Agent. *J. Mater. Civ. Eng.* **2018**, *30*, 04018291. [CrossRef]
93. Jin, F.; Abdollahzadeh, A.; Al-Tabbaa, A. Effect of Different MgO on the Hydration of MgO-Activated Granulated Ground Blastfurnace Slag Paste. 2013. Available online: http://www.claisse.info/Proceedings.htm (accessed on 22 June 2024).
94. Zhang, T.; Cheeseman, C.R.; Vandeperre, L.J. Development of low pH cement systems forming magnesium silicate hydrate (M-S-H). *Cement Concr. Res.* **2011**, *41*, 439–442. [CrossRef]
95. Brew, D.R.M.; Glasser, F.P. Synthesis and characterisation of magnesium silicate hydrate gels. *Cem. Concr. Res.* **2005**, *35*, 85–98. [CrossRef]
96. Liu, X.; Wu, Y.; Li, M.; Jiang, J.; Guo, L.; Wang, W.; Zhang, W.; Zhang, Z.; Duan, P. Effects of graphene oxide on microstructure and mechanical properties of graphene oxide-geopolymer composites. *Constr. Build. Mater.* **2020**, *247*, 118544. [CrossRef]
97. Mužek, M.N.; Zelić, J.; Jozić, D. Microstructural characteristics of geopolymers based on alkali-activated fly ash. *Chem. Biochem. Eng. Q.* **2012**, *26*, 89–95.
98. Swanepoel, J.C.; Strydom, C.A. Utilisation of fly ash in a geopolymeric material. *Appl. Geochem.* **2002**, *17*, 1143–1148. [CrossRef]
99. Álvarez-Ayuso, E.; Querol, X.; Plana, F.; Alastuey, A.; Moreno, N.; Izquierdo, M.; Font, O.; Moreno, T.; Diez, S.; Vázquez, E.; et al. Environmental, physical and structural characterisation of geopolymer matrixes synthesised from coal (co-) combustion fly ashes. *J. Hazard. Mater.* **2008**, *154*, 175–183. [CrossRef]
100. Kupwade-Patil, K.; Allouche, E.N. Impact of Alkali Silica Reaction on Fly Ash-Based Geopolymer Concrete. *J. Mater. Civ. Eng.* **2013**, *25*, 131–139. [CrossRef]
101. Mollah, M.Y.A.; Hess, T.R.; Cocke, D.L. Surface and bulk studies of leached and unleached fly ash using XPS, SEM, EDS and FTIR techniques. *Cem. Concr. Res.* **1994**, *24*, 109–118. [CrossRef]
102. Mozgawa, W.; Sitarz, M.; Rokita, M. Spectroscopic studies of different aluminosilicate structures. *J. Mol. Struct.* **1999**, *511–512*, 251–257. [CrossRef]
103. Sitarz, M.; Handke, M.; Mozgawa, W. Identification of silicooxygen rings in SiO_2 based on IR spectra. *Spectrochim. Acta Part A Mol. Biomol. Spectrosc.* **2000**, *56*, 1819–1823. [CrossRef]
104. Reig, L.; Soriano, L.; Borrachero, M.V.; Monzó, J.; Payá, J. Influence of calcium aluminate cement (CAC) on alkaline activation of red clay brick waste (RCBW). *Cem. Concr. Compos.* **2016**, *65*, 177–185. [CrossRef]

105. Wang, Y.; Liu, X.; Zhang, W.; Li, Z.; Zhang, Y.; Li, Y.; Ren, Y. Effects of Si/Al ratio on the efflorescence and properties of fly ash-based geopolymer. *J. Clean. Prod.* **2020**, *244*, 118852. [CrossRef]
106. Wan, Q.; Zhang, Y.; Zhang, R. Using mechanical activation of quartz to enhance the compressive strength of metakaolin-based geopolymers. *Cem. Concr. Compos.* **2020**, *111*, 103635. [CrossRef]

applied sciences

MDPI

Article

Performance Enhancement of Alkali-Activated Electric Arc Furnace Slag Mortars through an Accelerated CO_2 Curing Process

Dany Kassim, Ghandy Lamaa, Rui Vasco Silva * and Jorge de Brito

CERIS—Civil Engineering Research and Innovation for Sustainability, Instituto Superior Técnico, Universidade de Lisboa, Av. Rovisco Pais, 1049-001 Lisboa, Portugal; danykassim@tecnico.ulisboa.pt (D.K.); ghandy.lamaa@tecnico.ulisboa.pt (G.L.); jb@civil.ist.utl.pt (J.d.B.)
* Correspondence: rui.v.silva@tecnico.ulisboa.pt

Abstract: The use of electric arc furnace slag (EAFS) as sole precursor to produce alkali-activated mortars has been experimentally investigated. EAFS, a by-product of the steel recycling industry, is a coarse material with unevenly distributed and size-extensive particles. Milling of EAFS was required to achieve a cement-like sized powder before it could be used as precursor. Different combinations of sodium hydroxide (NaOH) and sodium silicate (Na_2SiO_3) were used, by varying the Na_2O/binder concentration (4%, 6%, 8%, 10%, 12%) and SiO_2/Na_2O ratio (0, 0.5, 1.0, 1.5, 2.0, 2.5) to maximize the mechanical performance. The alkaline solutions were prepared 24 h prior to mixing to unify temperatures for all mixes. The results showed that the SiO_2/Na_2O ratio and strength development are directly proportional. The maximum 28-day compressive strength obtained, after being subjected to an initial 24 h thermal curing at 80 °C, was 9.1 MPa in mixes with 4% Na_2O/binder and 2.5 SiO_2/Na_2O. However, after an additional 28 days of accelerated carbonation, the maximum compressive strength (i.e., 31 MPa compared to 3.9 MPa in uncarbonated mixes, corresponding to an 800% increase) was obtained in mixes with 12% and 1.0 for Na_2O/binder and SiO_2/Na_2O, respectively, thus showing an alteration in the optimal alkaline activator contents.

Keywords: alkali-activated materials; electric arc furnace slag; mechanical performance; carbonation; mortars

1. Introduction

The world's annual population is continuously increasing by 1.05% on average, with an increase of 81 million people in 2020 [1]. This growth seems to be proportionally linked to the rising demand for housing and other infrastructure that will simultaneously increase the need for construction materials, such as Portland cement. In 2016, the amount of cement produced worldwide reached 4.2 billion tonnes, 57% of which was produced by China alone [2,3]. It is a well-known fact that for each one tonne of cement produced, around 0.5–0.6 tonnes of CO_2 (0.59 according to IEA, 2020 [4]) are being released into the atmosphere, meaning that in 2016, the cement industry itself released nearly 2.5 billion tonnes of CO_2, which accounted for 8% of the global CO_2 emissions of the same year [5,6]. Solving this issue cannot be done by cutting the production of cement alone; other cement-replacing materials need to be found to meet the growing demand for this vital building material. Alkali-activated materials (AAM) are a relatively recent technology as integral cement replacement in the production of concrete and with reduced environmental impact and cost [7]. However, some of the precursors generally used for their production, such as ground granulated blast furnace slag (GGBFS) and fly ash (FA), are facing shortage in availability [8]. To maintain a constant flow in the supply chain, other aluminosilicate waste precursors, prone to little variation in yearly production over time, are needed. One contender is the electric arc furnace slag (EAFS), as the steel industry is shifting from new steelmaking to a recycling system, consequently reducing the amount of GGBFS. It has an aluminosilicate-rich chemistry giving it the possibility to be the next cement-replacing AAM used in concrete. In terms of availability, about 190–290 million tonnes of steel slag are generated each year, 15–20% of which are EAFS [9]. EAFS generation is based on the recycling of scrap metal and pig iron that can be limited at some locations [10]. It is being used mostly as an aggregate substitute for road base course layers and asphalt pavements [11,12], which is a downcycling process from what they can actually offer.

According to several studies [12–21], EAFS has led to notable improvements in mechanical strength when used as coarse and/or fine aggregates replacement. However, its richness in calcium oxide (free-CaO) and Ferric oxide (FeO) has pushed researchers to further explore the possibility of using EAFS as a cement substitute [22,23]. Xu and Deventer [24] found that any silica–alumina source material that has a pozzolanic component is suitable for dissolution in alkaline solutions. Since EAFS is rich in aluminium and silica, it is desirable to examine whether its activation in alkaline solutions can occur, considering this knowledge [25,26].

The use of EAFS as partial or full cement replacement has led to improvements in mechanical and microstructural performances [25,27]. Muhmood et al. [28] conducted an experimental study to understand the mechanical performance of partially replaced clinker with EAFS. The study concluded that a compressive strength of 50 MPa can be achieved with 30% clinker–EAFS replacement, compared to 58 MPa with 20% replacement and 58.6 MPa for the reference specimen with 100% cement. It has also been found that the substitution of untreated EAFS with treated EAFS can lead to an increase in strength [28]. Another study has recorded a 28-day compressive strength of 44.38 MPa using 25% EAFS as cement replacement compared to 59.34 MPa for the reference specimen with 100% cement [29]. Moreover, Zhao et al. [30] investigated the influence of cement's partial substitution with EAFS, focusing on the EAFS' particle distribution. The results revealed that concrete specimens containing properly ground EAFS showed better microstructure, higher compressive strength, lesser porosity and better durability performance compared to the reference blast furnace slag blended cement specimens. Other studies have reported similar results [29,31–33]. In addition, raising the fineness of EAFS particles, as well as integrating early thermal curing in the curing stages, led to better hydration performance and a faster early-age hydration [22,31,34,35]. Similar studies found that the compressive strength of alkali-activated concrete cured at 70 °C was higher than that of concrete cured at 30 °C [22,32,36]. Furthermore, Roslan et al. [36] reported later-age improvements in compressive, tensile and flexural strengths with a 20% cement–EAFS replacement.

The use of EAFS as a precursor for AAM production usually shows some shortcomings in performance in comparison with FA or GGBS. Ozturk et al. [25] studied the influence of the optimization of SiO_2/Na_2O ratio and Na_2O/binder percentage on the mechanical performance of mortars with 100% EAFS as a precursor. For a silicate modulus of 1, 1.5 and 2, the highest compressive strengths reported were 16.55 MPa, 17.11 MPa and 22.02 MPa, respectively. In addition, an increase in flexural strength was noticed with the increase in silicate modulus resulting in values of 2.85 MPa, 3.09 MPa and 4.18 MPa for SiO_2/Na_2O ratios of 1, 1.5 and 2, respectively. Moreover, the highest mechanical strengths were all reached under an 80 °C curing temperature and 12 h curing time. Nikolic et al. [37] investigated the influence of the NaOH and KOH concentration, solid to liquid ratio and temperature on the dissolution kinetics of Si and Al in EAFS. In the study, the increase in alkaline solution concentrations (NaOH or KOH) increased the dissolution of Si and Al of EAFS. Similarly, a decrease in the solid to liquid ratio and an increase in temperature also catalyzed this reaction. Furthermore, Turker et al. [35] studied the influence of thermal curing on the microstructural performance of alkali-activated EAFS mortars activated by a solution with 5% and 1.0 of Na_2O/binder and SiO_2/Na_2O ratios, respectively. A compressive strength of 25.6 MPa was reported for specimens cured at ambient temperature (21 °C), compared to 40.7 MPa for a similar sample cured at 60 °C for 6 h (59% increase). A similar study also demonstrated the benefits of early-age thermal curing on the strength development and shrinkage of alkali-activated EAFS mortars, after reporting an increase in compressive strength from 14 MPa to 21 MPa after a thermal curing stage at 70 °C with durations ranging between 3 h and 24 h, respectively [38]. Cesnovar et al. [39] studied the influence of different slag mixtures containing EAFS and ladle furnace basic slag (LS) with different mixing ratios and activated using potassium silicate with a fixed ratio to slag of 1/2. A compressive strength of 56.7 MPa was reported using an EAFS/LS ratio of 1/1. This optimal mixture was used to understand the influence of thermal curing on strength development. The results indicated that the specimens cured at 70 °C for 3 days attained similar compressive strength (~56 MPa) to that of specimens cured at room temperature for 28 days.

Despite the effort made, little is known on the use of EAFS as full binder replacement, and more experimental studies should take place to further explore this interesting research idea. In addition, the optimization of the alkali activators may pave the way toward more encouraging results, but the mechanical performance values recorded so far are not market competitive. This eagerness toward achieving better performance raised the idea of carbonation as a curing method. Exploiting the possibility of curing the EAFS concrete with CO_2 [40–42], by taking advantage of its CaO-rich chemistry, will not only lead to better mechanical performance but also an added value to CO_2-capturing revolutionary technologies. Monkman and Shao [40] assessed the carbonation behavior of six cementitious materials (cement, FA, GGBFS, EAFS, and hydrated lime) to determine and compare the strength development, $\%CO_2$ uptake and carbonation degree for each one of them. The samples were exposed to 100% CO_2 at a pressure of 5 bar for 2 h. Afterward, X-ray diffraction (XRD) and scanning electron microscopy (SEM-EDS) analyses were performed to determine the products of carbonation and to observe the morphology of the carbonation products. The results showed a 12% CO_2 uptake for EAFS, FA and cement, unlike lime and BFS that presented 40% and 7% CO_2 uptake, respectively. Similarly, cement and EAFS were comparable in terms of carbonation degree with a value of around 25%. The compressive strength (at 2 h) of EAFS reached 16.6 MPa, while that of FA was only 3.5 MPa.

In this paper, the use of alkali-activated electric arc furnace slag (AAEAFS) as a sole precursor in the production of mortars has been investigated. Its main objective is to reach the optimization of the activator's composition based on maximized mechanical performance. Twenty-one different mixes were produced to study the influence of different alkali activator concentrations and ratios, to find the best alkaline solution for the EAFS mortars. Different silicate modulus ratios (0, 0.5, 1.0, 1.5, 2.0, 2.5) and Na_2O/binder concentrations (4%, 6%, 8%, 10%, 12%) were selected for this study. Their effect on the performance was evaluated through comprehensive testing methodologies, including slump, shrinkage, flexural strength (FS), compressive strength (CS) and carbonation degree. A 28-day curing period, including a 7-day accelerated carbonation stage, followed. Finally, the correlations between the distinctive preparation conditions and the different tested parameters were graphically illustrated, serving the upcoming research studies toward the best future use of EAFS as a more sustainable cement-replacing material.

2. Materials and Methods

2.1. Electric Arc Furnace Slag (EAFS)

The EAFS used in this study is a by-product of steel recycling, collected from Siderurgia Nacional de Portugal and provided by HARSCO (Portugal). It presents an extensive and coarse particle size distribution that requires preparation and grinding before it can be used as a binder.

2.2. Alkaline Activator

The alkaline activator was prepared in the form of a liquid solution. Reactive grade anhydrous sodium hydroxide pellets (NaOH) from Crimolara (Portugal), with 98% purity and a density of 2.13 g/mL, then dissolved in tap water, complying with Directive 98/83/CE [43]. A commercial solution of sodium metasilicate (Na_2SiO_3) from Crimolara

(Portugal) was then added, containing 8 ± 0.6% of sodium oxide (Na_2O), 26.4 ± 1.5% of silicon oxide (SiO_2) and 65.6 ± 2% of water (H_2O). The Na_2SiO_3 has a relative density of 1.355 g/mL.

2.3. Fine Aggregate

The mortars were produced using two types of fine siliceous aggregates (i.e., coarse and fine river sands) to maximize compacity through an extensive particle size distribution. The particle size distribution showed maximum nominal sizes (NS) of 1 mm and 4 mm for fine and coarse aggregates, respectively. The 24 h water absorption (WA_{24}), including the values of the apparent (ρ_a), rodded-dry (ρ_{rd}), saturated surface-dry (ρ_{SSD}), and bulk (ρ_b) densities are shown in Table 1.

Table 1. Characterization of the aggregates used in the mixes.

Aggregates	NS	ρ_a	ρ_{rd}	ρ_{ssd}	ρ_b	WA_{24}
	mm	kg/m^3	kg/m^3	kg/m^3	kg/m^3	%
Fine sand	0/1	2652	2624	2637	1544	0.4
Coarse sand	0/4	2636	2601	2617	1556	0.5

2.4. Water-Reducing Admixture

The water-reducing admixture (WRA) used in this research was SikaPlast-717, consisting of a synthetic organic water-based naphthalene-based dispersant, with a density of 1.21 ± 0.03 kg/dm^3 and a pH of 10 ± 1.

2.5. Mortar Mix Design

The experimental research focused on optimizing the alkaline activator based on the mortar's mechanical performance. To achieve the optimum activator for EAFS, different mixing combinations were used. The Na_2O/precursor concentrations used were 4%, 6%, 8%, 10% and 12%, while the adopted SiO_2/Na_2O mass ratios were 0, 1.0, 1.5, 2.0 and 2.5 (Table 2). The amount of each constituent was calculated based on the volumetric and mass ratios of the components, in accordance with their densities. The binder/aggregate volumetric ratio (V_B/V_A) was 0.33, while the mass ratio of water/binder (M_W/M_B) was fixed at 0.3. The mass WRA/precursor ratio varied from 0% to 1% to achieve constant workability.

Table 2. Mix code of EAFS mixes.

Na_2O (%)	SiO_2/Na_2O					
	0	0.5	1	1.5	2	2.5
4	N4S0	N4S0.5	N4S1	N4S1.5	N4S2	N4S2.5
6	N6S0	N6S0.5	N6S1	N6S1.5	N6S2	-
8	N8S0	N8S0.5	N8S1	N8S1.5	-	-
10	N10S0	N10S0.5	N10S1	-	-	-
12	N12S0	N12S0.5	N12S1	-	-	-

2.6. Production Method

The production of mortars was based on EN 196-1 [44]. The alkaline solution was prepared by gradually dissolving the NaOH pellets in water and then leaving them to cool down for 24 h. The 40 × 40 × 160 mm^3 three-gang steel moulds were wrapped entirely with thin plastic film to demould the samples without the need for a release agent. The alkaline solution was poured first into the mixer followed by WRA and the precursor, before mixing for 3 min. After that, the mixer was paused to add the fine aggregates, and then mixing was resumed for another 2 min, followed by one additional minute of high-speed mixing. Next, the slump was tested using the slump table according to EN 1015-3 [45]. Afterward, the mix was moulded and covered with plastic film and immediately placed in the thermal curing chamber. Finally, the specimens were demoulded, and each specimen was sealed and left to cure in a dry chamber at a temperature of 23 ± 2 °C and relative humidity (RH) of 65%. The samples were left in the dry chamber until the testing day.

2.7. Curing Conditions and Testing Methods

As stated previously (Section 2.6), the moulded mix was thermally cured at 80 °C for 24 h. After that, the mortar specimens were demoulded and placed in their designated curing condition depending on the desired test method, as shown in Table 3. The loading rate for the flexural and compressive strength tests had a constant value of 30 N/s and 300 N/s, respectively, complying with its corresponding standard (Table 3). The compressive test was applied to both halves resulting from the flexural test. The shrinkage values of the specimens were measured using a mortar shrinkage apparatus following a certain measuring regime in accordance with EN 1015-13 [46].

Table 3. Curing conditions and testing methods assigned for the specimens.

Test	Standard	No. of Specimens	Curing Conditions
Flexural strength	EN 1015-11 [47]	6	Sealed specimens in a dry chamber until testing age.
Compressive strength			
Accelerated carbonation	EN 13295 [48]	4	14 days sealed + 14 days unsealed in the dry chamber; then placed in the carbonation chamber until testing age.
Shrinkage	EN 1015-13 [46]	2	Sealed specimens in a dry chamber after demoulding until the end of the test.

3. Results

3.1. Characterization of EAFS

This material presents an apparent density of 3770 kg/m^3. The oxide chemical composition of the raw material, obtained from X-ray fluorescence (XRF), is shown in Table 4. The studied EAFS contains 28.5% of Fe_2O_3, 28.2% of CaO, 17.7% of SiO_2, and 10.1% of Al_2O_3. The high amount of iron in EAFS could induce magnetic properties on AAEAFS concrete [49] and it was confirmed to be a strongly magnetic precursor with the use of neodymium magnets.

Table 4. Chemical composition of EAFS obtained from XRF (%).

Fe_2O_3	CaO	SiO_2	Al_2O_3	MgO	MnO_2	Cr_2O_3	TiO_2	P_2O_5	SO_3	Na_2O	BaO	K_2O	V_2O_5	CuO	ZnO
28.48	28.18	17.66	10.13	5.66	5.44	2.38	0.65	0.42	0.33	0.19	0.17	0.03	0.11	0.02	0.02

Figure 1 presents the results of the particle size distribution of the EAFS after the milling process. Ordinary Portland and fly ash (FA) from a coal power plant were tested as reference materials. EAFS showed a bimodal particle distribution curve similar to FA and cement. The highest peak of both the EAFS and cement curves indicated a similar particle diameter of 25 μm. However, the peak corresponding to cement's curve was sharper than the one of EAFS, showing a higher distribution percentage. EAFS and cement showed another peak at 0.35 μm, indicating the presence of very fine particles. In contrast, FA showed a wider peak at 20 μm and a lower distribution percentage at 0.35 μm compared to EAFS and cement.

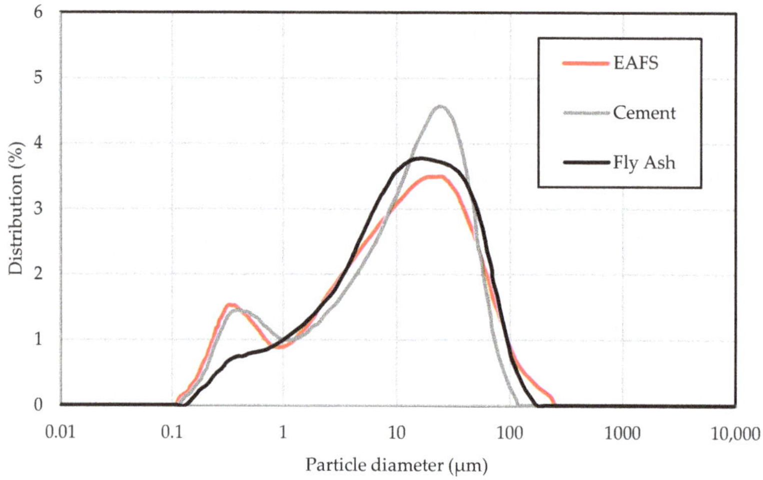

Figure 1. Particle size analysis.

In Figure 2a, the EAFS particles are irregular in shape, dispersed, and have angular morphology. In addition, finer particles were observed on the surface of the larger ones (Figure 2b). On a larger scale view, vitrified surfaces are seen resembling those of materials exposed to elevated temperatures (i.e., greater than 1600 °C) suggesting the possibility of pozzolanicity and reactivity to alkali activation (Figure 2c). Moreover, the EDS test was carried out on the area presented in Figure 2d and the results were graphically represented in Figure 3. The corresponding EAFS particle is composed of around 50.6% oxygen, 18.0% calcium, 7.1% silicon, 6.2% of iron, and other components. This primary composition matches the mineralogical composition of the EAFS XRD results (Section 3.1) showing high calcium content seen as white particles (Figure 2d), as well as silicon oxide and iron oxide.

(**a**)

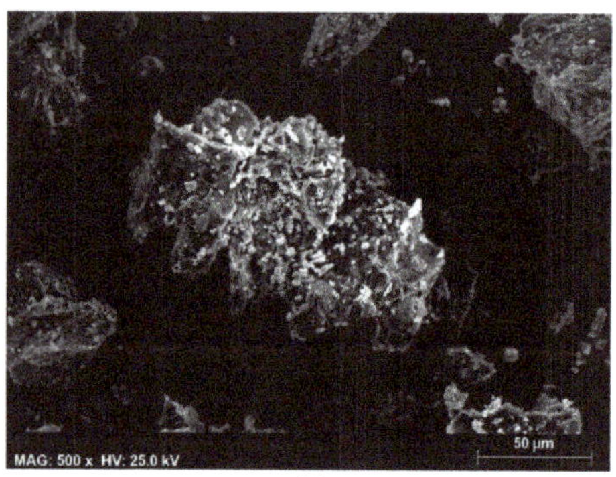

(**b**)

Figure 2. *Cont.*

(**c**)

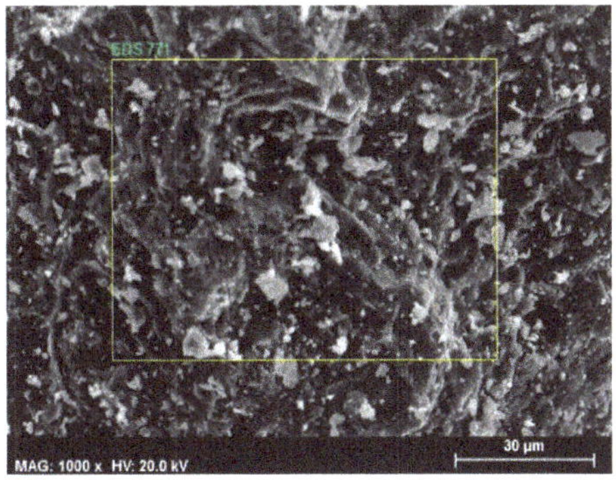

(**d**)

Figure 2. SEM micrographs of EAFS sample (magnification of (**a**) 100×; (**b**) 500×; (**c**) 5000×; (**d**) 1000×).

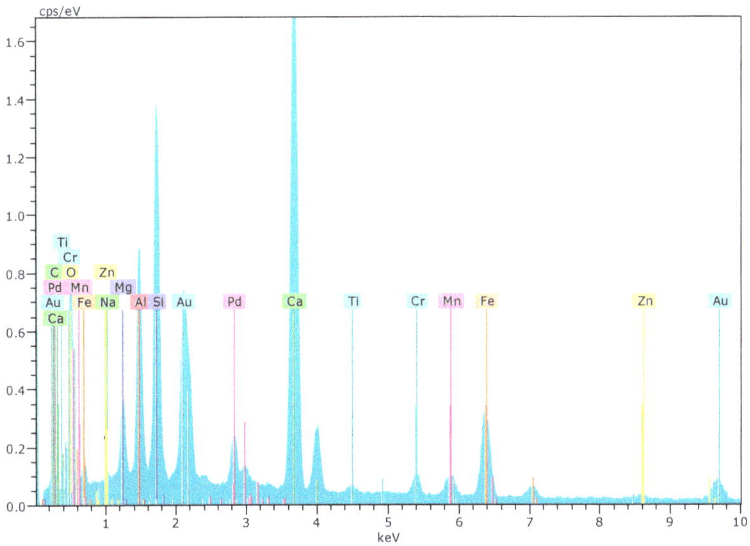

Figure 3. EDS coupled to SEM of EAFS sample.

Figure 4 shows the EAFS' mineralogical composition obtained by XRD. The results showed three fundamental phases: wustite (FeO), gehlenite ($Ca_2Al_2SiO_7$), and dicalcium silicate (Ca_2SiO_4) (Figure 4a). Other minor phases appeared as magnesioferrite. After heating the sample to 1000 °C in thermogravimetric analysis (TGA), the fundamental phases were gehlenite and magnesioferrite, as presented in Figure 4b; wustite was no longer present and hematite was formed as a result of the former's oxidation (mainly between 400 and 620 °C).

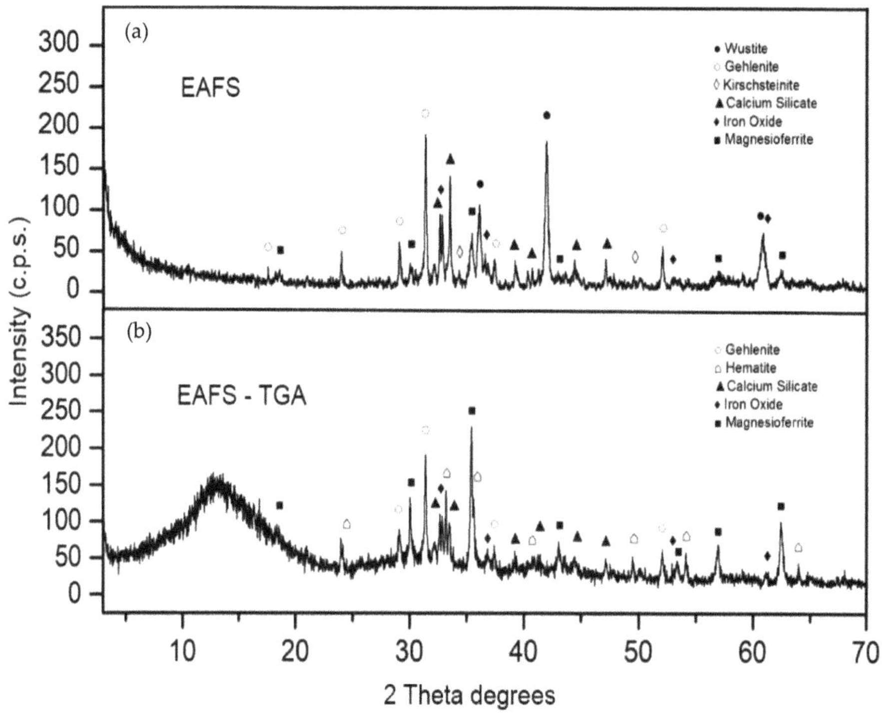

Figure 4. XRD pattern; (**a**) EAFS; (**b**) EAFS—TGA.

3.2. Fresh State Mortar Properties

The mortar's workability was evaluated by their slump according to the EN 1015-3 standard [45]. The target slump had been initially set at 140 ± 20 mm. The superplasticizer was used and adjusted accordingly between 0% and 1.0% depending on the Na_2O/binder and SiO_2/Na_2O ratios to achieve constant workability. Increasing the Na_2O/binder ratio led to a reduction in workability. This was due to the higher viscosity of the alkaline solution after the addition of a solid solute. It was also expected that the mixes would lose workability after a short time for higher SiO_2/Na_2O ratios due to flash setting; Ca^{2+} ions from the EAFS quickly react with the silicate ions from the solution, leading to the precipitation of an initial C-S-H, which is responsible for the setting [34]. Therefore, it can be stated that there is a linear correlation between the Na_2O/binder and SiO_2/Na_2O ratios and the workability. However, mixes N8S1, N10S1 and N12S1 presented a slightly lower slump between 109 and 117 mm (Figure 5a) due to the increase in viscosity of the alkaline activator. Therefore, the superplasticizer content was increased to 1.0% for mixes with silicate modulus ratios of 1.5 and higher, to adjust the workability and the slump which ranged between 130 and 147 mm (Figure 5b).

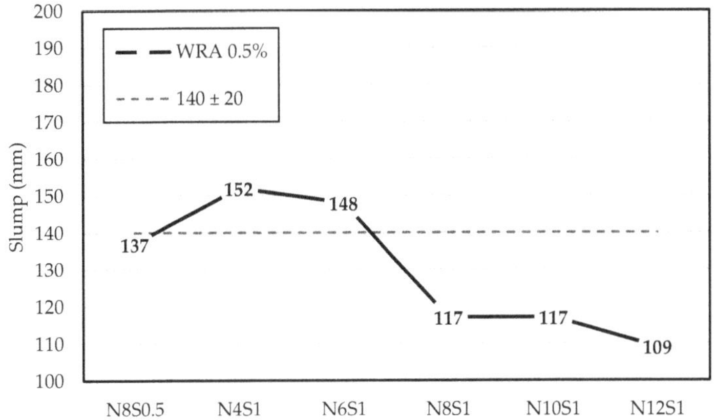

(a)

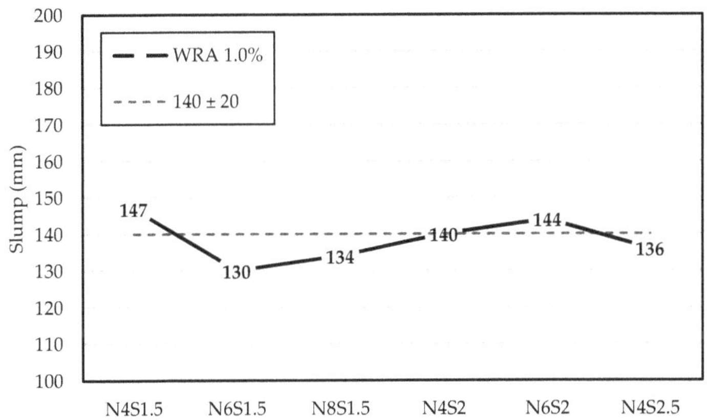

(b)

Figure 5. Slump values for mortar mixes (WRA content of (**a**) 0.5% and; (**b**) 1.0%).

3.3. Hardened State Mortar Properties

3.3.1. Compressive Strength

The average compressive strength of the AAEAFS uncarbonated samples is presented in Figure 6, which shows how the change in parameters affects the mechanical performance of the samples. The values presented in the figure have coefficients of variation ranging between 0.5 and 7.5%. Mixes N4S0.5, N6S0.5, N10S0.5, and N12S0.5 were eliminated due to inconsistent values and unexpected outcomes during the mixing process.

The alkaline solutions of mixes N10S0.5 and N12S0.5 showed precipitation, possibly due to the saturation of silica, during its 24 h of cooling before the mixing day. Thus, the solution must undergo different conditions to avoid any fluctuation in the results. Correspondingly, the alkali activator is the most important factor controlling compressive strength.

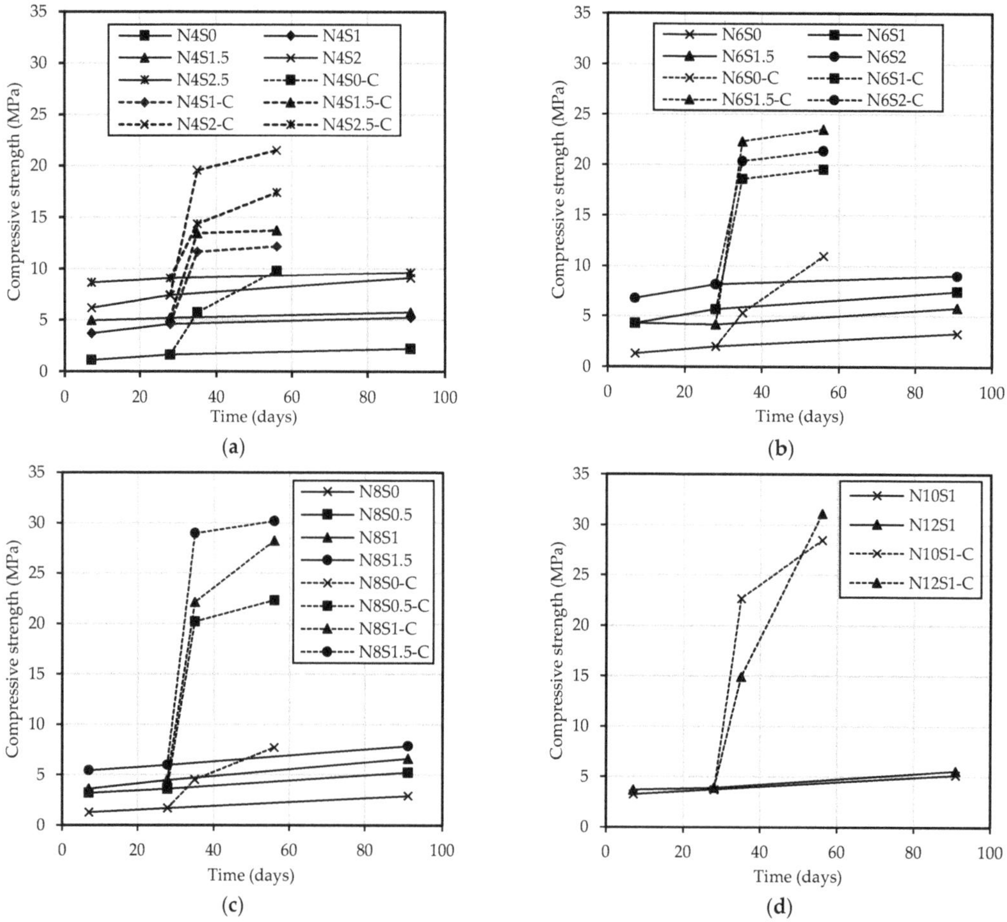

Figure 6. Average compressive strength: (**a**) 4% sodium concentration; (**b**) 6% sodium concentration; (**c**) 8% sodium concentration; (**d**) 10% and 12% sodium concentration (Mix code-C represents carbonated specimens).

Different studies on alkali-activated EAFS used at least one of the alkaline activators stated in this study; Turker et al. [35], Ozturk et al. [25], and Peys et al. [50] used both sodium hydroxide and sodium silicate to achieve compressive strength values of 40.7 MPa, 22.0 MPa, and 16.0 MPa, respectively. Abdollahnejad et al. [51] used only sodium hydroxide and recorded 27.0 MPa. This study presented maximum average compressive strength for uncarbonated specimens of 9.61 MPa for mixes with 4% and 2.5 of Na_2O/binder and SiO_2/Na_2O ratios, respectively. Mixes N10S0 and N12S0, which were specimens with insufficient stability, exhibited values close to zero and thus were not reported. This could be due to the excess amount of sodium hydroxide and the lack of sodium silicate. Increasing the amount of Na_2O to a given percentage increases the strength, after which the performance starts to deteriorate. Nevertheless, excess amount of OH^-, due to the $Ca(OH)_2$ across the particles of EAFS, reduces the interaction of Ca^{2+} ions from the surface of EAFS [34]. Therefore, it can be stated that the low strength results from the inadequate amount of C-S-H gels produced by the reaction of Ca^{2+} with Si^{4+}.

Sodium hydroxide and sodium silicate have a significant influence on the mechanical properties of AAEAFS. Si^{4+} and Al^{3+} ions within the EAFS dissolve much more with a high concentration of OH^- [52]. Song et al. [53] observed that increasing the concentration of the alkali activator increases the reaction rate as a result of a high alkali medium. Wang et al. [54] stated that sodium hydroxide and sodium silicate are directly proportional to compressive strength only to a given level (from 3–5% by weight of the mix). The authors also stated that, if the amount of sodium concentration increased over a given limit, it would cause efflorescence, as shown in Figure 7. This is due to the migration of Na^+ ions to the surface of the specimens, leading to the precipitation of sodium carbonate.

Another parameter affecting the compressive strength for all mixes was the curing time. The specimens in this study experienced 7-day, 28-days, and 91-days of curing in a dry chamber at 23 ± 2 °C and RH of 65%. Since all specimens were sealed, there was no exchange of humidity with the surrounding environment. The maximum compressive strength (9.61 MPa) reported for the uncarbonated specimen N4S2.5, was obtained at 91-days of curing age.

Figure 7. Efflorescence caused by Na_2CO_3 on the surface of a mortar specimen.

3.3.2. Flexural Strength

Figure 8 presents the results reported for the flexural strength of AAEAFS mortars. The values presented in the figure have coefficients of variation ranging between 0 and 11% except in one case that showed 17%. The maximum flexural strength for uncarbonated specimens reached 2.45 MPa for mix N12S1. The surface of the specimens presented some microcracks, possibly caused by the heat curing process and the expedited nature of the reactions at relatively high-temperature levels [55], thus causing a decline in strength. Although a higher SiO_2/Na_2O ratio is a good indicator of an improved performance [52,56–58], the low performance in the case of EAFS may be due to the low amount of amorphous phases present in the precursor, which did not react with the alkaline activator. Furthermore, even though one would expect to have improved performance from the interaction of Ca from EAFS with SiO_2 from the activator to produce C-S-H gels, it is possible that the Ca-bearing mineralogical phases were stable at high pH levels, thereby minimizing the dissolution of Ca^{2+} ions to the solution. In this context, flexural strength will only be enhanced when exposed to accelerated carbonation. This improvement in strength could be caused by the reaction of OH^- from the alkaline activator with the Ca^{2+} ions released from the decalcified phases of EAFS to generate $Ca(OH)_2$ and subsequently $CaCO_3$.

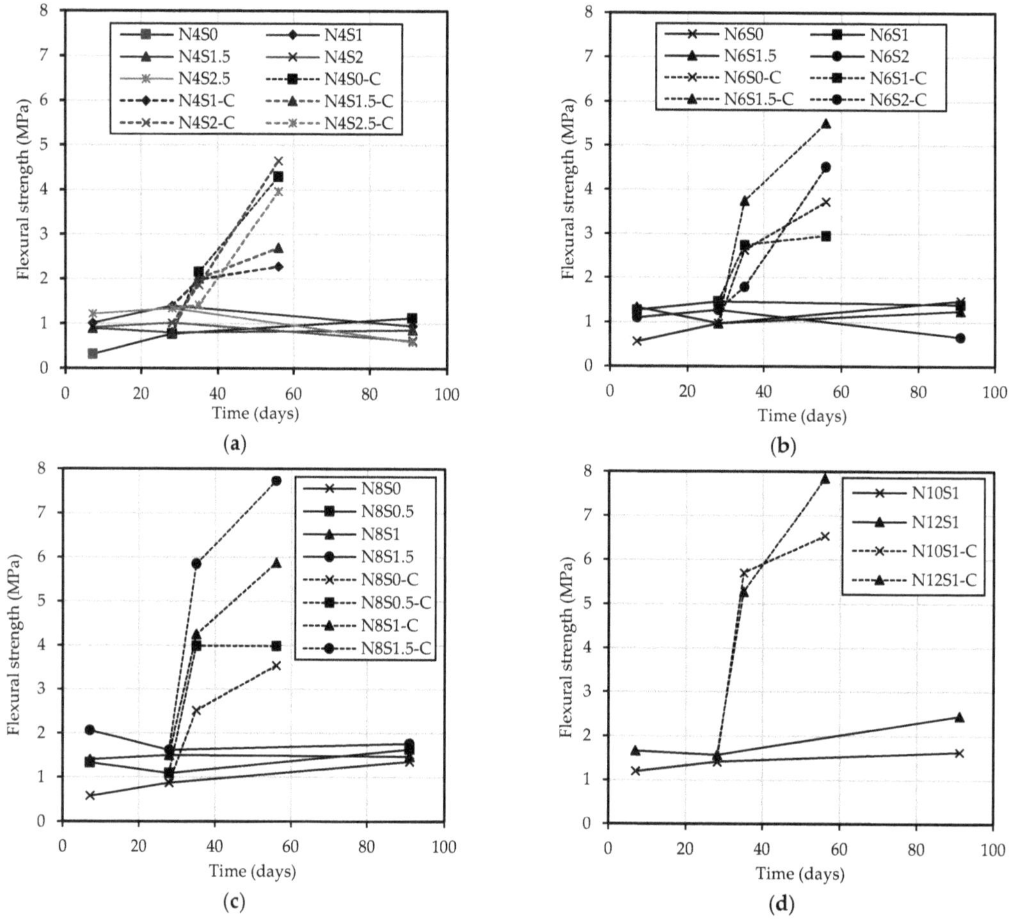

Figure 8. Average flexural strength: (**a**) 4% sodium concentration; (**b**) 6% sodium concentration; (**c**) 8% sodium concentration; (**d**) 10% and 12% sodium concentration (Mix code-C represents carbonated specimens).

3.3.3. Carbonation

The average compressive and flexural strength for the carbon-cured EAFS mortars are presented in Figures 6 and 8. The specimens in this study, after 28 days of curing in a dry chamber followed by 7 days of exposure to CO_2 (i.e., 35 days), were considered fully carbonated, as there was no indication of a pinkish hue from the phenolphthalein solution pH indicator, thereby making the CO_2 penetration impossible to read. Alkali-activated materials are known to present a fast decline in pH with ongoing polymerization reactions due to the consumption of the OH^- ions present in the pore solution. This phenomenon, in combination with the carbonation of Ca-bearing phases, led to an overall decline of the specimens' pH.

The average compressive and flexural strengths of the specimens subjected to accelerated carbonation were tested following EN 1015-11 [47]. A noticeable improvement in the mechanical performance of AAEAFS mortars was observed. The additional 28 days of accelerated carbonation following the 28 days in a dry chamber (i.e., 56 days) led to a near 800% and 500% increases in compressive (i.e., from 3.9 MPa to 31 MPa) and flexural (i.e., from 1.6 MPa to 7.85 MPa) strengths, respectively. The maximum mechanical performance was obtained in mixes with 12% and 1.0 for Na_2O/binder and SiO_2/Na_2O ratios, respectively, thus showing a shift in the optimal alkaline activator contents. It is likely that Ca^{2+} ions were released from the EAFS' Ca-bearing phases and reacted with CO_2 forced into the microstructure. This resulted in the precipitation of $CaCO_3$ polymorphs, which significantly densified the microstructure. The resulting SiO_2 gels from decalcified calcium silicate phases also led to a widespread polymerization thus a more effective binding of the microstructure. These processes resulted in enhanced performance [25,53].

The silicate modulus and sodium concentration also had an obvious effect on the mechanical performance. Contrary to that expected, the flexural strength of mixes with lower performance is often not correlated with the compressive strength, as presented in Figure 9.

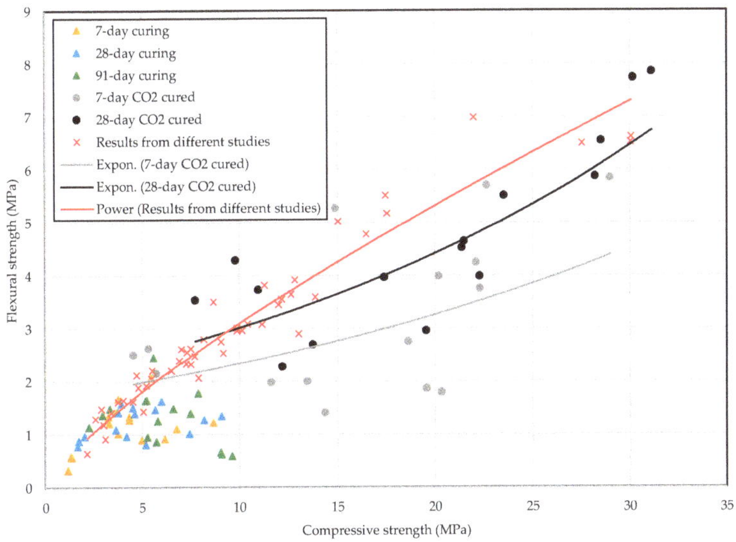

Figure 9. Flexural vs. compressive strengths of mortars from this study compared with other studies from the literature [59–66].

Carbonated specimens with enhanced performance, in which exponential relationships were observed between the flexural and compressive strength, suggested otherwise. The 28-day relationship of carbonated specimens was close to that obtained by standard 40 × 40 × 160 mm cement mortar prims from other studies [59–66]. The improvement in flexural strength between 7 days and 28 days is likely due to the greater polymerization of SiO_2 gels from the greater amount of decalcified calcium silicate phases (e.g., C-S-H). The isotropic growth of these phases, but especially of those growing perpendicularly to the loading vector, led to improved flexural strength.

3.3.4. Shrinkage

In Figure 10, the change in length (i.e., shrinkage) of sealed specimens (with little to no humidity exchange with the surrounding environment) was tested for 91 days. Most specimens presented considerable shrinkage, with one of them close to 3500 μm/m, which is three times what is typically observed in standard cement mortars. All specimens presented at least 45% of their total 91-day shrinkage in the first 28 days, except for mix N8S0.5. Mix N8S0.5 slightly expanded, with minor fluctuations, for the first 28 days and started to shrink later on to settle at 198 μm/m after 91 days. This behaviour also occurred in fly ash mortars studied by Atiş et al. [67]. The authors hypothesized that the expansion of mortars containing FA could be from the MgO and the high content of SO_3, which can result in long-term instability due to the formation of expansive calcium sulphate phases [67]. In Figure 10a, the shrinkage of specimens with a SiO_2/Na_2O ratio of 1.0 decreased with increasing Na_2O/binder ratio and somewhat correlated with their mechanical behaviour. It is likely that the greater dissolution of aluminosilicate phases from the higher pH level of the alkaline solution led to the formation of more strength-enhancing and densifying phases, thus leading to effective restrained shrinkage. However, the opposite was observed for mixes with a SiO_2/Na_2O ratio of 1.5. Further research is required to ascertain the trend of specimens with varying silicate modulus and Na_2O content.

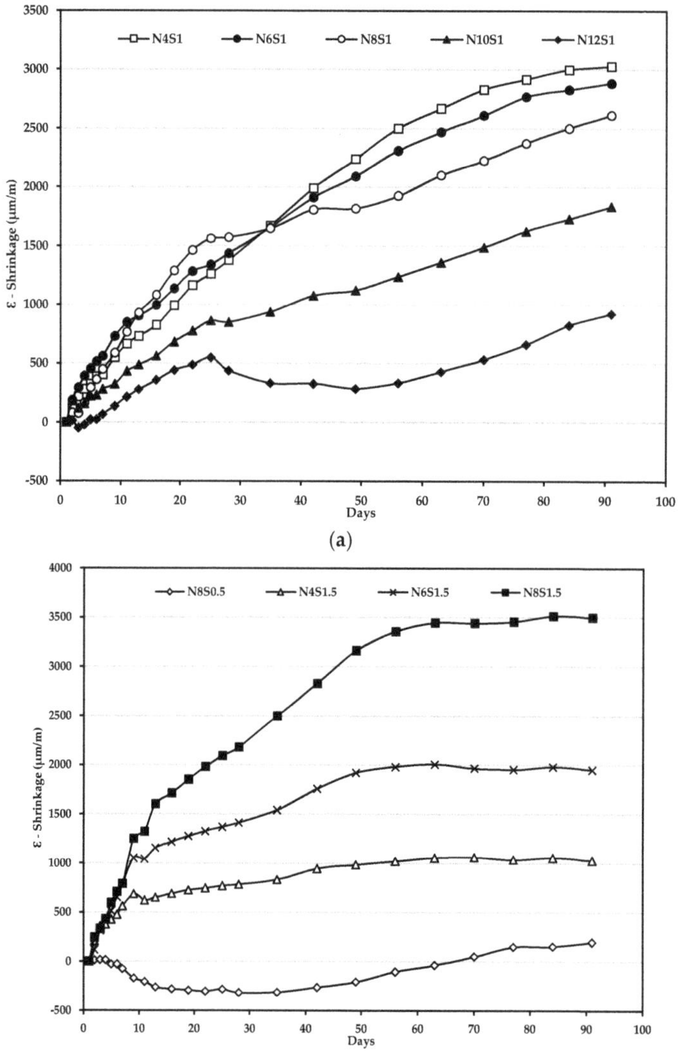

(a) (b)

Figure 10. Shrinkage for AAEAFS mortars: (**a**) values for mixes with silicate modulus of 1.0; (**b**) values for mixes with silicate modulus of 0.5 and 1.5.

4. Conclusions

In this study, electric arc furnace slag (EAFS) was studied as a potential full replacement for cement in the production of alkali-activated mortars. The results obtained in this study allowed a conclusion that EAFS as the sole precursor will result in mixes with relatively low performance. According to the XRD results, this is most likely due to the lower number of amorphous phases compared to other common aluminosilicate pozzolans.

The SiO_2/Na_2O ratio and compressive strength are generally directly proportional. The compressive strength was also affected by curing time. The continued reaction was observed after the 24 h thermal curing stage. The specimens in this study experienced 7 days, 28 days and 91 days of curing in a dry chamber at 23 ± 2 °C and RH of 65%. The maximum recorded compressive strength was obtained from specimens tested on day 91 corresponding to 4% and 2.5 of Na_2O/binder and SiO_2/Na_2O ratios, respectively.

Despite the shortcomings of EAFS as a sole precursor, the mechanical performance increased significantly after subjecting the specimens to an accelerated carbonation stage. After subjecting the specimens to a CO_2-enriched environment for 28 days, AAEAFS showed an average compressive strength increase of ~500%, with one case reaching 800% (i.e., from an initial 3.9 MPa in uncarbonated mixes to 31 MPa for carbonated ones). The maximum performance was observed in mixes with 12% and 1.0 for Na_2O/binder and SiO_2/Na_2O ratios, respectively, thus showing a shift in the optimal alkaline activator contents.

The sealed shrinkage test showed considerable dimensional variability over time. Great autogenous shrinkage may have occurred due to the continuous alkali activation reaction. Nevertheless, this phenomenon is still widely unknown and must be further researched.

The complete replacement of cement with alkali-activated aluminosilicate waste may translate into significant reductions in cost and minimal environmental impacts, especially with incorporating a forced carbonation curing stage using industrial CO_2-rich flue gases. Therefore, greater focus should be given to this curing technique in future research.

Author Contributions: Conceptualization, R.V.S.; methodology, R.V.S.; formal analysis, D.K.; investigation, D.K.; resources, R.V.S.; writing—original draft preparation, D.K. and G.L.; writing—review and editing, R.V.S. and J.d.B.; supervision, R.V.S. and J.d.B.; project administration, R.V.S.; funding acquisition, R.V.S. All authors have read and agreed to the published version of the manuscript.

Funding: This research was funded by FCT—Foundation for Science and Technology, through research projects PTDC/ECI-CON/29196/2017 (RInoPolyCrete) and EXPL/ECI-EGC/0288/2021 (ECO_2Alkrete).

Institutional Review Board Statement: Not applicable.

Informed Consent Statement: Not applicable.

Acknowledgments: The authors acknowledge the support of the CERIS Research Institute, IST, University of Lisbon. The authors would also like to acknowledge the support of HARSCO and SIKA for part of the materials provided for this experimental research.

Conflicts of Interest: The authors declare no conflict of interest.

References

1. Worldometer. Current World Population, World Population Sections. 2020. Available online: https://www.worldometers.info/world-population/ (accessed on 29 November 2021).
2. DBS Group Research. Production of cement in China from 1970 to 2020. DBS Group Research; CEIC; National Bureau of Statistics of China, Statista Research Department. 2018. Available online: https://www.statista.com/statistics/307647/china-production-volume-of-cement/ (accessed on 29 November 2021).
3. US Geological Survey. *Cement Production Worldwide from 1995 to 2020, USGS Mineral Commodity Summaries*; US Geological Survey: Reston, Virginia, 2021; p. 43.
4. IEA. *Cement*; IEA: Paris, France, 2021. Available online: https://www.iea.org/reports/cement (accessed on 10 December 2021).
5. Andrew, R.M. Global CO_2 emissions from cement production, 1928–2018. *Earth Syst. Sci. Data* **2019**, *11*, 1675–1710. [CrossRef]

6. Habert, G.; Roussel, N. Study of two concrete mix-design strategies to reach carbon mitigation objectives. *Cem. Concr. Compos.* **2009**, *31*, 397–402. [CrossRef]
7. Xiao, R.; Zhang, Y.; Jiang, X.; Polaczyk, P.; Ma, Y.; Huang, B. Alkali-activated slag supplemented with waste glass powder: Laboratory characterization, thermodynamic modelling and sustainability analysis. *J. Clean. Prod.* **2021**, *286*, 125554. [CrossRef]
8. Alberici, S.; de Beer, J.G.; van der Hoorn, I.; Staats, M. Fly Ash and Blast Furnace Slag for Cement Manufacturing. *BEIS Res. Pap.* **2017**, *19*, 1–34.
9. Teo, P.T.; Zakaria, S.K.; Salleh, S.Z.; Taib, M.A.A.; Mohd Sharif, N.; Abu Seman, A.; Mohamed, J.J.; Yusoff, M.; Yusoff, A.H.; Mohamad, M. Assessment of electric arc furnace (EAF) steel slag waste's recycling options into value added green products: A review. *Metals* **2020**, *10*, 1347. [CrossRef]
10. Fisher, L.V.; Barron, A.R. The recycling and reuse of steelmaking slags—A review. *Resour. Conserv. Recy.* **2019**, *146*, 244–255. [CrossRef]
11. Manso, J.M.; Gonzalez, J.J.; Polanco, J.A. Electric arc furnace slag in concrete. *J. Mater. Civ. Eng.* **2004**, *16*, 639–645. [CrossRef]
12. Autelitano, F.; Giuliani, F. Electric arc furnace slags in cement-treated materials for road construction: Mechanical and durability properties. *Constr. Build. Mater.* **2016**, *113*, 280–289. [CrossRef]
13. Abu-Eishah, S.I.; El-Dieb, A.S.; Bedir, M.S. Performance of concrete mixtures made with electric arc furnace (EAF) steel slag aggregate produced in the Arabian Gulf region. *Constr. Build. Mater.* **2012**, *34*, 249–256. [CrossRef]
14. Adegoloye, G.; Beaucour, A.-L.; Ortola, S.; Noumowé, A. Concretes made of EAF slag and AOD slag aggregates from stainless steel process: Mechanical properties and durability. *Constr. Build. Mater.* **2015**, *76*, 313–321. [CrossRef]
15. Alharbi, Y.R.; Abadel, A.A.; Elsayed, N.; Mayhoub, O.; Kohail, M. Mechanical properties of EAFS concrete after subjected to elevated temperature. *Ain Shams Eng. J.* **2021**, *12*, 1305–1311. [CrossRef]
16. Coppola, L.; Buoso, A.; Coffetti, D.; Kara, P.; Lorenzi, S. Electric arc furnace granulated slag for sustainable concrete. *Constr. Build. Mater.* **2016**, *123*, 115–119. [CrossRef]
17. Lam, M.N.-T.; Le, D.-H.; Jaritngam, S. Compressive strength and durability properties of roller-compacted concrete pavement containing electric arc furnace slag aggregate and fly ash. *Constr. Build. Mater.* **2018**, *191*, 912–922. [CrossRef]
18. Monosi, S.; Ruello, M.L.; Sani, D. Electric arc furnace slag as natural aggregate replacement in concrete production. *Cem. Concr. Compos.* **2016**, *66*, 66–72. [CrossRef]
19. Rooholamini, H.; Sedghi, R.; Ghobadipour, B.; Adresi, M. Effect of electric arc furnace steel slag on the mechanical and fracture properties of roller-compacted concrete. *Constr. Build. Mater.* **2019**, *211*, 88–98. [CrossRef]
20. San-José, J.T.; Vegas, I.; Arribas, I.; Marcos, I. The performance of steel-making slag concretes in the hardened state. *Mater. Des.* **2014**, *60*, 612–619. [CrossRef]
21. Santamaría, A.; Roji, E.; Skaf, M.; Marcos, I.; Gonzalez, J.J. The use of steelmaking slags and fly ash in structural mortars. *Constr. Build. Mater.* **2016**, *106*, 364–373. [CrossRef]
22. Lee, J.-Y.; Choi, J.-S.; Yuan, T.-F.; Yoon, Y.-S.; Mitchell, D. Comparing properties of concrete containing electric arc furnace slag and granulated blast furnace slag. *Materials* **2019**, *12*, 1371. [CrossRef]
23. Jiang, Y.; Ling, T.-C.; Shi, C.; Pan, S.-Y. Characteristics of steel slags and their use in cement and concrete—A review. *Resour. Conserv. Recy.* **2018**, *136*, 187–197. [CrossRef]
24. Xu, H.; Van Deventer, J. The geopolymerisation of alumino-silicate minerals. *Int. J. Miner. Process.* **2000**, *59*, 247–266. [CrossRef]
25. Ozturk, M.; Bankir, M.B.; Bolukbasi, O.S.; Sevim, U.K. Alkali activation of electric arc furnace slag: Mechanical properties and micro analyzes. *J. Build. Eng.* **2019**, *21*, 97–105. [CrossRef]
26. Hafez, H.; Kassim, D.; Kurda, R.; Silva, R.V.; de Brito, J. Assessing the sustainability potential of alkali-activated concrete from electric arc furnace slag using the ECO2 framework. *Constr. Build. Mater.* **2021**, *281*. [CrossRef]
27. Bernardo, G.; Marroccoli, M.; Nobili, M.; Telesca, A.; Valenti, G. The use of oil well-derived drilling waste and electric arc furnace slag as alternative raw materials in clinker production. *Resour. Conserv. Recy.* **2007**, *52*, 95–102. [CrossRef]
28. Muhmood, L.; Vitta, S.; Venkateswaran, D. Cementitious and pozzolanic behavior of electric arc furnace steel slags. *Cem. Concr. Res.* **2009**, *39*, 102–109. [CrossRef]
29. Parron-Rubio, M.E.; Perez-García, F.; Gonzalez-Herrera, A.; Rubio-Cintas, M.D. Concrete properties comparison when substituting a 25% cement with slag from different provenances. *Materials* **2018**, *11*, 1029. [CrossRef]
30. Zhao, J.; Wang, D.; Yan, P. Design and experimental study of a ternary blended cement containing high volume steel slag and blast-furnace slag based on Fuller distribution model. *Constr. Build. Mater.* **2017**, *140*, 248–256. [CrossRef]
31. Amin, M.N.; Khan, K.; Saleem, M.U.; Khurram, N.; Niazi, M.U.K. Influence of mechanically activated electric arc furnace slag on compressive strength of mortars incorporating curing moisture and temperature effects. *Sustainability* **2017**, *9*, 1178. [CrossRef]
32. Roslan, N.H.; Ismail, M.; Khalid, N.H.A.; Muhammad, B. Properties of concrete containing electric arc furnace steel slag and steel sludge. *J. Build. Eng.* **2020**, *28*, 101060. [CrossRef]
33. Traven, K.; Češnovar, M.; Ducman, V. Particle size manipulation as an influential parameter in the development of mechanical properties in electric arc furnace slag-based AAM. *Ceram. Int.* **2019**, *45*, 22632–22641. [CrossRef]
34. Bernal, S.A.; Provis, J.L.; Rose, V.; Mejía de Gutierrez, R. Evolution of binder structure in sodium silicate-activated slag-metakaolin blends. *Cem. Concr. Compos.* **2011**, *33*, 46–54. [CrossRef]
35. Türker, H.T.; Balçikanli, M.; Durmuş, İ.H.; Özbay, E.; Erdemir, M. Microstructural alteration of alkali activated slag mortars depend on exposed high temperature level. *Constr. Build. Mater.* **2016**, *104*, 169–180. [CrossRef]

36. Roslan, N.H.; Ismail, M.; Abdul-Majid, Z.; Ghoreishiamiri, S.; Muhammad, B. Performance of steel slag and steel sludge in concrete. *Constr. Build. Mater.* **2016**, *104*, 16–24. [CrossRef]
37. Nikolić, I.; Drinčić, A.; Djurović, D.; Karanović, L.; Radmilović, V.V.; Radmilović, V.R. Kinetics of electric arc furnace slag leaching in alkaline solutions. *Constr. Build. Mater.* **2016**, *108*, 1–9. [CrossRef]
38. Bakharev, T.; Sanjayan, J.G.; Cheng, Y.-B. Effect of elevated temperature curing on properties of alkali-activated slag concrete. *Cem. Concr. Res.* **1999**, *29*, 1619–1625. [CrossRef]
39. Češnovar, M.; Traven, K.; Horvat, B.; Ducman, V. The potential of ladle slag and electric arc furnace slag use in synthesizing alkali activated materials; the influence of curing on mechanical properties. *Materials* **2019**, *12*, 1173. [CrossRef]
40. Monkman, S.; Shao, Y. Assessing the carbonation behavior of cementitious materials. *J. Mater. Civ. Eng.* **2006**, *18*, 768–776. [CrossRef]
41. Pan, S.-Y.; Adhikari, R.; Chen, Y.-H.; Li, P.; Chiang, P.-C. Integrated and innovative steel slag utilization for iron reclamation, green material production and CO2 fixation via accelerated carbonation. *J. Clean. Prod.* **2016**, *137*, 617–631. [CrossRef]
42. Baciocchi, R.; Costa, G.; Di Bartolomeo, E.; Polettini, A.; Pomi, R. Carbonation of stainless steel slag as a process for CO_2 storage and slag valorization. *Waste Biomass Valoriz.* **2010**, *1*, 467–477. [CrossRef]
43. Europeia, U. Diretiva 98/83/CE do Conselho, de 3 de Novembro de 1998, Relativa à Qualidade da água Destinada ao Consumo Humano. 1998, Volume 330, pp. 32–54. Available online: https://eur-lex.europa.eu/legal-content/PT/TXT/PDF/?uri=CELEX:01998L0083-20151027&from=EN (accessed on 10 December 2021).
44. *EN-196-1*; Methods of Testing Cement—Part 1: Determination of Strength. European Committee for Standardization: Bruxelles, Belgium, 2005; p. 36.
45. *EN-1015-3*; Methods of Test for Mortar for Masonry—Part 3: Determination of Consistence of Fresh Mortar (by Flow Table). European Committee for Standardization: Bruxelles, Belgium, 1999; p. 10.
46. *EN-1015-13*; Methods of Test for Mortar for Masonry—Part 13: Determination of Dimensional Stability of Hardened Mortars. European Committee for Standardization: Bruxelles, Belgium, 1993; p. 20.
47. *EN-1015-11*; Methods of Test for Mortar for Masonry—Part 11: Determination of Flexural and Compressive Strength of Hardened Mortar. European Committee for Standardization: Bruxelles, Belgium, 1999; p. 12.
48. *EN-13295*; Products and Systems for the Protection and Repair of Concrete Structures. Test Methods. Determination of Resistance to Carbonation. European Committee for Standardization: Bruxelles, Belgium, 2004; p. 18.
49. Ozturk, M.; Akgol, O.; Sevim, U.K.; Karaaslan, M.; Demirci, M.; Unal, E. Experimental work on mechanical, electromagnetic and microwave shielding effectiveness properties of mortar containing electric arc furnace slag. *Constr. Build. Mater.* **2018**, *165*, 58–63. [CrossRef]
50. Peys, A.; Arnout, L.; Blanpain, B.; Rahier, H.; Van Acker, K.; Pontikes, Y. Mix-design parameters and real-life considerations in the pursuit of lower environmental impact inorganic polymers. *Waste Biomass Valori.* **2018**, *9*, 879–889. [CrossRef]
51. Abdollahnejad, Z.; Jesus, C.M.; Pacheco-Torgal, F.; Aguiar, J. One-part geopolymers versus Ordinary Portland Cement (OPC) mortars: Durability assessment. In Proceedings of the 2nd International Conference on Wastes: "Solutions, Treatments and Opportunities", Braga, Portugal, 11–13 September 2013; pp. 115–120.
52. Shearer, C.R.; Provis, J.L.; Bernal, S.A.; Kurtis, K.E. Alkali-activation potential of biomass-coal co-fired fly ash. *Cem. Concr. Compos.* **2016**, *73*, 62–74. [CrossRef]
53. Song, S.; Jennings, H.M. Pore solution chemistry of alkali-activated ground granulated blast-furnace slag. *Cem. Concr. Res.* **1999**, *29*, 159–170. [CrossRef]
54. Wang, S.-D.; Scrivener, K.L.; Pratt, P.L. Factors affecting the strength of alkali-activated slag. *Cem. Concr. Res.* **1994**, *24*, 1033–1043. [CrossRef]
55. Gu, Y.-m.; Fang, Y.-h.; You, D.; Gong, Y.-f.; Zhu, C.-h. Properties and microstructure of alkali-activated slag cement cured at below-and about-normal temperature. *Constr. Build. Mater.* **2015**, *79*, 1–8. [CrossRef]
56. Caijun, S.; Yinyu, L. Investigation on some factors affecting the characteristics of alkali-phosphorus slag cement. *Cem. Concr. Res.* **1989**, *19*, 527–533. [CrossRef]
57. Criado, M.; Fernández-Jiménez, A.; De La Torre, A.; Aranda, M.; Palomo, A. An XRD study of the effect of the SiO2/Na2O ratio on the alkali activation of fly ash. *Cem. Concr. Res.* **2007**, *37*, 671–679. [CrossRef]
58. Qureshi, M.N.; Ghosh, S. Effect of silicate content on the properties of alkali-activated blast furnace slag paste. *Arab. J. Sci. Eng.* **2014**, *39*, 5905–5916. [CrossRef]
59. Corinaldesi, V. Mechanical behavior of masonry assemblages manufactured with recycled-aggregate mortars. *Cem. Concr. Compos.* **2009**, *31*, 505–510. [CrossRef]
60. Corinaldesi, V.; Moriconi, G. Behaviour of cementitious mortars containing different kinds of recycled aggregate. *Constr. Build. Mater.* **2009**, *23*, 289–294. [CrossRef]
61. Vegas, I.; Azkarate, I.; Juarrero, A.; Frías, M. Design and performance of masonry mortars made with recycled concrete aggregates. *Mater. De Construcción* **2009**, *59*, 5–18. [CrossRef]
62. Jiménez, J.; Ayuso, J.; López, M.; Fernández, J.; De Brito, J. Use of fine recycled aggregates from ceramic waste in masonry mortar manufacturing. *Constr. Build. Mater.* **2013**, *40*, 679–690. [CrossRef]
63. Ledesma, E.F.; Jiménez, J.R.; Ayuso, J.; Fernández, J.M.; De Brito, J. Maximum feasible use of recycled sand from construction and demolition waste for eco-mortar production-Part-I: Ceramic masonry waste. *J. Clean. Prod.* **2015**, *87*, 692–706. [CrossRef]

64. Corinaldesi, V. Environmentally-friendly bedding mortars for repair of historical buildings. *Constr. Build. Mater.* **2012**, *35*, 778–784. [CrossRef]
65. Cuenca-Moyano, G.; Martín-Morales, M.; Valverde-Palacios, I.; Valverde-Espinosa, I.; Zamorano, M. Influence of pre-soaked recycled fine aggregate on the properties of masonry mortar. *Constr. Build. Mater.* **2014**, *70*, 71–79. [CrossRef]
66. Silva, R.; De Brito, J.; Dhir, R. Performance of cementitious renderings and masonry mortars containing recycled aggregates from construction and demolition wastes. *Constr. Build. Mater.* **2016**, *105*, 400–415. [CrossRef]
67. Atiş, C.D.; Kilic, A.; Sevim, U.K. Strength and shrinkage properties of mortar containing a nonstandard high-calcium fly ash. *Cem. Concr. Res.* **2004**, *34*, 99–102. [CrossRef]

applied sciences

MDPI

Article

Reduction in Drying Shrinkage and Efflorescence of Recycled Brick and Concrete Fine Powder–Slag-Based Geopolymer

Xiaoming Liu [1,*], Erping Liu [1] and Yongtong Fu [2]

1 School of Civil Engineering, Central South University, Changsha 410075, China
2 Shenzhen Municipal Design and Research Institute Co., Ltd., Shenzhen 518029, China
* Correspondence: 207076@csu.edu.cn

Abstract: It is an effective method to prepare geopolymer with recycled brick and concrete fine powder (RP) and slag as main materials for the resource utilization of construction waste. However, its hydration products have large drying shrinkage and high efflorescence risk under normal curing conditions. Until now, the durability of recycled brick and concrete fine powder–slag-based geopolymer (RPSG) has not been well documented, such as drying shrinkage and efflorescence. In this study, the effects of slag content, alkali equivalent and modulus on the durability properties of RPSG were evaluated. The results show: (1) Slag can significantly reduce the drying shrinkage and efflorescence of RPSG. (2) The potential for the efflorescence of RPSG increases with increasing alkali equivalent. The drying shrinkage of RPSG increases with the increase of alkali equivalent in the case of a low alkali equivalent (6 wt.% in this paper) and decreases with the increase of alkali equivalent in the case of a high alkali equivalent. (3) The drying shrinkage of RPSG increases with increasing modulus. In contrast, the degree of efflorescence decreases with increasing modulus. In this study, RP-S45-M1.3N6 (slag content: 45 wt.%; alkali equivalent: 6 wt.%; modulus: 1.3) is the best proportional design for RPSG with excellent durability. Compared to RP-S0-M1.3N6, the drying shrinkage of RP-S45-M1.3N6 is reduced by 76.32%, the capillary porosity is reduced by 60.9%, the visual efflorescence is significantly alleviated, and the early pH value is reduced by approximately 2.0. This paper systematically analyzed the drying shrinkage pattern and the efflorescence pattern of RPSG, which has a positive significance for promoting the recycling of RP from construction waste.

Keywords: recycled brick and concrete powder; slag; geopolymer; drying shrinkage; efflorescence

1. Introduction

With the rapid development of global urbanization, the emission of construction waste is increasing rapidly. According to the statistics, current construction waste accounts for about 30% of urban solid waste [1–5], and more than 10 billion tons of construction waste are generated worldwide every year, 80% of which are waste bricks and concrete [6,7]. At present, the treatment method for construction waste is mainly stacking and landfill, which not only wastes land but also seriously endangers the natural environment [8]. On the other hand, the exploitation of natural building materials in China is protected strictly. Therefore, the recycling of construction waste can not only protect the environment but also save resources and alleviate the current shortage of natural building materials.

In recent years, recycled aggregates produced from construction waste have been widely used in recycled concrete [9–11], recycled mortar [12,13], roadbase [14] and other fields. In the process of producing recycled aggregate from construction waste, a large amount of fine powder with a size of less than 0.16 mm, which is called recycled fine powder, is produced. Compared with untreated construction waste, the long-term accumulation of recycled fine powder will not only occupy land but also produce dust, pollute water resources and do greater harm to the environment [15].

Geopolymers have many advantages, such as faster hardening, higher strength, excellent acid and alkali resistance, lower carbon emission, and are considered to be the most potential low-carbon green cementitious material to replace ordinary Portland cement (OPC). The reaction mechanism is that the silicon oxygen bond and aluminum oxygen bond in the geopolymer precursor materials break in an alkaline environment to form AlO_4 and SiO_4 tetrahedral monomers, and then polymerize and recombine to form a three-dimensional network structure [16,17]. The preparation of geopolymers from some wastes, such as metakaolin, slag, fly ash, rice husk ash, etc., with aluminosilicate as the main component is a research frontier [16,17]. As a material mainly composed of silica and alumina, recycled fine powder has the potential to develop geopolymers. At present, the use of recycled fine powder as a precursor material of geopolymers has been studied [18–23]. Meanwhile, the mechanical properties of recycled fine powder-based geopolymers have been widely qualified and have great potential to replace OPC [22–25].

Drying shrinkage is one of the main reasons that durability and applicability of concrete structures are reduced [26]. The shrinkage of concrete will cause tensile stress due to the restraint of itself and external conditions. When the shrinkage is large, it will cause harmful crack formation and eventually lead to serious crack damage of concrete [27]. Compared with OPC, the geopolymer will produce greater dry shrinking under dry conditions, restricting its actual engineering applications [28–30]. Previous studies [31–33] have shown that the shrinkage of geopolymer concrete is 2–4 times that of ordinary concrete. Many scholars have studied how to reduce the drying shrinkage of alkali-activated binders in terms of alkali equivalents, activator types, mineral admixtures, chemical additives and curing conditions [31]. Duxson et al. [34,35] found that Si/Al and gel structure are two main parameters of sodium-based geopolymers, which have a direct impact on shrinkage. Deb et al. [36] showed that both incorporating a small amount of fly ash and reducing the modulus of modified sodium silicate reduced the drying shrinkage of geopolymer concrete cured at room temperature. Ma et al. [37] also found that lowering the modulus of activators can effectively reduce the drying shrinkage of geopolymer concrete. Additives are one of the effective means to reduce the drying shrinkage of geopolymer concrete. There were popular additives, such as magnesium oxide [38], calcium oxide [39], gypsum [40], expansion agent [41] and absorbent polymer materials [42]. Hardjito et al. [43] found that increasing the curing temperature is another measure to reduce the drying shrinkage of geopolymers. Palacios et al. [44] found that the drying shrinkage of geopolymer can also be reduced by increasing the curing humidity.

However, the chemical nature of recycled fine powders differs significantly from that of other types of precursor materials, and it remains unknown whether these strategies will be effective in reducing the drying shrinkage of recycled fine powder-based geopolymers.

In geopolymer systems, efflorescence is mainly caused by alkali metal cations freeing to the surface of the material and reacting with carbon dioxide to form substances such as sodium carbonate and sodium bicarbonate [45,46]. The highly alkaline reaction conditions of geopolymers will inevitably result in a much higher degree of efflorescence than that of OPC. Efflorescence not only affects the esthetics of the building material, but also reduces the strength of the material and affects the durability of the structure. The efflorescence of alkali-activated binders severely limits their practical engineering applications [47–49]. Therefore, it is crucial to limit the efflorescence of geopolymers. Zhang et al. [50] found that the addition of 20% slag to fly ash-based cementitious materials were able to reduce the rate of efflorescence in specimens, while efflorescence could also be inhibited by high-temperature curing. A similar result was found by Kani et al. [51], who showed that the addition of alumina-rich precursors or the addition of slag was effective in inhibiting alkali efflorescence. In contrast, Xiao et al. [52] found that the addition of less than 50% mass of slag to fly ash did not significantly change the degree of efflorescence in the specimens, which showed higher levels of efflorescence with slag content above 50%. Therefore, the relationship between the amount of slag and efflorescence is not entirely positive. Zhang et al. [50] found that soluble silica reduced the porosity of the specimens to inhibit efflorescence at the same alkali equivalent, and also pointed out that the efflorescence degree of NaOH-activated geopolymers was lower than that of sodium silicate-activated geopolymers. With regard to the effect of alkali equivalent on the efflorescence properties, Saha et al. [53] concluded that reducing the alkali equivalent would reduce the degree of efflorescence, but when the alkali equivalent was less, it would again affect the strength of the specimen, so a relative compromise was needed. Wu et al. [54] suggested that water evaporation is an important factor affecting the efflorescence of fly ash-based geopolymers. Tan et al. [55] first investigated efflorescence mitigation in recycled fine powder-based geopolymers and demonstrated that both slag and metakaolin could mitigate the degree of efflorescence in recycled fine powder based-geopolymers.

As can be seen from the above, recycled fine powder-based geopolymers have great potential to replace OPC. However, little research has been reported on the durability of recycled fine powder-based geopolymers in terms of drying shrinkage and efflorescence, which restricts the promotion and application of recycled fine powder-based geopolymers.

In this paper, recycled brick and concrete fine powder–slag-based geopolymer (RPSG) was prepared by using recycled brick and concrete fine powder (RP) and slag. The effects of different slag contents, alkali equivalents and moduli on drying shrinkage and efflorescence were studied. X-ray diffraction (XRD), Fourier transform infrared spectroscopy (FTIR), thermogravimetry (TG) and scanning electron microscope (SEM) were used to analyze the hydration products and their microstructure. The reaction mechanism of RP and slag was discussed, and the drying shrinkage and efflorescence patterns of the RPSG were systematically analyzed.

2. Research Significance

Firstly, this paper systematically investigates the drying shrinkage pattern and the efflorescence of RPSG, which is a useful addition to the research related to the durability performance of recycled fine powder-based geopolymers. The research results of this paper are beneficial to promote the practical application of recycled fine powder-based geopolymers. As a low-carbon green cementitious material, RPSG could replace OPC in part of the engineering field, which has great practical significance to the sustainable development of cement industry.

3. Materials and Methods

3.1. Materials

Recycled brick and concrete fine powder (RP): it was obtained from recycled brick and concrete fine aggregate (0–4.75 mm), mainly including clay bricks, mortar, tiles, ceramics, etc. It was dried to constant weight at 105 °C, and then passed through a standard sieve with 2.36 mm. Then, it was crushed and sieved by a small laboratory crusher to obtain fine powder with a particle size less than 0.16 mm. The process is shown in Figure 1. Considering the crushing efficiency, the crushing time was 40 s, and the powder yield was 56.41%. The main chemical composition of RP can be seen in Table 1. SiO_2 and Al_2O_3 are the two main components. The XRD results, morphology and particle size distribution are shown in Figures 2–4, respectively.

Slag: S95 grade granulated blast furnace slag with a density of 2.9 kg/cm^3, specific surface area of 412 m^2/kg and 28-day activity index of 95.5% was used. The main chemical components can be seen in Table 1.

Sodium silicate solution: The liquid sodium silicate solution with a modulus of 3.2 kg/cm^3 and water content of 65% was used in this test.

Sodium hydroxide: NaOH used in the test was flake NaOH with a purity of 99.0%.

Water (W): Except for the pH value test, which used purified water, the water used for the tests was laboratory tap water.

Figure 1. The preparation process of RP.

Table 1. The chemical compositions of RP and slag.

	SiO_2	Al_2O_3	CaO	Fe_2O_3	K_2O	MgO	Na_2O	TiO_2	Others
RP	49.4	20.2	17.29	4.7	2.1	1.4	1.3	0.6	3.0
S	30.3	14.2	39.3	0.7	0.4	7.1	0.4	0.5	7.1

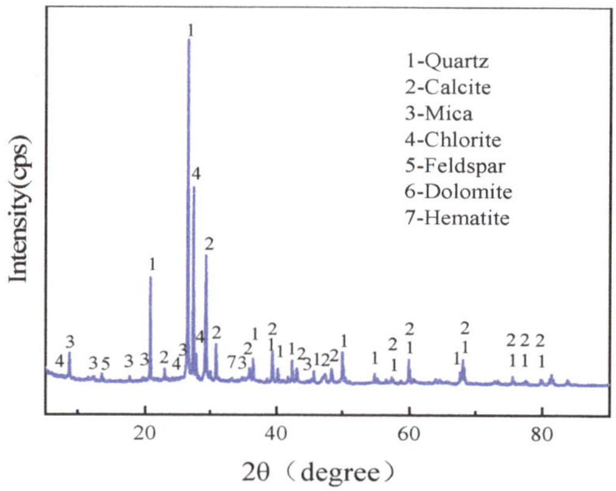

Figure 2. The XRD patterns of RP.

Figure 3. The SEM images of RP [56].

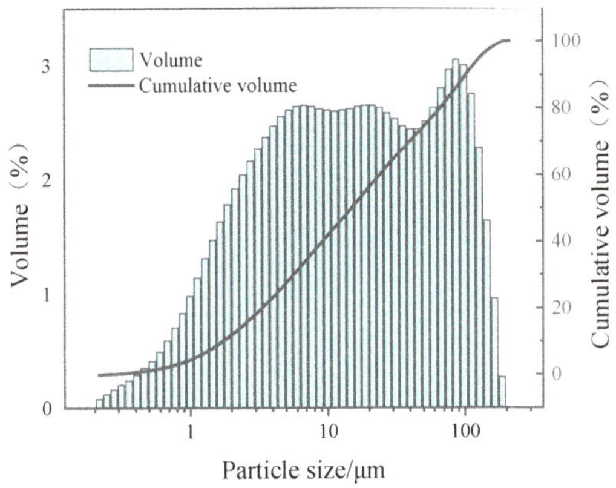

Figure 4. The particle size distribution of RP.

3.2. Mixture Proportion Design

In order to study the influence of slag content on the properties of RPSG, the specimens were prepared with slag contents of 0 wt.%, 15 wt.%, 30 wt.% and 45 wt.% (calculated by the total mass of RP and slag), and the fixed modulus and alkali equivalent were 6 wt.% and 1.3, respectively. In order to study the effect of modulus and alkali equivalent on the performance of RPSG, the fixed slag content was 30 wt.%, and the moduli were adjusted to 0.9, 1.1, 1.3, 1.5 and 1.7, and the alkali equivalents were 3 wt.%, 6 wt.%, 9 wt.%, 12 wt.% and 15 wt.%, respectively. The mixture proportions are shown in Table 2. The specimens were numbered according to the slag content, modulus and alkali equivalent. For example, for RP-S0-M1.3N6, RP represents recycled brick and concrete fine powder, S0 represents 0 wt.% slag content, M1.3 represents 1.3 modulus of alkali activator solution, and N6 represents 6 wt.% alkali equivalent of composite cementitious system. The water binder ratio in this test was all 0.35. In order to investigate the hydration products and microstructure of RPSG, specimens RP-S0-M1.3N6 and RP-S30-M1.3N6 were selected for further micro-analysis.

Table 2. The mixture proportions of RPSG.

Mix ID	[a] W/[b] B	RP	Slag	[c] W	NaOH	Na_2SiO_3
RP-S0-M1.3N6	0.35	500.00	0.00	107.95	23.09	142.42
RP-S15-M1.3N6	0.35	425.00	75.00	107.95	23.09	142.42
RP-S30-M1.3N6	0.35	350.00	150.00	107.95	23.09	142.42
RP-S45-M1.3N6	0.35	275.00	225.00	107.95	23.09	142.42
RP-S30-M0.9N6	0.35	350.00	150.00	132.75	27.90	98.60
RP-S30-M1.1N6	0.35	350.00	150.00	120.35	25.49	120.51
RP-S30-M1.5N6	0.35	350.00	150.00	95.55	20.69	164.33
RP-S30-M1.7N6	0.35	350.00	150.00	83.15	18.28	186.24
RP-S30-M1.3N3	0.35	350.00	150.00	141.48	11.54	71.21
RP-S30-M1.3N9	0.35	350.00	150.00	74.43	34.63	213.63
RP-S30-M1.3N12	0.35	350.00	150.00	40.91	46.18	284.84
RP-S30-M1.3N15	0.35	350.00	150.00	7.38	57.72	356.06

[a] W (Water content) = water content of Na_2SiO_3 solution + [c] W. [b] B = RP + S + solid content of Na_2SiO_3 solution.
[c] W—additional water.

3.3. Test Method

(1) Dry shrinkage test: The dry shrinkage test was conducted in accordance with the Test Method for Dry Shrinkage of Cement Mortar in China (JC/T 603-2004), and the mix proportions are shown in Table 2. The modified sodium silicate solution with the appropriate modulus and alkali equivalent was prepared 24 h in advance, then the

modified sodium silicate solution was mixed with RP and slag and poured into the cement mortar mixer. After stirring on low speed for 3 min, a scraper was used to scrape up the slurry from the bottom of the mortar mixing pot and stir manually for 2 min, then stirring continued on high speed for 3 min. The mixed slurry was rapidly poured into a 25 mm × 25 mm × 280 mm mold with embedded metal nails at both ends. After 20 s of vibrating, it was cured with plastic film at room temperature for 48 h, then stripped and transferred to 20 ± 2 °C and 50 ± 5% relative humidity for 28 days of dry shrinkage curing. BC-300 cement length meter was used to measure the size change, and the length was measured once every 24 h. Each group contained 3 specimens, and the average value of the results was analyzed.

(2) Capillary porosity test: According to Table 2, the rectangular specimens with 40 mm × 40 mm × 40 mm were wrapped by film under standard conditions for 28 days, and then crushed into pieces. The capillary porosity of the pieces was determined using the saturated water method [45].

(3) Rapid efflorescence test: The specimen preparation method was the same as that of the capillary porosity test. The specimens were wrapped in plastic film and maintained for 28 days under standard conditions, then placed in plastic containers with an appropriate amount of tap water so that the bottom of the specimen was immersed in water for approximately 5 mm. A camera was used to record the images of the specimens over time. The temperature and relative humidity during the experiment were 25 °C and 55%, respectively. During the test, the proper amount of water was added to the plastic container every 5 h to maintain a fixed water level at the bottom of the specimen. After 7 days, the addition of water to the container was stopped, and the water in the container underwent complete evaporation. The white crystals and broken epidermis on the surface of the specimens were scraped out with a spatula and then weighed.

(4) PH value test: The specimens were crushed into granules from 28 days of film curing. Solid particles with a size range of 0.6–1.18 mm were selected, and then immersed in purified water. The pH value of the immersion solution was measured every 10 min for the first hour. After 1 h, measurements were taken every 1 h until the pH value of the immersion solution did not modify significantly and then stopped. In this experiment, the ratio of solid to liquid was 1:50.

(5) XRF: 3 g powder specimens were prepared. An X-ray fluorescence spectrometer was used, and the test mode was in the form of oxide.

(6) XRD: The specimen at the specified age was first crushed into small pieces. The pieces < 1.18 mm were taken and immersed in anhydrous alcohol for 7 days to terminate the hydration process, then dried at 50 °C to constant weight. Finally, the specimens were ground into powder, and 1 g was selected for the XRD test after passing 200 mesh sieve. X-ray diffractometer with scanning speed of 2°/min and scanning angle range of 5–90° was adopted.

(7) FTIR: The specimen preparation method was the same as that of XRD. The test mode was conventional powder tablet. The wavenumber range was 400–4000 cm^{-1}.

(8) TG: The specimen preparation method was the same as that of XRD. 20 mg powder specimen was selected and tested with a thermogravimetric analyzer. The test temperature range was room temperature ~1000 °C, the test gas atmosphere was nitrogen, and the heating rate was 10 °C/min.

(9) SEM: Block specimens with thickness less than 1 cm, diameter ≤ 1 cm and relatively flat fracture surface were selected and dried at 50 °C to constant weight. A scanning electron microscope was used to test the morphology.

4. Results and Discussion

4.1. Drying Shrinkage

4.1.1. Effect of Slag Content on Drying Shrinkage

Figure 5 shows the drying shrinkage of RPSG with different slag contents within 28 days. It can be seen that the drying shrinkage decreases with the increase of slag content.

The 28-day drying shrinkage of RP-S0-M1.3N6, RP-S15-M1.3N6, RP-S30-M1.3N6 and RP-S45-M1.3N6 are 2.98%, 2.57%, 0.93% and 0.70%, respectively. The drying shrinkage of RPSG with slag content of 15 wt.%, 30 wt.% and 45 wt.% are 13.53%, 68.69% and 76.32% lower than that of RPSG without slag. The drying shrinkage value of RPSG decreases with the increase of Ca/Si, which is contrary to the traditional conclusion that the drying shrinkage value of alkali-activated cementitious materials increases with the increase of Ca/Si [57], such as alkali-activated fly ash/slag, Kaolin/slag, silica fume/slag and other cementitious systems. It is noted that the RPSG with 60 wt.% slag content had a large drying shrinkage value, and cracking occurred on the third day after forming.

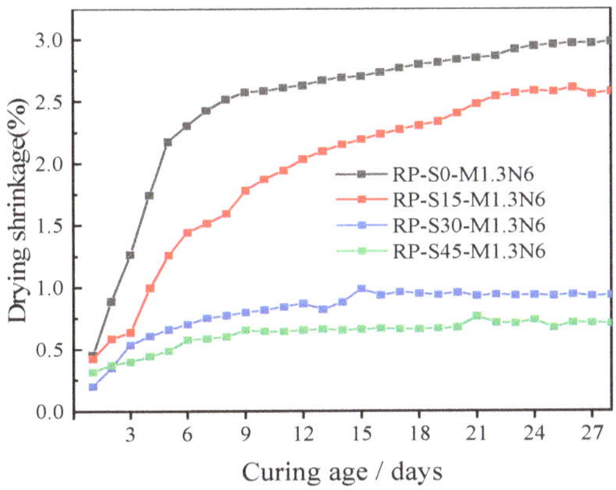

Figure 5. Drying shrinkage results of RPSG with different slag contents.

The capillary pressure theory shows that the essence of shrinkage is that the free water evaporates from the saturated mesoporous pores in a dry environment, and the surface tension in the pore solution forms capillary pressure. Thus, the curved moon surface of liquid–gas was formed, resulting in the reduction of slurry volume. This mechanism is applicable to alkali-activated cementitious materials [58]. The pore size distribution is the main factor affecting its drying shrinkage. However, further analysis is required for the cementitious system with two or more precursors, and a variety of factors should be considered [59]. According to the Mackenzie Bentz equation, the parameters affecting the drying shrinkage of RPSG mainly include bulk modulus, bulk modulus of solid material, pore saturation and capillary tension of liquid phase [60]. Capillary tension is mainly composed of surface tension, contact angle and Kelvin radius. The surface tension and contact angle of the pore solution are the same for different mixtures under the same humidity and are mainly affected by the pore diameter. According to the above theory, the effective factors of drying shrinkage caused by capillary tension are pore diameter, pore saturation and bulk modulus [2].

Slag can promote the production of hydrophilic products such as NASH and CASH, making the slurry denser, so as to improve the bulk modulus of the slurry and the ability of the slurry to resist drying shrinkage and deformation. Slag can refine pore structure, reduce pore size and increase capillary tension. The particle size of slag particles is smaller than that of RP. Slag particles are dispersed to the surface of RP particles during mixing. After hydration, the products can fill pores and form closed pores, improving pore saturation [59]. It can be seen from the above analysis that the drying shrinkage result is the compound action of the bulk modulus, pore diameter and pore saturation. The results show that bulk modulus is the dominant factor affecting the drying shrinkage of RPSG when the slag content is small. Therefore, slag can reduce the drying shrinkage of RPSG. When the slag content is high, the dominant factors affecting the drying shrinkage are pore diameter and pore saturation [61].

The activity of RP under alkaline conditions is much lower than that of slag. After slag replaces RP, the consumed water through chemical reaction and the active SiO_2 in sodium silicate solution increase, and the residual water and active SiO_2 in the slurry decrease. Relevant studies [30] show that silica gel with high water content is easy to dehydrate under dry conditions, resulting in larger shrinkage, while the addition of slag can reduce the water content in the slurry and consume some sodium silicate, thus reducing the drying shrinkage. In addition, RP can be used as an "internal curing agent" to reduce the drying shrinkage of alkali-activated slag cementitious systems due to its internal porous and water absorption [2,62]. Therefore, slag can significantly reduce the drying shrinkage of RPSG.

4.1.2. Effect of Alkali Equivalent on Drying Shrinkage

Figure 6 shows the drying shrinkage of RPSG with different alkali equivalents within 28 days. The test results show that the drying shrinkage does not have a certain positive correlation with the alkali equivalent. The 28-day drying shrinkage of RP-S30-M1.3N3, RP-S30-M1.3N6, RP-S30-M1.3N9, RP-S30-M1.3N12 and RP-S30-M1.3N15 are 0.40%, 0.93%, 0.83%, 0.57% and 0.18%, respectively. When the alkali equivalent is small (In this paper, the alkali equivalent is less than 6 wt.%), drying shrinkage increases with the increase of alkali equivalent. This is consistent with some current studies on alkali-activated binders. It can be explained that the increase of alkali concentration increases the proportion of pores and gel pores in the slurry, reduces the content of large pores [62] and increases the capillary tensile stress. In addition, a certain alkaline environment is the basis of slag hydration, and an appropriate increase in alkali equivalent can promote slag hydration [63]. Therefore, for the low alkali equivalent specimens, slag hydration is incomplete, and part of the slag that is not hydrated fills the large-sized pores of the recycled brick mix microfine, which slows down water evaporation. At the same time, slag that is not hydrated has a "microskeleton" effect, which can also reduce the drying shrinkage performance of the specimen. These reasons together contribute to the low alkali equivalent specimen drying shrinkage being small. When the alkali equivalent is large, drying shrinkage decreases with the increase of alkali equivalent. The specimens with high alkali equivalent are unstable, as more alkali will promote the precipitation of $Ca(OH)_2$ and the formation of low Ca/Si gel, reduce mesoporous proportion of the matrix and reduce drying shrinkage of the specimen. For example, the drying shrinkage of RP-S30-M1.3N9, RP-S30-M1.3N12 and RP-S30-M1.3N15 decreased by 11.33%, 39.27% and 80.71%, respectively, compared with RP-S30-M1.3N12. In addition, the porosity of the RPSG system is higher than that of the traditional alkali-activated cementitious system, and there are more macropores in the slurry. Although it has higher activity under high alkali equivalent, it cannot actually consume the excessive alkali component in the system, and the excessive alkali will generate white crystals in the macropores in the slurry and on the surface of the specimen under a dry environment. After crystallization is generated in the internal pores of the slurry, a certain amount of crystallization pressure will be generated on the pore wall to offset the capillary tensile stress caused by the mesoporous pores so that the drying shrinkage value of the specimen is reduced and even expanded with the extension of the curing age. It can be well verified from the drying shrinkage results of the specimen RP-S30-M1.3N15. It can be seen from Figure 6 that the specimen RP-S30-M1.3N15 with 15 wt.% alkali equivalent has a large shrinkage in the early stage, but the shrinkage value decreases with the increase of age. It can be observed that white alkali crystals are separated from the slurry of RPSG, and some visible cracks appear on the surface of the specimen. These cracks are not shrinkage cracks caused by capillary tension but expansion cracks caused by pressure formed by free alkali crystallization. However, specimens with high alkali equivalents are very unstable and prone to large cracks under ambient conditioning conditions, contributing to a reduction in specimen strength. Therefore, increasing the alkali equivalent is not a good measure to reduce the drying shrinkage of RPSG.

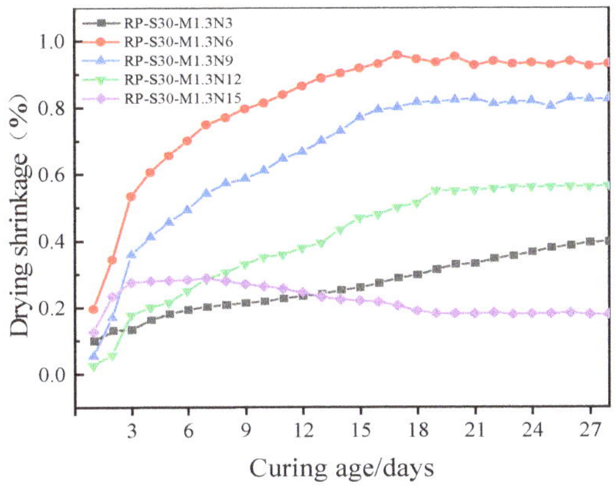

Figure 6. Drying shrinkage results of RPSG with different alkali equivalents.

4.1.3. Effect of Modulus on Drying Shrinkage

Figure 7 shows the drying shrinkage of RPSG with different moduli within 28 days. The test results show that the drying shrinkage has a certain positive correlation with the modulus. When the alkali equivalent is constant, drying shrinkage increases with the increase of modulus, which is consistent with the conclusions drawn by other studies [29]. The 28-day drying shrinkage of RP-S30-M0.9N6, RP-S30-M1.1N6, RP-S30-M1.3N6, RP-S30-M1.5N6 and RP-S30-M1.7N6 are 0.42%, 0.82%, 0.93%, 1.17% and 1.32%, respectively. For the RPSG with fixed alkali equivalent, the main factor affecting drying shrinkage is the pore distribution in the binders. When the alkali equivalent is constant, the content of sodium silicate increases relatively with the increase in modulus. Generally speaking, the increase of sodium silicate will lead to the decrease of the total porosity of the slurry and the increase of mesopore ratio, while the increase of mesopore volume will lead to the increase of capillary stress, thus increasing the drying shrinkage of the specimen. In addition, excessive sodium silicate will not react completely, and it is very easy to lose water under dry conditions, resulting in dry shrinkage of the slurry.

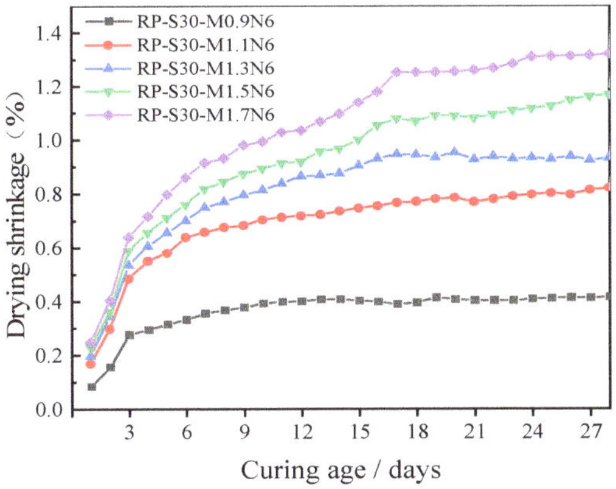

Figure 7. Drying shrinkage results of RPSG with different moduli.

From the above analysis, it is clear that RPSG has large drying shrinkage, with maximum drying shrinkage reaching 2.98% after 28 days, which is far higher than the drying shrinkage of metakaolin-based polymer (approx. 0.4–0.6%) [64]. However, the drying shrinkage of RPSG could be reduced by optimizing the mix ratio, such as the specimen RP-S30-M1.3N3, whose drying shrinkage is only 0.40%

4.2. Capillary Porosity

4.2.1. Effect of Slag Content on Capillary Porosity

The permeability of the specimen is one of the important factors affecting its efflorescence, and capillary porosity is an important indicator for evaluating the permeability of the specimen. Figure 8 shows the capillary porosity of RPSG with different slag contents. It can be seen from Figure 8 that the capillary porosity of RPSG decreases with the increase of slag content. The capillary porosity of the specimens RP-S0-M1.3N6, RP-S15-M1.3N6, RP-S30-M1.3N6, RP-S45-M1.3N6 are 0.33%, 0.24%, 0.18% and 0.13%, respectively. Compared with the specimens without slag, the capillary porosity of the specimens with 15 wt.%, 30 wt.% and 45 wt.% slag content decreased by 27.3%, 45.5% and 60.6%, respectively. This is because the incorporation of slag can introduce more active calcium components into the matrix, which significantly increases the content of active components, thereby promoting the formation of hydration products such as CSH, CASH and NASH. The generated hydration products can fill the pores of the RPSG and improve the density of the slurry, which in turn leads to a decrease in the capillary porosity of the specimen. In addition, unreacted slag can also fill the pores of the RPSG, improve the density of the slurry and reduce the capillary porosity of the slurry.

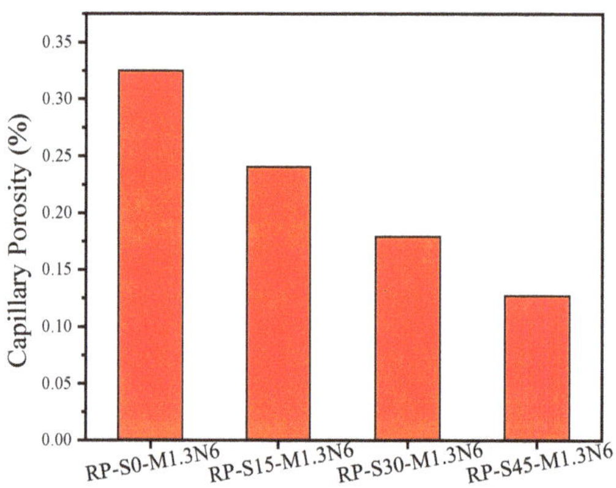

Figure 8. Capillary porosity of RPSG with different slag contents.

4.2.2. Effect of Alkali Equivalent on Capillary Porosity

Figure 9 shows the capillary porosity of RPSG with different alkali equivalents. As can be seen from Figure 9, the capillary porosity of RPSG does not have a simple linear relationship with the alkali equivalent. The capillary porosity of specimen RP-S30-M1.3N6 was reduced by 48.90% compared to that of specimen RP-S30-M1.3N3, while the capillary porosity of specimens RP-S30-M1.3N9 and RP-S30-M1.3N12 remained essentially unchanged compared to that of specimen RP-S30-M1.3N6. Meanwhile, the capillary porosity of RP-S30-M1.3N15 increased by 50.5% compared with that of RP-S30-M1.3N12. The silica and aluminum components in RP need to be dissolved and polymerized in an alkaline environment to form NASH hydration gel. The active CaO in slag also needs an alkaline environment to dissolve and polymerize to form CSH and CASH gels. Properly increasing the alkali equivalent and improving the alkaline environment of the geopolymer reaction can promote the formation of more NASH, CSH and CASH gels, thereby improving the compactness of the specimen. When the alkali equivalent exceeds a certain threshold, the strong alkali environment will not only inhibit the dissolution of the silicon–alumina components in the RP, but also guide the active CaO in slag to form precipitation and adhere to the particle surface, hindering the progress of the RPSG hydration reaction [65]. Therefore, continuously increasing the alkali equivalent cannot continuously improve the compactness of the specimen and may even damage the compactness of the specimen.

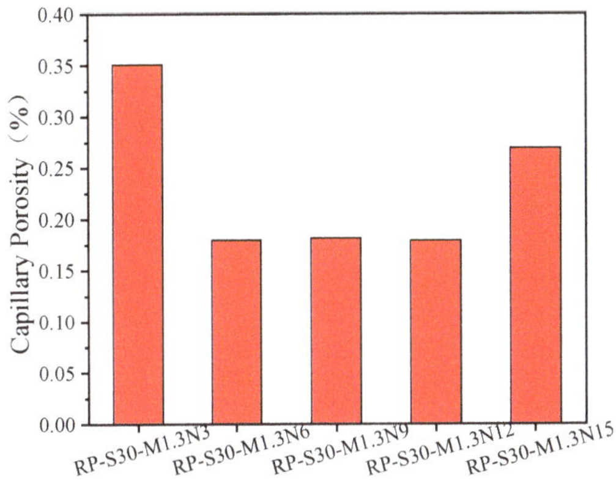

Figure 9. Capillary porosity of RPSG with different alkali equivalents.

4.2.3. Effect of Modulus on Capillary Porosity

Figure 10 shows the capillary porosity of RPSG with different moduli. It can be seen from Figure 10 that within a certain range, the capillary porosity of RPSG decreases with the increase of the modulus. Compared to the capillary porosity of specimen RP-S30-M0.9N6, the capillary porosity of specimens RP-S30-M1.1N6, RP-S30-M1.3N6, RP-S30-M1.5N6 and RP-S30-M1.7N6 was reduced by 4.9%, 10.8%, 19.4% and 19.4%, respectively. This is because the high modulus modified sodium silicate solution contains more active SiO_2, which is conducive to the formation of a denser gel and thus reduces the capillary porosity of the specimens [66]. Meanwhile, the hydration reaction of RPSG requires a certain alkaline environment, and the weak alkaline environment is not conducive to the formation of hydration and gelation products such as CSH, CASH and NASH. Therefore, high modulus RPSG may have higher capillary porosity.

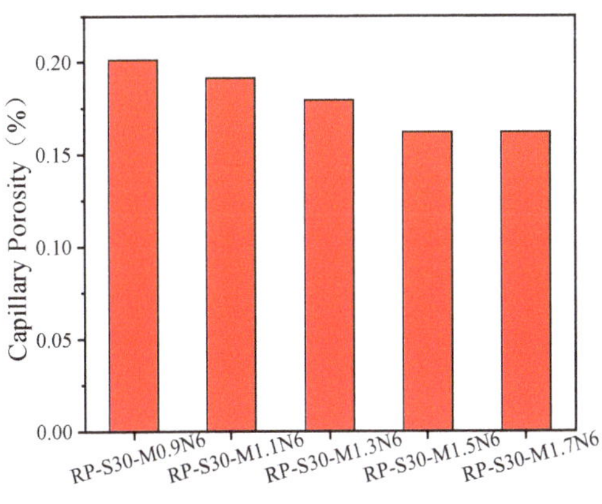

Figure 10. Capillary porosity of RPSG with different moduli.

4.3. Visual Efflorescence

4.3.1. Effect of Slag Content on Visual Efflorescence

Visual efflorescence can intuitively reflect the degree of efflorescence of the specimen. Figure 11 shows the visual efflorescence changes of RPSG with different slag contents at 1 day, 3 days and 7 days. It can be seen from Figure 11 that after 1 day of immersion, the white crystalline substance first appears around the top of the specimen RP-S0-M1.3N6 without slag addition, while there is no obvious change around the top of the specimens RP-S15-M1.3N6, RP-S30-M1.3N6 and RP-S45-M1.3N6. After 3 days of immersion, white crystalline substances precipitated around and at the top of the specimen RP-S0-M1.3N6, while the specimens RP-S15-M1.3N6 and RP-S45-M1.3N6 only showed a small circle of crystalline substances at the bottom above the immersed part. However, crystallization appeared on the top of the specimen RP-S30-M1.3N6 with a slag content of 30 wt.%. This is due to the joint porosity between the top and bottom of the specimen, which causes the abnormal phenomenon that the surface of the specimen crystallizes prematurely. After 7 days of immersion, the surface of specimen RP-S0-M1.3N6 starts peeling on the surface of the crystallization pressure gauge, and part of the paste swells or falls off. This can be attributed to the fact that crystals are formed when alkaline-rich solutions inside or on the surface of the specimen pores react with CO_2 and continue to migrate with the moisture. The crystals continue to grow, generating a crystallization pressure. When the pressure exceeds the internal tensile strength of the specimen, cracks and rupture occurs [67]. The degree of efflorescence of other specimens is not significantly different from that of 3 days. It can be concluded from the test phenomena that the addition of slag can significantly reduce the visual efflorescence of RPSG and reduce the visual efflorescence rate.

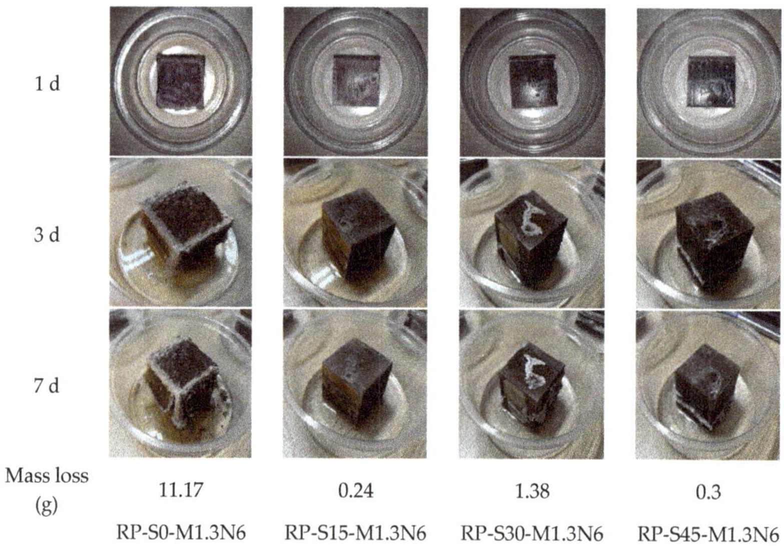

Figure 11. Visual efflorescence of RPSG with different slag contents.

Visual efflorescence can intuitively show the phenomenon of efflorescence when the specimen is immersed in water at the bottom, but this test method cannot quantify the degree of efflorescence in the slurry. Figure 11 shows the mass loss results of RPSG with different slag contents due to efflorescence. The mass loss of RP-S0-M1.3N6 was the highest, reaching 11.17 g, and the results were in good agreement with visual efflorescence. Unlike the decrease in capillary porosity with increasing slag content as shown in Figure 8, specimen RP-S15-M1.3N6 showed minimal visual efflorescence and mass loss. At the same time, there is an obvious macro-crack above the submerged part at the bottom of the specimen. This is because alkali-activated materials can occur not only on the surface of the specimen but also in the larger capillary pores on the surface or cracks in the interior of the specimen. Efflorescence will produce crystallization pressure and force cracks to expand further, which will further enhance the phenomenon of efflorescence [45]. At the same time, the appearance of efflorescence substances will also be different due to the speed of evaporation, so the efflorescence substances cannot be completely scraped out during the test. In addition, the visual image of efflorescence will also have certain differences due to the different shapes of the crystalline substances.

4.3.2. Effect of Alkali Equivalent on Visual Efflorescence

Figure 12 shows the visual efflorescence changes of RPSG with different alkali equivalents at 1 day, 3 days and 7 days. It can be seen from Figure 12 that after 1 day of immersion, the specimens RP-S30-M1.3N3 and RP-S30-M1.3N6 did not produce obvious crystalline substances, while the specimens RP-S30-M1.3N9, RP-S30-M1.3N12 and RP-S30-M1.3N15 have a little crystalline material on the surface. After 3 days of immersion, crystalline substances appeared on the surface and surrounding of all the specimens, and the most crystalline substances appeared in the specimens RP-S30-M1.3N3 and RP-S30-M1.3N15. After 7 days of immersion, visual efflorescence of the specimens was more obvious. The activity of RP and slag is weaker at a low alkali equivalent, and the specimen has smaller strength and larger capillary porosity (as shown in Figure 9), so the rate of efflorescence of specimen RP-S30-M1.3N3 is faster. However, the specimen RP-S30-M1.3N15 with a high alkali equivalent also had an obvious efflorescence phenomenon after immersion in water for 1 day, and the degree of efflorescence was high. The reason for this is that specimens with a high alkali equivalent are not stable and the crystallization pressure generated by the excess residual alkali in the pores causes cracks to form in the top and bottom of the specimen. This leads to rapid and severe efflorescence, which eventually leads to the rupture of the specimen. According to the mass loss results in Figure 12, it can be seen that the relationship between efflorescence and alkali equivalent is not an absolute positive correlation. At lower alkali equivalents, the degree of efflorescence decreases as the alkali equivalent increases. For the specimens with high alkali equivalents, the higher the alkali equivalent, the higher the degree of efflorescence. In this paper, the 6 wt.% alkali equivalent specimen has the least efflorescence. From the results of the analysis based on Figure 9, it can be seen that the capillary porosity of the specimens RP-S30-M1.3N6, RP-S30-M1.3N9 and RP-S30-M1.3N12 is basically the same, but the relationship between the degree of efflorescence is: RP-S30-M1.3N12 > RP-S30-M1.3N9 > RP-S30-M1.3N6. This is because the modified sodium silicate solution with high alkali equivalent has a higher concentration of alkali metal ions.

Figure 12. Visual efflorescence of RPSG with different alkali equivalents.

4.3.3. Effect of Modulus on Visual Efflorescence

Figure 13 shows the visual efflorescence changes of RPSG with different moduli at 1 day, 3 days and 7 days. It can be seen from Figure 13 that after immersion in water for

1 day, no obvious crystallization appeared in all the specimens. After immersion in water for 3 days, flower-like white crystals appeared about 5 mm above the immersed part of the bottom of all specimens, and this phenomenon was more obvious after immersion in water for 7 days. The mass losses of the specimens are all at small values, while the mass losses of the high modulus specimens RP-S30-M1.5N6 and RP-S30-M1.7N6 are significantly smaller than those of the low modulus specimens RP-S30-M0.9N6, RP-S30-M1.1N6 and RP-S30-M1.3N6. This is because the modified sodium silicate solution with high modulus contains more active SiO_2, which facilitates the formation of denser gels in the specimens, thereby improving the degree of efflorescence of the specimens [64]. This is well demonstrated by the analysis in Figure 5 above.

Figure 13. Visual efflorescence of RPSG with different moduli.

4.4. PH Value

4.4.1. Effect of Slag Content on pH Value

Figure 14 shows the pH values of RPSG with different slag contents. As shown in Figure 14, slag delayed the increase in pH value of the specimens. After 72 h of immersion, the pH values of the individual specimens gradually reach their maximum values with minimal differences. It suggests that most of the alkali metal ions in the specimens will be released from the specimens as the immersion time is extended. However, at 2 h of immersion, the pH values of the different specimens showed significant differences. The pH value of specimen RP-S0-M1.3N6 was the largest, and the pH value of specimen RP-S45-M1.3N6 was the smallest. It shows that the efflorescence rate of the specimens becomes slower with increasing slag content.

4.4.2. Effect of Alkali Equivalent on pH Value

Figure 15 shows the pH values of RPSG with different alkali equivalents. As shown in Figure 15, the pH value of specimen RP-S30-M1.3N6 was the smallest (pH $\approx$ 9.4), and the pH value of specimen RP-S30-M1.3N15 was the largest (pH $\approx$ 11.4) at the 2 h immersion time. The results were validated against each other with the visual efflorescence results in Figure 12. After 24 h of immersion, all specimens reached a pH value above 10.0 and gradually increased in the later stages. Meanwhile, as can be observed in Figure 15, the pH value of the specimens increased with increasing alkali equivalents. It suggests that the efflorescence potential of the specimens becomes higher with increasing alkali equivalents. However, this does not indicate the actual degree of efflorescence in relation to the alkali equivalents. According to the results in Figures 9 and 12, the efflorescence of the specimen is also influenced by the denseness of the matrix.

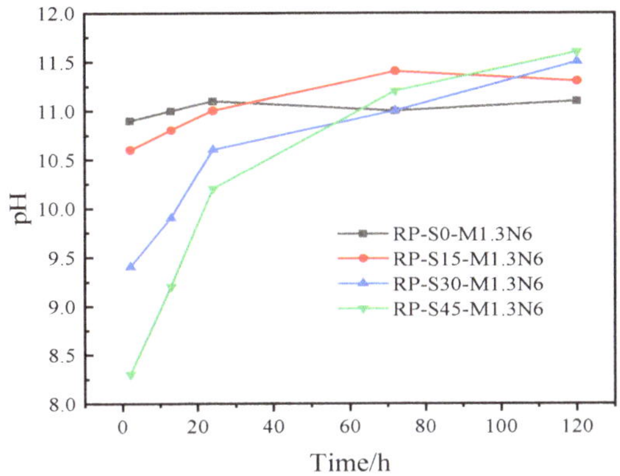

Figure 14. PH values of RPSG with different slag contents.

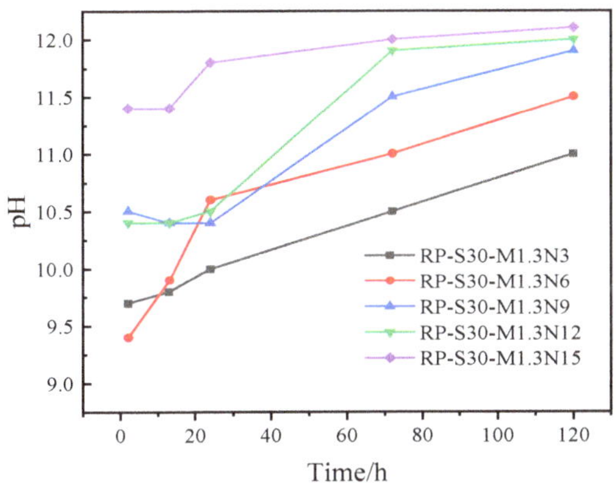

Figure 15. PH values of RPSG with different alkali equivalents.

4.4.3. Effect of Modulus on pH Value

Figure 16 shows the pH values of RPSG with different moduli. It can be observed from Figure 16 that the pH value of the lower modulus specimens is greater than that of the higher modulus specimens. It shows that increasing the modulus is an effective means of improving the efflorescence of the matrix. Meanwhile, it can be observed that the pH value of the specimens is slightly influenced by the modulus. It suggests that the modulus has a lower influence on the degree of efflorescence of RPSG.

4.5. Micro-Analysis

Based on the analysis results of drying shrinkage and efflorescence above, it is clear that the relationship of each factor on the durability performance of RPSG is: slag content > alkali equivalent > modulus. Meanwhile, the changes in alkali equivalent and modulus influence the durability performance of RPSG by inhibiting or promoting the hydration reaction of RP and slag. Therefore, the specimens with different slag contents were selected for further micro-analysis in this paper.

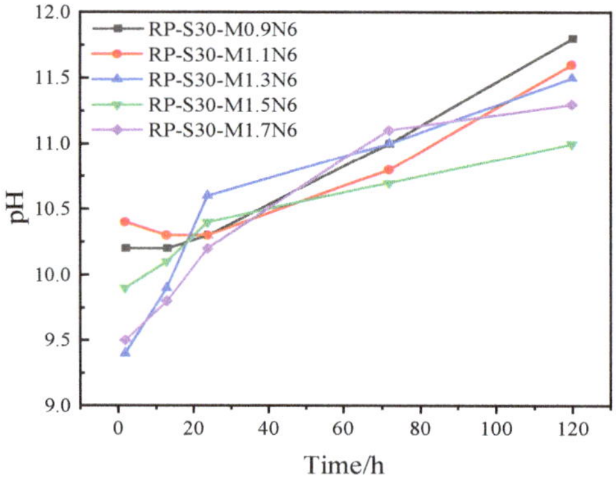

Figure 16. PH values of RPSG with different moduli.

4.5.1. XRD

Figure 17 shows the XRD diffraction patterns of RPSG. According to Figure 17, compared with the specimen RP, the diffraction peaks of chlorite, mica and feldspar in RP-S0-M1.3N6 and RP-S30-M1.3N6 are significantly lower. This shows that the alkaline environment provided by the modified sodium silicate can dissolve the layered silicoaluminate minerals of chlorite in the RP. The modified sodium silicate can provide reaction conditions for the RPSG. The specimen RP-S30-M1.3N6 showed some broader diffraction peaks between 20° and 40°, but the phenomenon was not significant. This is because the slag prompted the generation of poorly crystalline hydration products CASH and NASH from the RPSG. The amorphous nature of these hydration products, coupled with their overlap with the diffraction peaks of the more crystalline minerals in the XRD diffraction pattern, means that they do not have a distinct diffraction peak in Figure 17. In addition, specimen RP-S30-M1.3N6 has a weakened diffraction peak in the calcite phase compared to specimen RP-S0-M1.3N6, which has a significant effect on drying shrinkage of the calcite relative to the slurry [68]. It should be noted that the diffraction peak intensity of quartz decreases with the addition of activator or slag, and this is mostly caused by the dilution effect [69]. In fact, the amount of crystalline quartz dissolved in an alkaline environment is so small that it has a negligible effect on the variation of its diffraction intensity.

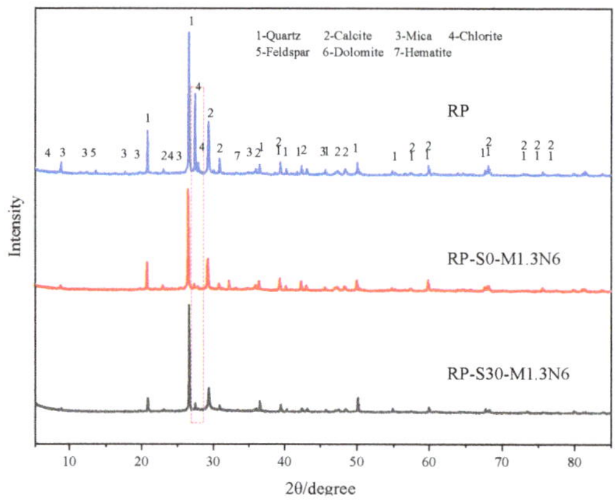

Figure 17. RD patterns of RPSG with different slag contents.

4.5.2. FTIR

Figure 18 shows the FTIR images of RPSG. As shown in Figure 18, significant variations can be observed around wave numbers 400 to 800 cm^{-1} and 1000 cm^{-1}. For the spectrograms in the wave number range of 400 to 800 cm^{-1}, the wave peak of specimen RP-S30-M1.3N6 is significantly smaller than that of specimen RP-S0-M1.3N6. This wavenumber band corresponds to the vibrational band produced by SiO_4 and AlO_4 tetrahedra, suggesting that the addition of slag promotes the dissolution of the silicon aluminum component of the RP [20]. Around wave number 1000 cm^{-1}, the peak intensity of specimen RP-S30-M1.3N6 with the addition of slag was significantly weaker compared to that of specimen RP-S0-M1.3N6. At the same time, the wave peak of specimen RP-S30-M1.3N6 was shifted towards a lower wavenumber compared to that of specimen RP-S0-M1.3N6. The spectral band near 1000 cm^{-1} corresponds to the stretching vibration of Si-O-Si and the asymmetric stretching vibration of T-O-Si (T is Si or O), indicating that the addition of slag promotes the dissolution of silica–aluminates and the hydration of low polymerization gel formation [70,71].

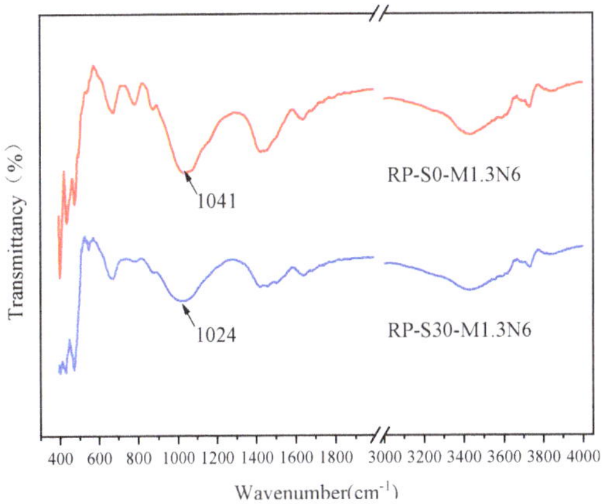

Figure 18. FTIR results of RPSG with different slag contents.

4.5.3. TG

Figure 19 shows the TG/DTG curves of RPSG. As can be seen from Figure 19a,b, the specimens have three main regions of mass loss corresponding to three temperature ranges: (1) room temperature ~ 200 °C, (2) 200–600 °C and (3) 600–1000 °C. Higher temperature results are not covered in this paper. The mass loss of the specimens below 200 °C is mainly the loss of free water adsorbed in the voids, while 200 to 600 °C is mainly the loss of bound water, corresponding to the dehydroxylation process of the hydration products NASH and CASH [72]. The mass loss of specimen RP-S30-M1.3N6 in this range is greater than that of specimen RP-S0-M1.3N6. This is because the incorporation of slag increased the internal hydrated gel content of the specimens, and more free water was adsorbed on the surface of the gel pores, while slag promoted the generation of CASH and NASH gels (corroborating the analysis in Figure 17). The mass loss of the specimen at 600–1000 °C is mainly the loss of calcium carbonate, which is closely related to the carbonation resistance of the cementitious material [15]. The mass loss rate of specimen RP-S30-M1.3N6 in this range is less than that of specimen RP-S0-M1.3N6, indicating that slag can improve the carbonation resistance of RPSG.

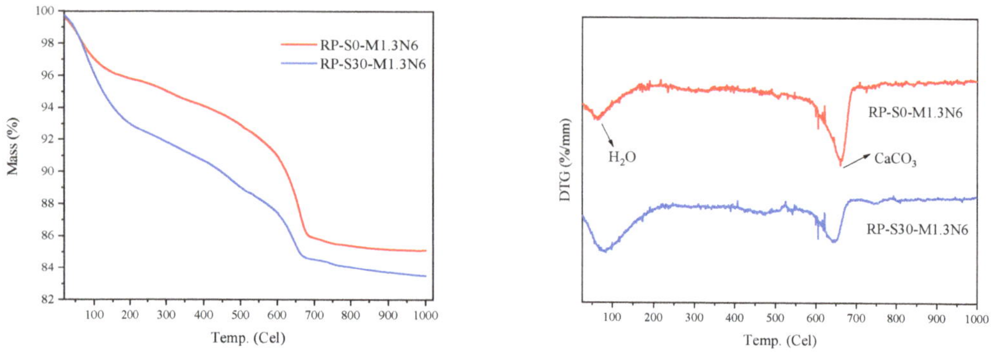

Figure 19. TG/DTG curves of RPSG with different slag contents.; (**a**) Mass loss curves of RPSG; (**b**) DTG curves of RPSG.

4.5.4. SEM

Figure 20 shows the SEM photos of RPSG with different slag contents after standard curing for 28 days under the condition of fixed alkali equivalent and modulus. It can be observed that the SEM photos of RP-S30-M1.3N6 show smaller pores, and the structure is denser. In combination with the analyses in Figures 17–19, it is clear that slag promotes the dissolution of the silica–aluminum component of the RPSG and the formation of hydrated gels such as NASH and CASH with low polymerization, resulting in a denser structure of the specimen RP-S30-M1.3N6. Some unbound loose stroma can be seen in the lower right corner of RP-S0-M1.3N6, while no similar result is observed in RP-S30-M1.3N6. This indicates that the slag particles in RP-S30-M1.3N6 and RP particles have been fully bonded after 28 days. This is because slag rapidly dissolves a large number of calcium ions in the alkaline environment, which rapidly reacts with active SiO_2 dissolved from the RP to form geopolymer gel with a network tetrahedron structure, making the matrix denser [73,74].

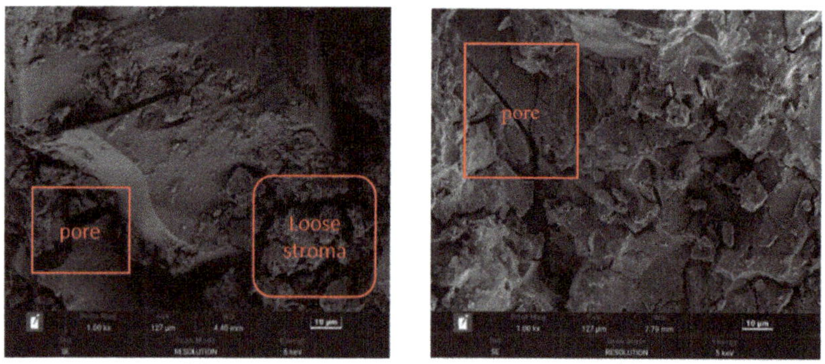

Figure 20. SEM images of RPSG with different slag contents [68,75]; (**a**) RP-S0- M1.3N6 (1000×); (**b**) RP-S30- M1.3N6 (1000×).

4.6. Discussion

Currently, the international community does not have a uniform rule for measuring the degree of efflorescence of geopolymers [50,76,77]. This paper selected capillary porosity, visual efflorescence and PH value as indexes of efflorescence degree to ensure the conclusion is more reliable. Before the mix proportion design, the group conducted exploratory tests to ensure that the formed specimens had good mechanical and workability properties.

Meanwhile, the conclusions are valid within the limits of the paper. However, further study is needed to determine whether the conclusions continue to be valid if the limited range is exceeded, such as modulus higher than 1.7. In addition, the recycled micronized powder originated from construction waste disposal plants and mainly contained mortar, clay bricks, ceramics and tiles without further screening.

5. Conclusions

In this paper, the effects of slag content, alkali equivalent and modulus on drying shrinkage and efflorescence of RPSG under normal temperature curing were studied. The drying shrinkage pattern and efflorescence pattern of RPSG were systematically analyzed. The main conclusions of this paper are as follows:

(1) Slag significantly reduces the drying shrinkage and efflorescence of RPSG. Slag can promote RPSG to generate more hydration products, which causes the matrix to become denser and reduces the capillary porosity, thus significantly improving its drying shrinkage and efflorescence performance.

(2) When the modulus is constant, the potential for the efflorescence of RPSG increases with increasing alkali equivalent. However, the relationship between drying shrinkage and alkali equivalent is not a simple linear relationship. At low alkali equivalents (6% in this paper), the drying shrinkage of RPSG increases with increasing alkali equivalents and decreases with increasing alkali equivalents at high alkali equivalents.

(3) The drying shrinkage of RPSG increases with increasing modulus when the alkali equivalent is constant. In contrast, the degree of efflorescence decreases with increasing modulus.

(4) Micro-analysis has shown that slag promotes the dissolution of the silica–aluminum composition in RPSG and the formation of hydrated gels such as NASH and CASH with low polymerization. This results in a slurry with smaller pores and a denser structure.

The research in this paper shows that the effects of slag content, alkali equivalent and modulus on the drying shrinkage and efflorescence of RPSG are not identical. In this study, RP-S45-M1.3N6 is the best proportional design for RPSG. The drying shrinkage properties and efflorescence performance of RP-S45-M1.3N6 were significantly reduced compared with RP-S0-M1.3N6 (drying shrinkage was reduced by 76.32%, visual efflorescence was significantly alleviated, capillary porosity was reduced by 60.9% and early pH value was reduced by approximately 2.0). The research has demonstrated that geopolymers with excellent durability can be prepared by using slag and RP, which has a beneficial impact on promoting the resource utilization of RP.

The main materials used in the prepared RPSG are solid wastes, which have complex and variable chemical compositions. The effects of the content of the main chemical components, such as CaO, Al_2O_3 and SiO_2, on the properties of RPSG should be further investigated.

Author Contributions: Conceptualization, X.L.; methodology, X.L.; resources, X.L.; writing—original draft, X.L. and E.L.; supervision, X.L.; validation, E.L. and Y.F.; investigation, E.L. and Y.F.; data curation, E.L. and Y.F.; writing—review and editing, E.L. and Y.F. All authors have read and agreed to the published version of the manuscript.

Funding: This research received no external funding.

Institutional Review Board Statement: Not applicable.

Informed Consent Statement: Not applicable.

Data Availability Statement: The data presented in this study are available on request from the corresponding author.

Acknowledgments: This research did not receive any specific grant from funding agencies in the public, commercial, or not-for-profit sectors.

Conflicts of Interest: The authors declare no conflict of interest.

References

1. Liu, J.; Yi, Y.; Wang, X. Exploring factors influencing construction waste reduction: A structural equation modeling approach. *J. Clean. Prod.* **2020**, *276*, 123185. [CrossRef]
2. Liang, G.; Liu, T.; Li, H.; Wu, K. Shrinkage mitigation, strength enhancement and microstructure improvement of alkali-activated slag/fly ash binders by ultrafine waste concrete powder. *Compos. Part B-Eng.* **2022**, *231*, 109570. [CrossRef]
3. Frappa, G.; Pauletta, M.; Di Marco, C.; Russo, G. Experimental tests for the assessment of residual strength of r.C. Structures after fire—Case study. *Eng. Struct.* **2022**, *252*, 113681. [CrossRef]
4. Wagih, A.M.; El-Karmoty, H.Z.; Ebid, M.; Okba, S.H. Recycled construction and demolition concrete waste as aggregate for structural concrete. *HBRC J.* **2013**, *9*, 193–200. [CrossRef]
5. Rao, A.; Jha, K.N.; Misra, S. Use of aggregates from recycled construction and demolition waste in concrete. *Constr. Build. Mater.* **2022**, *330*, 71–81. [CrossRef]
6. Ye, T.; Xiao, J.; Duan, Z.; Li, S. Geopolymers made of recycled brick and concrete powder—A critical review. *Constr. Build. Mater.* **2022**, *330*, 127232. [CrossRef]
7. Wu, H.; Zuo, J.; Zillante, G.; Wang, J.; Yuan, H. Status quo and future directions of construction and demolition waste research: A critical review. *J. Clean. Prod.* **2019**, *240*, 118163. [CrossRef]
8. Chen, X. Experimental study on waste concrete recycled micro-powder as supplementary cementitious material. Master's Thesis, Qinghai University, Qinghai, China, 2021.
9. Premkumar, R.; Hariharan, P.; Rajesh, S. Effect of silica fume and recycled concrete aggregate on the mechanical properties of ggbs based geopolymer concrete. *Mater. Today Proc.* **2022**, *60*, 211–215. [CrossRef]
10. Zhao, M.Z.; Geng, Y.; Wang, Y.Y.; Hu, J.X. Compounding effect and an expanded theoretical model for recycled coarse and fine aggregate concretes under uniaxial loading. *Constr. Build. Mater.* **2022**, *320*, 126226. [CrossRef]
11. Afroughsabet, V.; Biolzi, L.; Monteiro, P.J.M.; Gastaldi, M.M. Investigation of the mechanical and durability properties of sustainable high performance concrete based on calcium sulfoaluminate cement. *J. Build. Eng.* **2021**, *43*, 102656. [CrossRef]
12. Murthi, P.; Krishnamoorthi, S.; Poongodi, K.; Saravanan, R. Development of green masonry mortar using fine recycled aggregate based on the shear bond strength of brick masonry. *Mater. Today Proc.* **2021**, *61*, 413–419. [CrossRef]
13. Sahin, F.; Uysal, M.; Canpolat, O.; Aygormez, Y.; Cosgun, T.; Dehghanpour, H. Effect of basalt fiber on metakaolin-based geopolymer mortars containing rilem, basalt and recycled waste concrete aggregates. *Constr. Build. Mater.* **2021**, *301*, 124113. [CrossRef]
14. Zhu, Z.; Gu, S.; Tang, Z.; Song, L. Experimental study on road base material of geopolymer recycled concrete. *J. Test. Eval.* **2021**, *49*, 1747–1762. [CrossRef]
15. Liu, C. Research on Preparation and Application of Alkali Activated Recycled Cementtitious Material. Master's Thesis, Yangzhou University, Yangzhou, China, 2021.
16. Alhawat, M.; Ashour, A.; Yildirim, G.; Aldemir, A.; Sahmaran, M. Properties of geopolymers sourced from construction and demolition waste: A review. *J. Build. Eng.* **2022**, *50*, 104104. [CrossRef]
17. Cong, P.; Cheng, Y. Advances in geopolymer materials: A comprehensive review. *J. Traffic Transp. Eng. Engl. Ed.* **2021**, *8*, 283–314. [CrossRef]
18. Xu, J.; Kang, A.; Wu, Z.; Gong, Y.; Xiao, P. The effect of mechanical-thermal synergistic activation on the mechanical properties and microstructure of recycled powder geopolymer. *J. Clean. Prod.* **2021**, *327*, 129477. [CrossRef]
19. Tan, J.; Cai, J.; Li, X.; Pan, J.; Li, J. Development of eco-friendly geopolymers with ground mixed recycled aggregates and slag. *J. Clean. Prod.* **2020**, *256*, 120369. [CrossRef]
20. Huo, W.W.; Zhu, Z.D.; Chen, W.; Zhang, J.; Kang, Z.; Pu, S.; Wan, Y. Effect of synthesis parameters on the development of unconfined compressive strength of recycled waste concrete powder-based geopolymers. *Constr. Build. Mater.* **2021**, *292*, 123264. [CrossRef]
21. Tan, J.; Cai, J.; Huang, L.; Yang, Q.; Mao, M.; Li, J. Feasibility of using microwave curing to enhance the compressive strength of mixed recycled aggregate powder based geopolymer. *Constr. Build. Mater.* **2020**, *262*, 120897. [CrossRef]
22. Ulugol, H.; Gunal, M.F.; Yaman, I.O.; Yildirim, G.; Sahmaran, M. Effects of self-healing on the microstructure, transport, and electrical properties of 100% construction- and demolition-waste-based geopolymer composites. *Cem. Concr. Compos.* **2021**, *121*, 104081. [CrossRef]
23. Mahmoodi, O.; Siad, H.; Lachemi, M.; Dadsetan, S.; Sahmaran, M. Development of normal and very high strength geopolymer binders based on concrete waste at ambient environment. *J. Clean. Prod.* **2021**, *279*, 123436. [CrossRef]
24. Dadsetan, S.; Siad, H.; Lachemi, M.; Sahmaran, M. Construction and demolition waste in geopolymer concrete technology: A review. *Mag. Concr. Res.* **2019**, *71*, 1232–1252. [CrossRef]
25. Frappa, G.; Miceli, M.; Pauletta, M. Destructive and non-destructive tests on columns and cube specimens made with the same concrete mix. *Constr. Build. Mater.* **2022**, *349*, 128807. [CrossRef]
26. Gao, X.; Yu, Q.L.; Brouwers, H.J.H. Assessing the porosity and shrinkage of alkali activated slag-fly ash composites designed applying a packing model. *Constr. Build. Mater.* **2016**, *119*, 175–184. [CrossRef]
27. Frappa, G.; Pauletta, M.; Russo, G. Failure analysis of three rigid block assemblies-a real collapse resulting in death. *Eng. Fail. Anal.* **2023**, *145*, 107001. [CrossRef]

28. Collins, F.; Sanjayan, J.G. Effect of pore size distribution on drying shrinking of alkali-activated slag concrete. *Cem. Concr. Res.* **2000**, *30*, 1401–1406. [CrossRef]
29. Thomas, R.J.; Lezama, D.; Peethamparan, S. On drying shrinkage in alkali-activated concrete: Improving dimensional stability by aging or heat-curing. *Cem. Concr. Res.* **2017**, *91*, 13–23. [CrossRef]
30. Krizan, D.; Zivanovic, B. Effects of dosage and modulus of water glass on early hydration of alkali-slag cements. *Cem. Concr. Res.* **2002**, *32*, 1181–1188. [CrossRef]
31. Zhang, B.; Zhu, H.; Cheng, Y.; Huseien, G.F.; Shah, K.W. Shrinkage mechanisms and shrinkage-mitigating strategies of alkali-activated slag composites: A critical review. *Constr. Build. Mater.* **2022**, *318*, 125993. [CrossRef]
32. Cartwright, C.; Rajabipour, F.; Radlińska, A. Shrinkage characteristics of alkali-activated slag cements. *J. Mater. Civ. Eng.* **2014**, *27*, B4014007. [CrossRef]
33. Rashad, A.M.; Morsi, W.M.; Khafaga, S.A. Effect of limestone powder on mechanical strength, durability and drying shrinkage of alkali-activated slag pastes. *Innov. Infrastruct. Solut.* **2021**, *6*, 127. [CrossRef]
34. Duxson, P.; Fernández-Jiménez, A.; Provis, J.L.; Lukey, G.C.; Palomo, A.; Deventer, J.S. Geopolymer technology: The current state of the art. *J. Mater. Sci.* **2007**, *42*, 2917–2933. [CrossRef]
35. Duxson, P.; Lukey, G.C.; van Deventer, J.S.J. Evolution of Gel Structure during Thermal Processing of Na-Geopolymer Gels. *Langmuir* **2006**, *22*, 8750–8757. [CrossRef] [PubMed]
36. Deb, P.S.; Nath, P.; Sarker, P.K. Drying shrinkage of slag blended fly ash geopolymer concrete cured at room temperature. In *5th Euro Asia Civil Engineering Forum (EACEF)*; Petra Christian University: Surabaya, Indonesia, 2015; pp. 594–600.
37. Ma, Y.; Ye, G. The shrinkage of alkali activated fly ash. *Cem. Concr. Res.* **2015**, *68*, 75–82. [CrossRef]
38. Yang, Y.; Li, Z.; Zhang, T.; Yin, S.; Wei, J.; Chen, X.; Yu, Q. Mechanism of active magnesium oxide compensating the shrinkage of inorganic polymer pastes. *J. South China Univ. Technol.* **2017**, *45*, 102–109.
39. Liu, Q. Research on Preparation of Cements and Alkali-Activated Cementitious Materials with Calcium Silicate Slag. Ph.D. Thesis, China Academy of Building Materials Science, Beijing, China, 2015.
40. Melo Neto, A.A.; Cincotto, M.A.; Repette, W. Mechanical properties, drying and autogenous shrinkage of blast furnace slag activated with hydrated lime and gypsum. *Cem. Concr. Compos.* **2010**, *32*, 312–318. [CrossRef]
41. Yuan, X.H.; Chen, W.; Lu, Z.A.; Chen, H. Shrinkage compensation of alkali-activated slag concrete and microstructural analysis. *Constr. Build. Mater.* **2014**, *66*, 422–428. [CrossRef]
42. Jiang, D.; Li, X.; Lv, Y.; Li, C.J.; Jiang, W.G.; Liu, Z.L.; Xu, J.S.; Zhou, Y.; Dan, J.M. Autogenous shrinkage and hydration property of alkali activated slag pastes containing superabsorbent polymer. *Cem. Concr. Res.* **2021**, *149*, 106581. [CrossRef]
43. Hardjito, D.; Wallah, S.E.; Sumajouw, D.M.J.; Rangan, B.V. On the development of fly ash-based geopolymer concrete. *Aci Mater. J.* **2004**, *101*, 467–472.
44. Palacios, M.; Puertas, F. Effect of shrinkage-reducing admixtures on the properties of alkali-activated slag mortars and pastes. *Cem. Concr. Res.* **2007**, *37*, 691–702. [CrossRef]
45. Tang, D.; Yang, C.; Li, X.; Zhu, X.; Yang, K.; Yu, L. Mitigation of efflorescence of alkali-activated slag mortars by incorporating calcium hydroxide. *Constr. Build. Mater.* **2021**, *298*, 123873. [CrossRef]
46. De Oliveira, L.B.; De Azevedo, A.R.G.; Marvila, M.T.; Pereira, E.C.; Fediuk, R.; Vieira, C.M.F. Durability of geopolymers with industrial waste. *Case Stud. Constr. Mater.* **2022**, *16*, e00839. [CrossRef]
47. Skvara, F.; Kopecky, L.; Kyskova, L.; Smilauer, V.; Alberovska, L.; Vinsova, L. Aluminosilicate polymers–influence of elevated temperatures, efflorescence. *Ceram. -Silik.* **2009**, *53*, 276–282.
48. Zhang, Z.; Provis, J.L.; Ma, X.; Reid, A.; Wang, H. Efflorescence and subflorescence induced microstructural and mechanical evolution in fly ash-based geopolymers. *Cem. Concr. Compos.* **2018**, *92*, 165–177. [CrossRef]
49. Longhi, M.A.; Zhang, Z.; Rodríguez, E.D.; Kirchheim, A.; Wang, H. Efflorescence of Alkali-Activated Cements (Geopolymers) and the Impacts on Material Structures: A Critical Analysis. *Front. Mater.* **2019**, *6*, 89. [CrossRef]
50. Zhang, Z.; Provis, J.L.; Reid, A.; Wang, H. Fly ash-based geopolymers: The relationship between composition, pore structure and efflorescence. *Cem. Concr. Res.* **2014**, *64*, 30–41. [CrossRef]
51. Kani, E.N.; Allahverdi, A.; Provis, J.L. Efflorescence control in geopolymer binders based on natural pozzolan. *Cem. Concr. Compos.* **2012**, *34*, 25–33. [CrossRef]
52. Yao, X.; Yang, T.; Zhang, Z. Compressive strength development and shrinkage of alkali-activated fly ash–slag blends associated with efflorescence. *Mater. Struct.* **2015**, *49*, 2907–2918. [CrossRef]
53. Saha, S.; Rajasekaran, C. Enhancement of the properties of fly ash based geopolymer paste by incorporating ground granulated blast furnace slag. *Constr. Build. Mater.* **2017**, *146*, 615–620. [CrossRef]
54. Wu, B.; Li, L.; Deng, H.; Zheng, Z.; Xiang, Y.; Li, Y.; Ma, X. Characteristics and mechanism of efflorescence in fly ash-based geopolymer mortars under quasi-natural condition. *J. Build. Eng.* **2022**, *55*, 104708. [CrossRef]
55. Tan, J.; Cizer, Ö.; Vandevyvere, B.; De Vlieger, J.; Dan, H.; Li, J. Efflorescence mitigation in construction and demolition waste (cdw) based geopolymer. *J. Build. Eng.* **2022**, *53*, 105001. [CrossRef]
56. Lingzhu, Y.; Jian, L. Basic principle and application of scanning electron microscope. *Exp. Sci. Technol.* **2019**, *17*, 85–93. [CrossRef]
57. Wang, G.; Ma, Y. Drying shrinkage of alkali-activated fly ash/slag blended system. *J. Sustain. Cem. Mater.* **2018**, *7*, 203–213. [CrossRef]

58. Huang, D.; Chen, P.; Peng, H.; Yang, Y.W.; Yuan, Q.M.; Su, M. A review and comparison study on drying shrinkage prediction between alkali-activated fly ash/slag and ordinary Portland cement. *Constr. Build. Mater.* **2021**, *305*, 124760. [CrossRef]
59. Hojati, M.; Radlińska, A. Shrinkage and strength development of alkali-activated fly ash-slag binary cements. *Constr. Build. Mater.* **2017**, *150*, 808–816. [CrossRef]
60. Bentz, D.P.; Quenard, D.A.; Garboczi, E.J. Modelling drying shrinkage in reconstructed porous materials: Application to porous vycor glass. *Model. Simul. Mater. Sci. Eng.* **1998**, *6*, 211. [CrossRef]
61. Chindaprasirt, P.; Thaiwitcharoen, S.; Kaewpirom, S.; Rattanasak, U. Controlling ettringite formation in FBC fly ash geopolymer concrete. *Cem. Concr. Compos.* **2013**, *41*, 24–28. [CrossRef]
62. Lee, N.; Jang, J.; Lee, H. Shrinkage characteristics of alkali-activated fly ash/slag paste and mortar at early ages. *Cem. Concr. Compos.* **2014**, *53*, 239–248. [CrossRef]
63. Shi, D.; Ye, J.; Zhang, W. Effects of activator content on properties, mineralogy, hydration and microstructure of alkali-activated materials synthesized from calcium silicate slag and ground granulated blast furnace slag. *J. Build. Eng.* **2020**, *32*, 101791. [CrossRef]
64. Yang, T.; Zhu, H.; Zhang, Z. Influence of fly ash on the pore structure and shrinkage characteristics of metakaolin-based geopolymer pastes and mortars. *Constr. Build. Mater.* **2017**, *153*, 284–293. [CrossRef]
65. Li, C. Research on the Glass Phase of Slag, High Calcium Fly Ash and Low Calcium Fly Ash and Their Hydration Mechanism. Ph.D. Thesis, Tsinghua University, Beijing, China, 2011.
66. Longhi, M.A.; Rodríguez, E.D.; Walkley, B.; Zhang, Z.; Kirchheim, A. Metakaolin-based geopolymers: Relation between formulation, physicochemical properties and efflorescence formation. *Compos. Part B-Eng.* **2020**, *182*, 107671. [CrossRef]
67. Scherer, G.W. Stress from crystallization of salt. *Cem. Concr. Res.* **2004**, *34*, 1613–1624. [CrossRef]
68. Taghvayi, H.; Behfarnia, K.; Khalili, M. The Effect of Alkali Concentration and Sodium Silicate Modulus on the Properties of Alkali-Activated Slag Concrete. *J. Adv. Concr. Technol.* **2018**, *16*, 293–305. [CrossRef]
69. Das, S.K.; Shrivastava, S. Influence of molarity and alkali mixture ratio on ambient temperature cured waste cement concrete based geopolymer mortar. *Constr. Build. Mater.* **2021**, *301*, 124380. [CrossRef]
70. Wong, C.L.; Mo, K.H.; Alengaram, U.J.; Yap, S. Mechanical strength and permeation properties of high calcium fly ash-based geopolymer containing recycled brick powder. *J. Build. Eng.* **2020**, *32*, 101655. [CrossRef]
71. Hwang, C.L.; Yehualaw, M.D.; Duy-Hai, V.; Trong-Phuoc, H.; Largo, A. Performance evaluation of alkali activated mortar containing high volume of waste brick powder blended with ground granulated blast furnace slag cured at ambient temperature. *Constr. Build. Mater.* **2019**, *223*, 657–667. [CrossRef]
72. Yang, Z.; Liu, Z.; Li, F.; Lu, Y.; Li, S. Effect of multi-walled carbon nanotubes on durability of high-strength slag-based geopolymer. *Ceram. Int.* **2022**, *(in press)*. [CrossRef]
73. Puligilla, S.; Mondal, P. Role of slag in microstructural development and hardening of fly ash-slag geopolymer. *Cem. Concr. Res.* **2013**, *43*, 70–80. [CrossRef]
74. Wu, P.; Lu, X.J.; Hu, S.G.; Zhang, L. Research progress on the activation of gelling properties of granulated blast furnace slag. *Met. Mine* **2012**, *10*, 157–161. [CrossRef]
75. Golnaz, S.; Kiachehr, B.; Mohammad, T. Drying shrinkage of one-part alkali-activated slag concrete. *J. Build. Eng.* **2022**, *51*, 104263.
76. Weng, T.-L.; Lin, W.-T.; Cheng, A. Effect of Metakaolin on Strength and Efflorescence Quantity of Cement-Based Composites. *Sci. World J.* **2013**, *2013*, 606524. [CrossRef] [PubMed]
77. Peruchi, A.B.R.; Zuchinali, F.F.; Bernardin, A.M. Development of a water-based acrylic paint with resistance to efflorescence and test method to determine the appearance of stains. *J. Build. Eng.* **2020**, *35*, 102005. [CrossRef]

applied sciences

MDPI

Article

Effect of Elastomeric Expandable Additive on Compressive Strength and Linear Expansion of Fly-Ash-Based Strength-Enhanced Geopolymer Cement for Shrinkage-Resistant Oil-Well Cementing

Siti Humairah Abd Rahman [1,*], Syed Ahmad Farhan [2,3], Yon Azwa Sazali [1], Luqmanul Hakim Shafiee [1], Nadzhratul Husna [4], Afif Izwan Abd Hamid [2], Nasir Shafiq [2], Nurul Nazmin Zulkarnain [1] and Mohd Firdaus Habarudin [1]

1 PETRONAS Research Sdn. Bhd., Bandar Baru Bangi 43000, Malaysia; yonazwa.sazali@petronas.com (Y.A.S.); hakim.shafiee@petronas.com (L.H.S.); nazmin.zulkarnain@petronas.com (N.N.Z.); mfirdaus.habarudin@petronas.com (M.F.H.)
2 Institute of Self-Sustainable Building for Smart Living, Universiti Teknologi PETRONAS, Seri Iskandar 32610, Malaysia; syed.af_g02626@utp.edu.my (S.A.F.); afif_g03455@utp.edu.my (A.I.A.H.); nasirshafiq@utp.edu.my (N.S.)
3 Department of Civil Engineering, Indian Institute of Technology Madras, Chennai 600036, India
4 Institute of Autonomous System for Autonomous Facilities, Universiti Teknologi PETRONAS, Seri Iskandar 32610, Malaysia; husna.puad@utp.edu.my
* Correspondence: humairah.rahman@petronas.com

Abstract: The present study aimed to investigate the effect of an expandable additive on the compressive strength and linear expansion of geopolymer cement, which is an alternative to ordinary Portland cement, for oil-well cementing. Fly-ash-based geopolymer cement samples, with the addition of slag cement as a strength enhancer, were prepared by using an elastomeric expandable additive (*R*-additive), which consists of styrene–butadiene rubber with a specific gravity of 0.945, at concentrations of 10%, 15%, 20% and 25% by weight of the solid blend, and cured in a water bath at 60 °C and atmospheric pressure, and a curing chamber at 90 °C and 3000 psi, or approximately 20.68 MPa. Mixability, amount of free water and slurry density were studied, and the effects of the concentration of *R*-additive on the compressive strength (F) and linear expansion ($\Delta l/l_0$) of the samples were analyzed. When cured at 60 °C and atmospheric pressure, the highest F of 15.01 MPa was obtained when the concentration of *R*-additive was 10%, while the highest $\Delta l/l_0$ of 0.9985% was obtained when the concentration of *R*-additive was 25%. An increase in the curing temperature and pressure to 90 °C and 3000 psi (≈20.68 MPa) resulted in the reduction of F from 15.01 to 14.62 MPa and from 10.33 to 9.61 MPa, and the increase in $\Delta l/l_0$ from 0.52% to 0.63%, and from 0.99% to 1.32%, when the concentrations of *R*-additive were 10% and 25%, respectively. The findings suggest that the formulations adopted, which contain *R*-additive at concentrations ranging from 10% to 25%, fulfilled the requirements of the oil and gas industry.

Keywords: compressive strength; expandable additive; fly ash; geopolymer cement; linear expansion; ordinary Portland cement; volumetric shrinkage

1. Introduction

During the construction of oil wells, cementing is performed primarily to establish zonal isolation. Apart from that, cementing is essential to support the casing, as well as to shield the casing from corrosion [1,2]. It has to be designed to possess a long-term structural integrity and be serviceable at a wide range of temperatures, from as low as below freezing temperatures, as in permafrost zones, to as high as above 500 °C, as in geothermal wells [3]. Conventionally, ordinary Portland cement (OPC), in compliance with the requirements of Class G cement, as specified by the American Petroleum Institute (API), is adopted to perform cementing for various well operations [3–5]. Class G well cement satisfies the exigent specifications stated in API Specification 10A [5], inclusive of fluid loss control, low free fluid, low viscosity, predictable thickening time and strength. Notwithstanding the ubiquity of OPC as the material employed for well cementing, its adoption coincides with drawbacks that limit its application when exposed to critical conditions that are confronted by acid-rich, deep-water and geothermal wells [3]. At critical levels of temperature and pressure, as well as in the presence of high amounts of carbon dioxide (CO_2), OPC endures degradation that results in strength retrogression in tandem with exacerbations of permeability and porosity [6–9]. The degradation will eventually lead to the loss of zonal isolation, owing to the failure of the cement sheath [1] and, hence, compromising the structural integrity of the well [10]. Failure mechanisms associated with OPC-based well cement are, among others, the formation of channels through the cement matrix, micro-annuli at cement interfaces and radial cracks within the cement sheath [11,12].

Volumetric shrinkage of cement during hydration has always been a concern in oil-well cementing [13–15]. It instigates the formation of micro-annulus cracks due to contraction of the external dimensions of the cement [16,17] and de-bonding between the cement, casing and formation [18,19]. Elevation in the casing pressure is then induced as a result of the migration of gas and liquid from the formation [20,21] that, in due course, may lead to failure of the cement sheath.

The presence of gaps inside the cement sheath, between the casing and cement, and between the casing and formation facilitates the migration of fluid from the formation [22], where the fluid flows upward to the surface via flow paths. Sustained casing pressure (SCP) may occur owing to the accumulation of fluid underneath the wellhead [23] that can potentially lead to the loss of hydrocarbon reserves and pollution of the aquifer and sea [17]. Accordingly, wells that are subjected to SCP require work-over jobs that include the replacement of corroded tubes and remedial squeeze cementing. The process of de-hydration would be employed to perform the replacement, which involves the injection of cement slurry into the leak paths [17]. Moreover, work-over jobs have been previously reported to be costly, with exorbitant rates that can reach USD 100,000 per well by virtue of the low success rate of earlier attempts prior to successfully attaining an adequate seal that is serviceable [24].

Volumetric shrinkage of cement transpires due to the absorption of water into crystals of ettringite ($Al_2Ca_6H_{12}O_{24}S_3$), which is formed in the hydrated OPC during the initial phase of cementing. The formation of $Al_2Ca_6H_{12}O_{24}S_3$ occurs as a result of the chemical reaction between calcium aluminate ($CaAl_2O_4$) and calcium sulfate ($CaSO_4$), which are both present in OPC [9,25,26]. It can be obstructed by means of adding chemicals that can induce expansion in the cement [18,19], namely expandable additives, which can be mixed into the slurry, and function as a shape-memory agent that expands prior to setting of the cement [1]. Accordingly, the expansion that occurs after hardening allows for the sealing of micro-annulus cracks to be performed [27].

Contemporarily, geopolymer cement (GPC) has been accentuated in previous research as an alternative to OPC for oil-well cementing, with findings that substantiate its potential for various applications in the industry and field of research pertaining to civil engineering [28]. Geopolymers possess long-range, covalently bonded and amorphous networks that are produced via the activation of a source material, such as, among others, fly ash, metakaolin and slag, with an alkaline-activator solution. Prevalently, a combination of sodium hydroxide (NaOH) and sodium silicate (Na_2SiO_3) [29,30] is adopted as the solution, which forms inorganic polymers with the empirical formula, as shown in Equation (1) [31]. Chains of geopolymer are present, mainly in the forms of polysialate (Al–O–Si), polysialate siloxo (Al–O–Si–Si) and polysialate disiloxo (Al–O–Si–Si–Si) [32].

$$M_n[-(SiO_2)z - AlO_2]n.wH_2O \quad (1)$$

where M is an alkali cation, such as calcium (Ca^{2+}), potassium (K^+) or sodium (Na^+) [31,33]; z is either 1, 2 or 3; and n is the degree of polycondensation.

The synthesis of GPC entails dissolution, reorientation and solidification reactions as part of the mechanism of geopolymerization [20,33], where forming of aluminosilicate gels takes place owing to the presence of alkali, which does not necessitate the absorption of water. In view of the mechanism for activation of OPC that entails the absorption of water to enable cement hydration to occur, volumetric shrinkage in GPC is anticipated to be less than that of OPC [9]. Furthermore, the calcium content of GPC is considerably less than that of OPC, owing to the low calcium content of the aluminosilicate source. Conversely, the high calcium content in OPC leads to, in the course of time, elevations in porosity and permeability and, consequently, loss in mechanical strength and structural integrity of the cement sheath. Expenditure of calcium ions that are present in calcium-silicate-hydrate (C-S-H) occurs due to chemical reactions between the calcium ions and carbonic acid (H_2CO_3) to form calcium carbonate ($CaCO_3$). Subsequent to the depletion of the calcium ions in C-S-H, further chemical reactions between H_2CO_3 and $CaCO_3$ will occur [2]. Accordingly, the lower calcium content of GPC in comparison to OPC signifies that, under high exposure to CO_2 in CO_2-rich well environments, GPC will be more resistant toward carbonation than OPC [7,34,35].

Apart from low shrinkage and resistance toward carbonation, a vast array of other advantages pertaining to the material properties of GPC in comparison to OPC has been reported in previous research; the advantages include, but are not limited to, high mechanical strength; high pumpability; low permeability toward the flow of water and gas, with the potential for further reduction when exposed to higher pressures, as in oil-well cementing; low Young's modulus; resistance to acid attacks; resistance to alkali-aggregate reaction; resistance to freeze–thaw cycles; stability at high temperatures; and tolerance to contamination with oil-based mud [1,33,36–41]. Moreover, the adoption of GPC as an alternative to OPC presents a cost- and energy-efficient solution, as the manufacturing process of GPC consumes less energy with a carbon emission of only about 0.184 ton of CO_2 for every ton of GPC produced, as compared to 1 ton of CO_2 for every ton of OPC produced, with, coincidentally, less cost [1,41–45].

GPC has been accentuated in previous research as an alternative to OPC for oil-well cementing [1,17,20]. Previous research on the application of geopolymer for other civil engineering applications is wide-ranging. Contrarily, research on the application of GPC for oil-well cementing is scarce, and, accordingly, studies that attempt to address the concern of volumetric shrinkage of cement in oil-well cementing are inadequate. Therefore, the present study aims to investigate the effect of an expandable additive on the compressive strength and linear expansion of GPC for the oil-well cementing application. Fly-ash-based strength-enhanced GPC samples, with the addition of slag cement as a strength enhancer, were prepared by using an elastomeric expandable additive (R-additive), which consists of styrene–butadiene rubber with a specific gravity of 0.945, at concentrations of 10%, 15%, 20% and 25% by weight of the solid blend, and cured in a water bath at 60 °C and atmospheric pressure, and a curing chamber at 90 °C and 3000 psi, or approximately 20.68 MPa. Even though the adoption of styrene–butadiene as an additive in the geopolymer matrix has been explored in Lee et al. [46] and Ekinci et al. [47], the studies did not focus on the conditions for oil-well cementing. Mixability, amount of free water and slurry density were studied. Subsequently, the effects of the concentration of R-additive on the compressive strength and linear expansion of the samples were analyzed. Findings of the present study will contribute toward addressing, in general, the inadequacy of data on the application of GPC for oil-well cementing and, in particular, formulation of shrinkage-resistant oil-well cementing based on GPC.

2. Materials and Methods

2.1. Preparation of Geopolymer Cement (GPC) Samples

GPC samples were prepared by using fly ash as the aluminosilicate source. The fly ash was obtained from the Tanjung Bin power plant in Johor, Malaysia, which is a 2100-MW coal-fired power plant [48] that implements denitrification technology. Figure 1 presents the particle size distributions of the fly ash sample obtained. According to the American Society for Testing Materials (ASTM), the fly ash can be classified as a Class F fly ash with a maximum calcium oxide (CaO) content of 6.72%, as per ASTM C618-19 [49]. The chemical composition of the fly ash, which was determined by using X-ray fluorescence of model S8 Tiger developed by Bruker (Billerica, MA, USA), is presented in Table 1. Micrographs of the fly ash were produced by using a scanning electron microscope of model Evo LS15 VPSEM developed by Zeiss (Jena, Germany), at magnifications of 1000× and 10,000×, as disclosed in Figure 2.

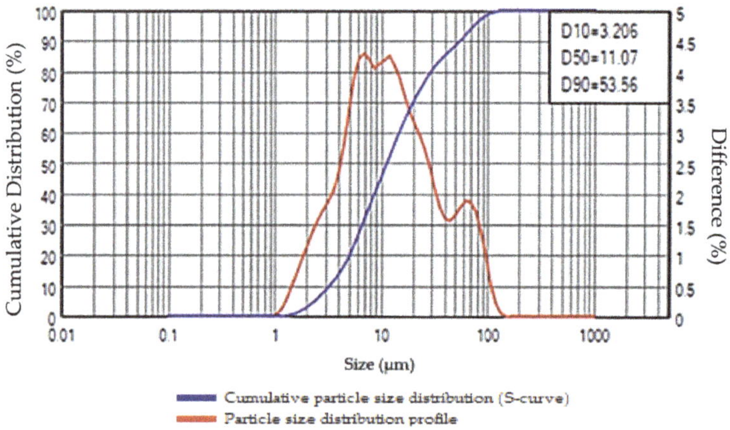

Figure 1. Particle size distribution of the fly ash obtained from the Tanjung Bin power plant.

Table 1. Chemical composition of the fly ash employed in the present study.

Element/Parameter	Weight (%)
Silicon Dioxide (SiO_2)	46.47
Aluminum Oxide (Al_2O_3)	25.95
Iron (III) Oxide (Fe_2O_3)	8.31
Calcium Oxide (CaO)	6.88
Magnesium Oxide (MgO)	4.95
Potassium Oxide (K_2O)	2.11
Sodium Oxide (Na_2O)	1.72
Titanium (IV) Oxide (TiO_2)	1.16
Sulfur Trioxide (SO_3)	0.63
Chlorine (Cl)	<0.1
Moisture	0.11
Loss of Ignition	1.61

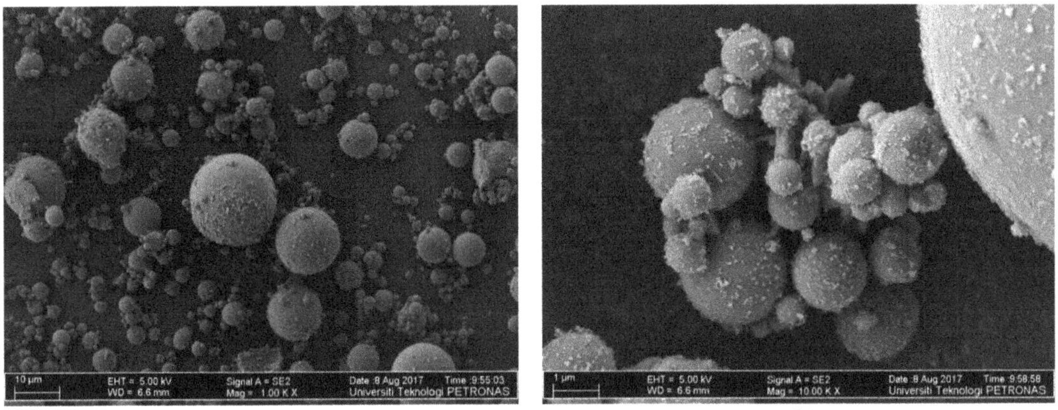

(a) (b)

Figure 2. Micrographs of the fly ash employed in the present study at magnifications of (**a**) 1000× and (**b**) 10,000×.

In order to enhance the strength of the cement to obtain the desired strength range for oil-well cementing, slag cement with a strength grade of 32.5 [50] was added to the mix as a strength enhancer, as implemented in previous studies [51–53], with a concentration of 10% of the weight of the fly ash and slag cement. Furthermore, *R*-additive, which is an elastomeric expandable additive that consists of styrene–butadiene rubber with a specific gravity of 0.945, was incorporated into the mix to induce expansion in the cement in an attempt to obstruct volumetric shrinkage. Four (4) mix formulations of the samples, namely R10, R15, R20 and R25, were specified, which were composed of, by weight of the mix, 65% of a solid blend that comprises fly ash, slag cement and *R*-additive, and 35% of an alkaline-activator solution that comprises 8M NaOH and Na_2SiO_3 with a Na_2SiO_3-to-NaOH molar ratio of 0.25. Concentration of the *R*-additive was increased from 10% to 15%, 20% and 25% by weight of the solid blend for formulations R10, R15, R20 and R25, respectively. Preliminary tests were conducted on trial mix designs prior to selecting the concentration range of *R*-additive, which was specified with the aim of attaining cement expansions that are acceptable to perform cementing for various well operations. Concentrations of the fly ash and slag cement, also by weight of the solid blend, were adjusted accordingly, as clarified in Table 2, with the percentage of solid blend fixed at 65% by weight of the mix.

Table 2. Concentrations of each constituent of the solid blend for each mix formulation.

Mix Formulation	Concentration (% of Solid Blend)		
	R-Additive	Fly Ash	Slag Cement
R10	10	81.0	9.0
R15	15	76.5	8.5
R20	20	72.0	8.0
R25	25	67.5	7.5

Cement slurry with a volume of up to 600 mL was prepared for each mix. Mixing of slurry was performed by using a constant speed mixer of Model 3260, which was developed by Ametek Chandler Engineering (Tulsa, OK, USA), at an initial rotational speed of 4000 rpm for 15 s, which was subsequently increased to a final rotational speed of 12,000 rpm for 35 s after the solid constituents were completely poured into the mixer. Mixing was performed in accordance with API RP 10B-2 [54], which was also adopted to perform free-water and rheology tests on the slurry.

Slurry density is pivotal to modulate the pumping of slurry through the wellbore, where, during the operation, the equivalent circulating density (*ECD*) is adopted, which is the coalescence of the slurry density and annular pressure loss, as presented in Equation (2). In the present research, the slurry density (ρ) of each sample was measured in g/cm^3, using a pressurized mud balance.

$$ECD = \frac{P_{annular} \times 100}{9.81 \times TVD} + \rho \tag{2}$$

where *ECD* is the equivalent circulating density measured in kg/m^3, $P_{annular}$ is the annular pressure loss measured in kg/m^2, *TVD* is the true vertical depth in m and ρ is the slurry density in kg/m^3.

The free-water test was conducted by pouring the slurry into a 250-mL measuring cylinder and then leaving the slurry undisturbed for two (2) hours. Subsequently, the water that was present on top of the cement was collected to determine the amount of free water.

Rheological properties of the slurry were measured using an atmospheric rheometer of Model 35 manufactured by Fann Instrument Company (Houston, TX, USA). Kinematic viscosity (v) was recorded at rotational speeds of 100 and 300 rpm, and averages of five readings were taken for each rotational speed to calculate plastic viscosity (*PV*) and yield point (*YP*), as per Equations (3) and (4).

$$PV = (v_{300} - v_{100}) \times 1.5 \tag{3}$$

$$YP = v_{300} - PV \tag{4}$$

where *PV* is the plastic viscosity measured in cP; *YP* is the yield point measured in N/m^2; and v_{100} and v_{300} are the kinematic viscosities at 100 and 300 rpm, respectively, measured in cP.

2.2. Measurement of Compressive Strength and Linear Expansion of GPC Samples

Cube samples of GPC with dimensions of 50 mm × 50 mm × 50 mm for each mix formulation were casted and cured in a water bath at 60 °C and atmospheric pressure, and a curing chamber at 90 °C and 3000 psi, which is approximately 20.68 MPa. Measurements of compressive strength and linear expansion were performed on the samples, which were cured for 1 day, 14 days, 30 days and 60 days. Measurement of compressive strength was conducted in accordance with API SPEC 10A [5], using a digital compressive strength tester of Model 4207D manufactured by Ametek Chandler Engineering (Tulsa, OK, USA). The compressive strength (*F*) of the samples was determined based on Equation (5).

$$F = \frac{P}{A} \tag{5}$$

where *F* is the compressive strength measured in psi, *P* is the compressive load at the point of failure measured in lbf and *A* is the cross-sectional surface area of the sample measured in inch^2.

Measurement of linear expansion was performed based on the procedure described in API RP10B-5 [55], using an expansion cell based on Equation (6). The slurry was subjected to conditioning for 30 min and then poured into the expansion cell. The distance between two steel balls of the cell before expansion was recorded as the initial length (L_i). Subsequently, the cell was placed into the water bath that was preheated to 60 °C and cured for a predetermined curing time. After curing, the cell was removed from the water bath, and the distance between the two steel balls after expansion was recorded as the final length (L_f) [17].

$$\frac{\Delta l}{l_0} = \frac{\left(L_f - L_i\right) \times 0.358}{L_i} \times 100\,\% \tag{6}$$

where $\Delta l/l_0$ is the linear expansion measured by percentage, L_f is the final length measured in mm and L_i is the initial length measured in mm.

The process flow of the research methodology employed in the present study is illustrated in Figure 3.

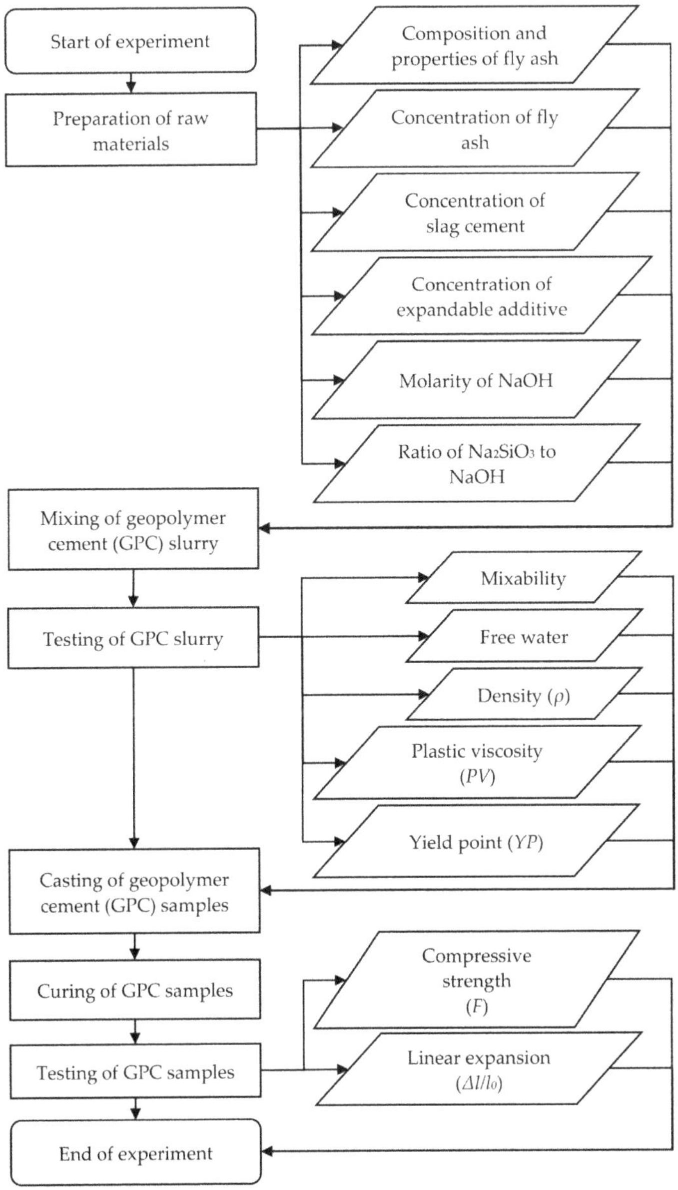

Figure 3. Process flow of the research methodology employed in the present study.

3. Results and Discussion

3.1. Mixability, Amount of Free Water and Density of GPC Slurry (ρ)

All formulations exhibited stability during mixing; the mixability and homogeneity were adequate; and the amount of free water collected was zero. Measurements of ρ revealed that the increase in concentration of *R*-additive from 10%, to 15%, 20% and 25% led to the reduction in slurry density from 1.76 to 1.74 g/cm^3 and from 1.66 and 1.64 g/cm^3, respectively, as shown in Figure 4. The rate of decline in ρ was the highest when the concentration of *R*-additive was increased from 15% to 20%, where ρ was reduced from 1.74 to 1.66 g/cm^3.

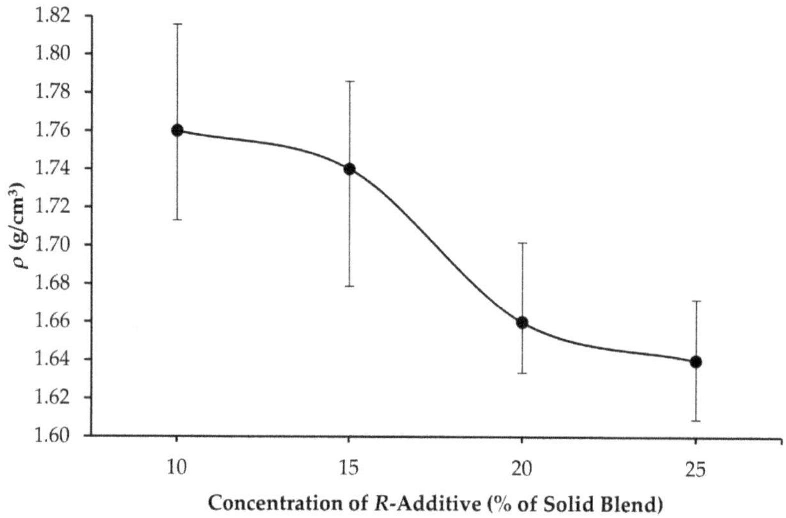

Figure 4. Density (ρ) of geopolymer cement (GPC) slurry at varying concentrations of elastomeric expandable additive (*R*-additive).

3.2. Rheological Properties of GPC Slurry

The results reveal that the increase in the concentration of *R*-additive has led to the increase in rheological properties of the slurry, *PV* and *YP*. As shown in Figure 5, as the concentration of *R*-additive was raised from 10% to 15%, 20% and 25%, *PV* increased from 48 cP to 74, 83 and 104 cP, respectively. In view of the difficulties experienced during the pumping of slurry with *PV* that exceeds 100 cP through the wellbore, as elaborated by Igbani et al. [56] and Zahid et al. [57], the addition of *R*-additive at concentrations of not more than 20% is recommended. In parallel with the impact on *PV*, as presented in Figure 6, as the concentration of *R*-additive was raised from 10% to 15%, 20% and 25%, *YP* also increased from 3.8 N/m^2 to 6.7, 10.1 and 12.3 N/m^2, respectively. The increase in *PV* and *YP* with respect to the increase in concentration of *R*-additive is in agreement with the findings of Richhariya et al. [21], which employed dual-coated polyacrylamide (DPAM) as the additive, where it was deduced that the higher *PV* was obtained as a result of the gelation characteristics of the slurry, and the concentration of DPAM of 16% resulted in optimal rheological characteristics, as further additions of DPAM beyond 16% did not exhibit further variations of *PV*.

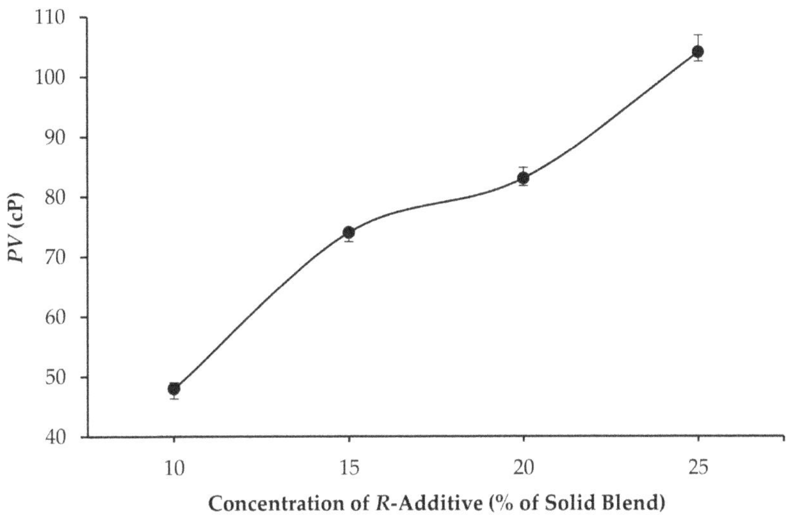

Figure 5. Plastic viscosity (*PV*) of GPC slurry at varying concentrations of *R*-additive.

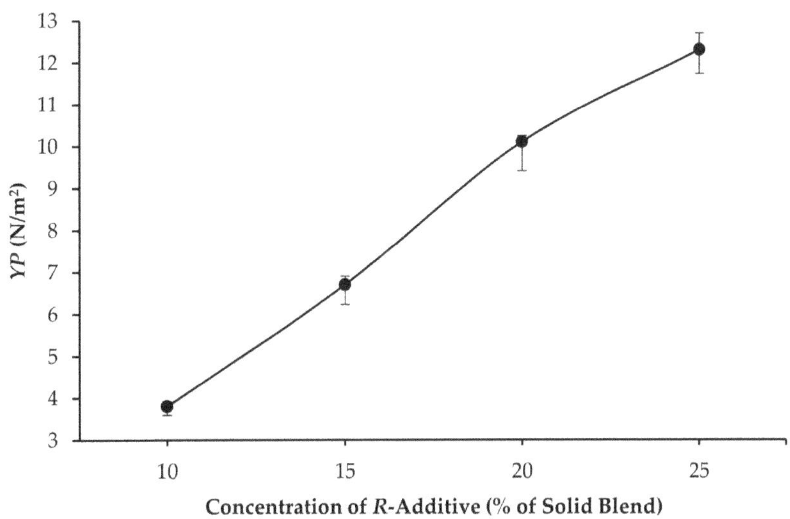

Figure 6. Yield point (*YP*) of GPC slurry at varying concentrations of *R*-additive.

3.3. Compressive Strength and Linear Expansion of GPC Samples

In essence, the findings indicate that, as the curing time was extended, F and $\Delta l/l_0$ increased with gradients that vary from the curing times of 1 day to 14, 30 and 60 days. For GPC samples that were cured at 60 °C, at atmospheric pressure, the highest F of 15.01 MPa was obtained by R10, while the highest $\Delta l/l_0$ of 0.9985% was obtained by R25.

As the curing time increased from 1 day to 60 days, F and $\Delta l/l_0$ of each formulation increased as follows:

- R10: F increased from 4.71 to 15.01 MPa, and $\Delta l/l_0$ increased from 0.3508% to 0.5213%, as per Figure 7.
- R15: F increased from 6.01 to 10.49 MPa, and $\Delta l/l_0$ increased from 0.1432% to 0.3101%, as per Figure 8.

- R20: *F* increased from 5.14 to 10.82 MPa, and $\Delta l/l_0$ increased from 0.7625% to 0.9903%, as per Figure 9.
- R25: *F* increased from 1.72 to 10.33 MPa, and $\Delta l/l_0$ increased from 0.7625% to 0.9985%, as per Figure 10.

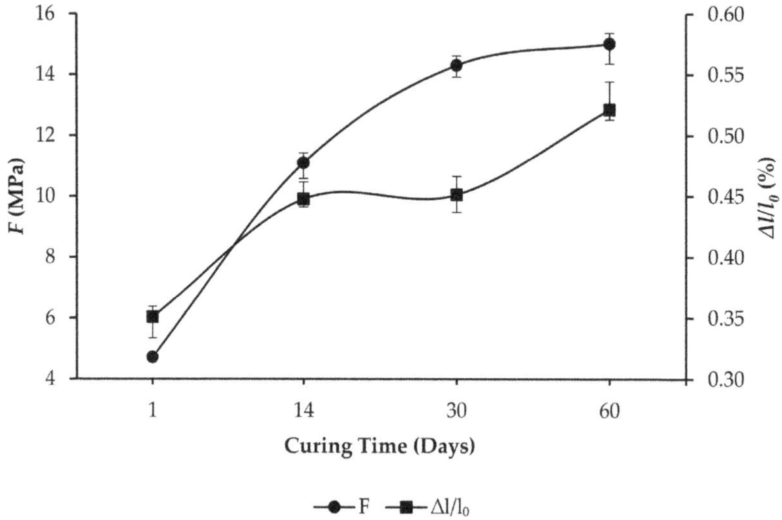

Figure 7. Compressive strength (*F*) and linear expansion ($\Delta l/l_0$), at varying curing times, of GPC samples with R10 formulation that were cured at 60 °C, at atmospheric pressure.

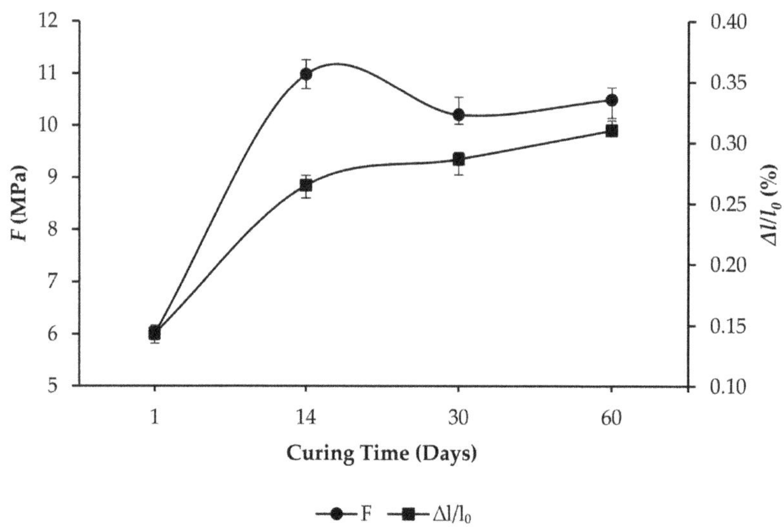

Figure 8. *F* and $\Delta l/l_0$, at varying curing times, of GPC samples with R15 formulation that were cured at 60 °C, at atmospheric pressure.

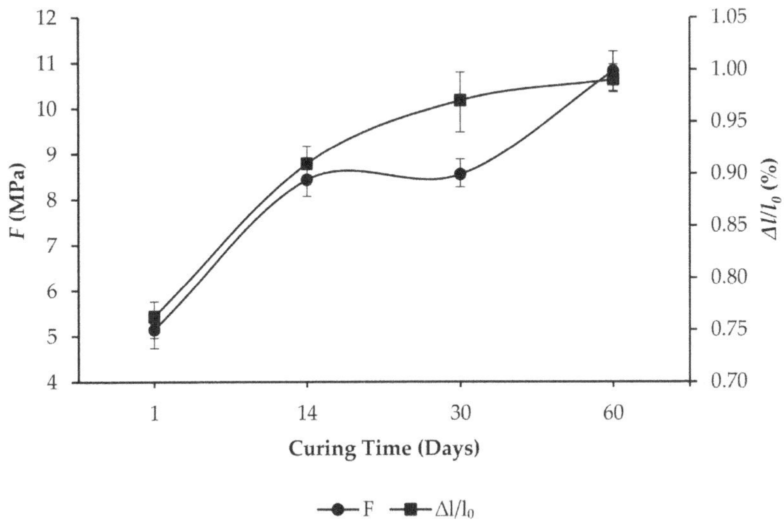

Figure 9. F and $\Delta l/l_0$, at varying curing times, of GPC samples with R20 formulation that were cured at 60 °C, at atmospheric pressure.

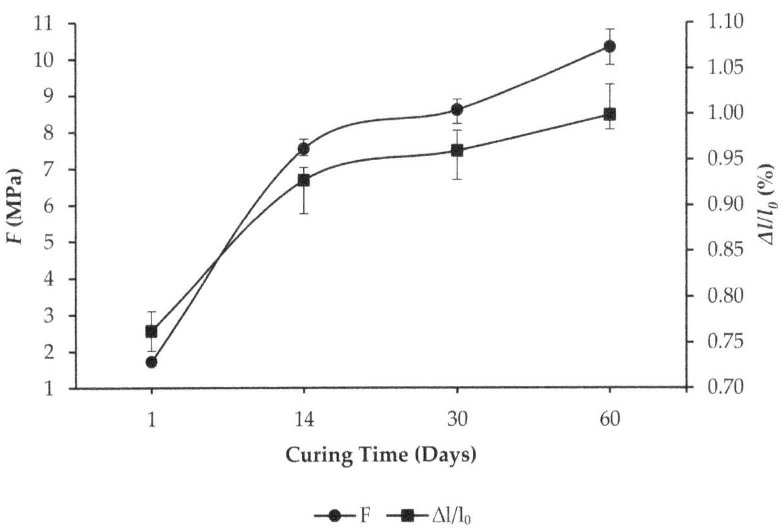

Figure 10. F and $\Delta l/l_0$, at varying curing times, of GPC samples with R25 formulation that were cured at 60 °C, at atmospheric pressure.

The impacts of increasing the concentration of R-additive on F and $\Delta l/l_0$ are elucidated in Figures 11–14 for samples that were cured for 1, 14, 30 and 60 days, respectively. Principally, as the concentration of R-additive was raised, F decreased with gradients that vary from the concentrations of R-additive of 10% to 15%, 20% and 25%. When the concentration of R-additive was increased from 10% to 15%, F increased for samples that were cured for 1 day, as shown in Figure 11; exhibited a trivial change for samples that were cured for 14 days, as shown in Figure 12; and, conversely, declined with a high gradient for samples that were cured for 30 and 60 days, as shown in Figures 12 and 13, respectively. The findings are in agreement with Lee et al. [46] and Ekinci et al. [47], who highlighted that styrene–

butadiene lowers the pH of the alkaline-activator solution and, in consequence, that of the geopolymer matrix, thus impeding the development of compressive strength. Accordingly, the addition of styrene–butadiene at higher concentrations led to further reduction in the pH of the geopolymer matrix and, as a consequence, further reduction in the compressive strength. Furthermore, Ekinci et al. [47] suggested that styrene–butadiene hinders the reaction between the raw material and alkaline-activator solution as it covers the interfaces of the raw material. Moreover, the addition of *R*-additive as an expandable additive led to the increase in bulk volume of the internal cement as explained in Baumgarte et al. [58]. Furthermore, based on Sofi [59], it can be added that higher concentrations of *R*-additive resulted in further reduction of *F* by virtue of the reduction in density. *R*-additive is an oil-swellable particle, and hence, as explained in Barlet-Gouédard et al. [60], it can counteract the formation of cracks, as it swells upon contact with oil; this can potentially address any issues associated with the formation of micro-annulus and micro-cracks that adversely enable formation fluids to flow through.

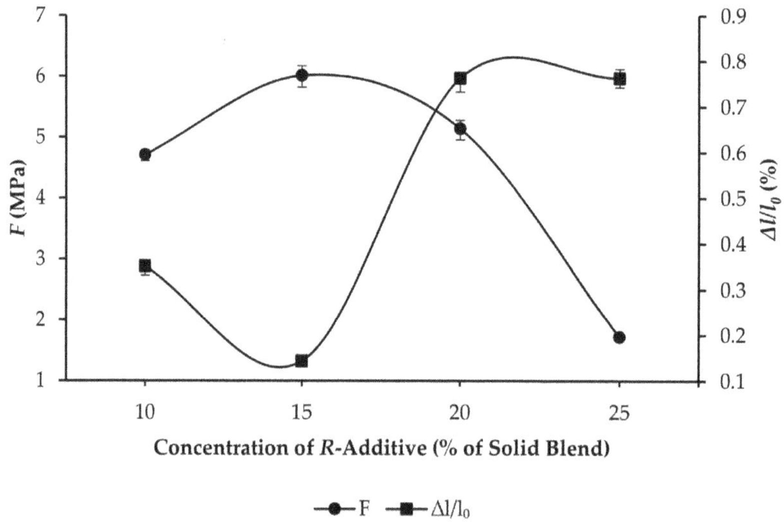

Figure 11. *F* and $\Delta l/l_0$, at varying concentrations of *R*-additive, of GPC samples that were cured at 60 °C, at atmospheric pressure, after 1 day of curing.

On the other hand, results reveal that, as the concentration of *R*-additive was raised from 10% to 15%, $\Delta l/l_0$ decreased. Subsequently, as the concentration of *R*-additive was raised further from 15% to 20%, $\Delta l/l_0$ increased with a higher gradient as compared to the prior decline in $\Delta l/l_0$. Subsequent change in $\Delta l/l_0$, owing to the further addition of *R*-additive from 20% to 25%, is inconsequential. Abd Rahman et al. [17] highlighted that, in contrast to OPC, GPC can expand by itself when exposed to water, even without the presence of expandable additives. The findings of Abd Rahman et al. [17] revealed that the amount of expansion of geopolymer cement was approximately 2.5 times higher than that of OPC at all curing durations. According to their findings, Abd Rahman et al. [17] suggested that the addition of expandable additives in geopolymer binders that are of the same types as those that are commercially available for OPC are compatible. Hence, in view of the findings of Abd Rahman et al. [17], the present study suggests the adoption of GPC coupled with the addition of an optimum concentration of expandable additives to further increase $\Delta l/l_0$.

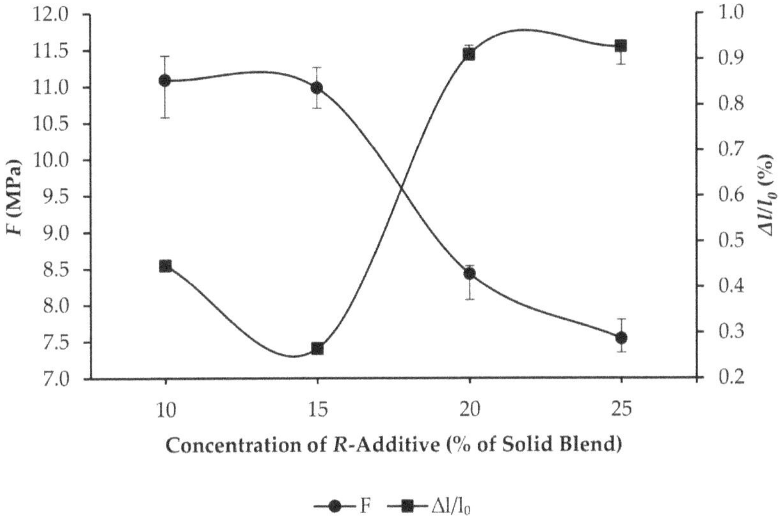

Figure 12. *F* and $\Delta l/l_0$, at varying concentrations of *R*-additive, of GPC samples that were cured at 60 °C, at atmospheric pressure, after 14 days of curing.

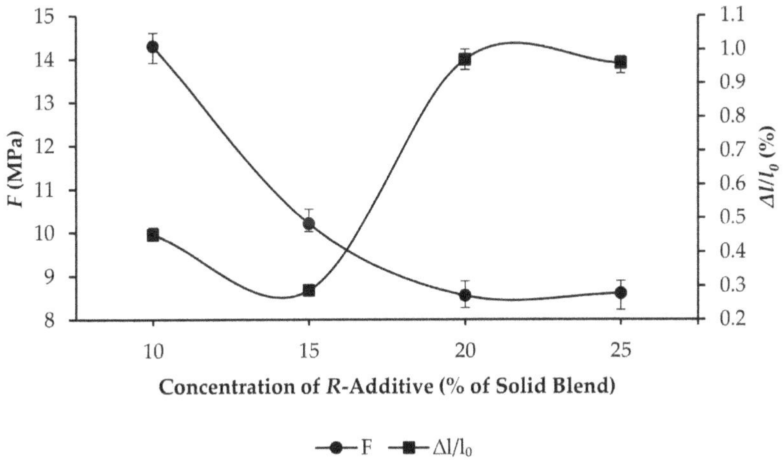

Figure 13. *F* and $\Delta l/l_0$, at varying concentrations of *R*-additive, of GPC samples that were cured at 60 °C, at atmospheric pressure, after 30 days of curing.

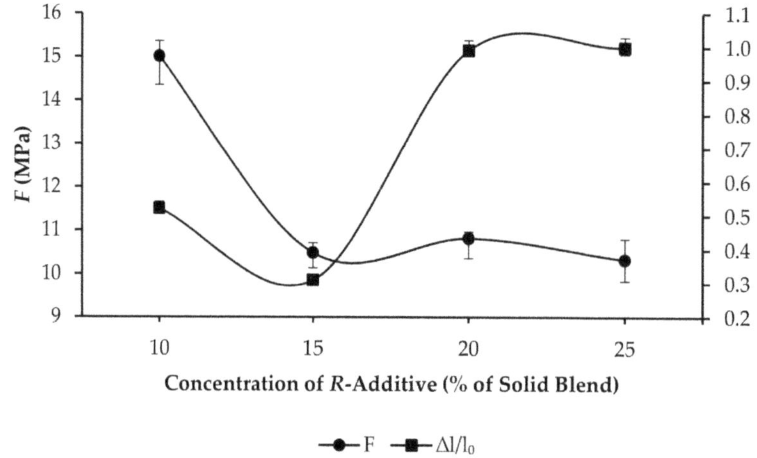

Figure 14. *F* and $\Delta l/l_0$, at varying concentrations of *R*-additive, of GPC samples that were cured at 60 °C in atmospheric pressure after 60 days of curing.

The effects of curing time on *F* and $\Delta l/l_0$ of GPC samples with R10 and R25 formulations that were cured at 90 °C and 3000 psi (≈20.68 MPa) are presented in Figures 15 and 16, respectively. When the R10 formulation was adopted, as the curing time increased from 1 day to 14, 30 and 60 days, *F* increased from 3.77 MPa to 11.69, 14.47 and 14.62 MPa, while $\Delta l/l_0$ increased from 0.04% to 0.10%, 0.45% and 0.63%, respectively. On the other hand, when the R25 formulation was adopted, *F* increased from 2.90 MPa to 8.79, 9.53 and 9.61 MPa, while $\Delta l/l_0$ increased from 0.19% to 0.21%, 1.17% and 1.32%, respectively. The profiles of *F* reveal that the increase in concentration of *R*-additive consistently led to lower values of *F* throughout the curing period, with an increasing difference between *F* of the R10 and R25 formulations as curing time increased.

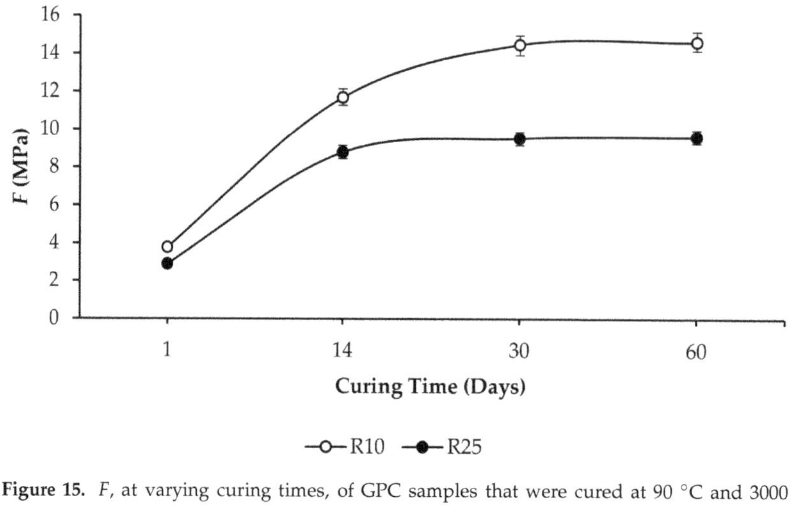

Figure 15. *F*, at varying curing times, of GPC samples that were cured at 90 °C and 3000 psi (≈20.68 MPa).

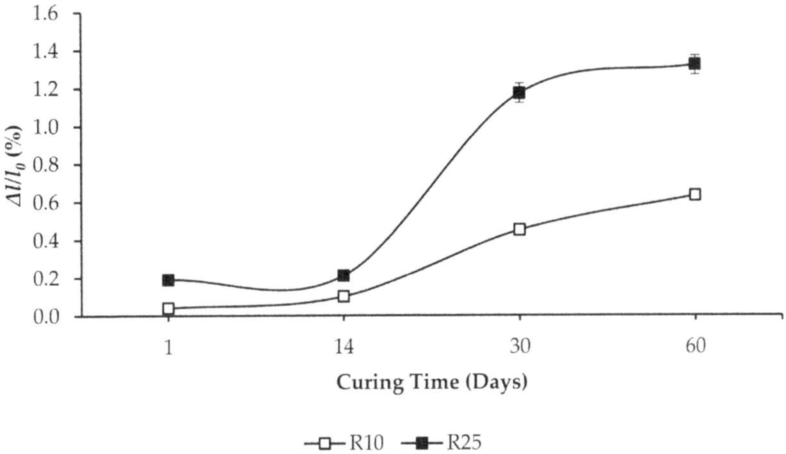

Figure 16. $\Delta l/l_0$, at varying curing times, of GPC samples that were cured at 90 °C and 3000 psi (≈20.68 MPa).

As revealed in Figure 17, the increase in curing temperature and pressure from 60 °C and atmospheric pressure to 90 °C and 3000 psi (≈20.68 MPa), respectively, resulted in the reduction in *F* from 15.01 to 14.62 MPa and from 10.33 to 9.61 MPa when the concentrations of *R*-additive were 10% and 25%, respectively. The rise in temperature resulted in the loss of the initial properties of the *R*-additive within the cement matrix, which, as a consequence, led to the reduction in *F* of the GPC sample [61]. On the other hand, as revealed in Figure 18, an increase in $\Delta l/l_0$ from 0.52% to 0.63% and from 0.99% to 1.32% occurred due to the faster movement of molecules in the elastomer chain of the *R*-additive and an increase in its diffusion coefficient owing to the increase in temperature and pressure [62].

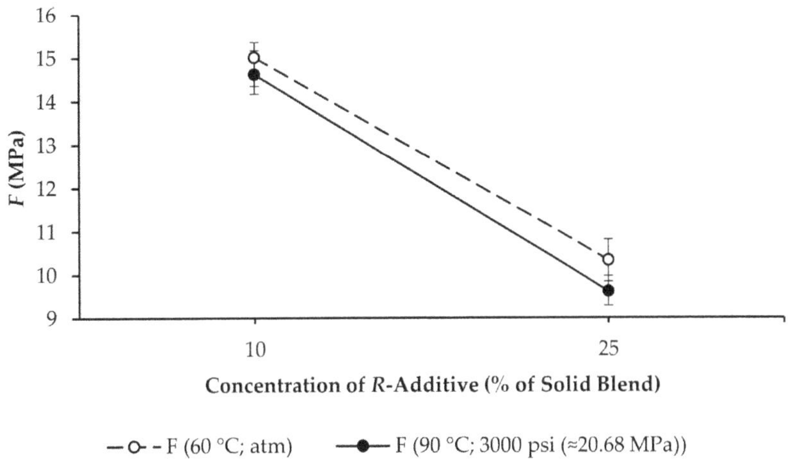

Figure 17. *F*, at 10% and 25% concentrations of *R*-additive, of GPC samples after 60 days of curing.

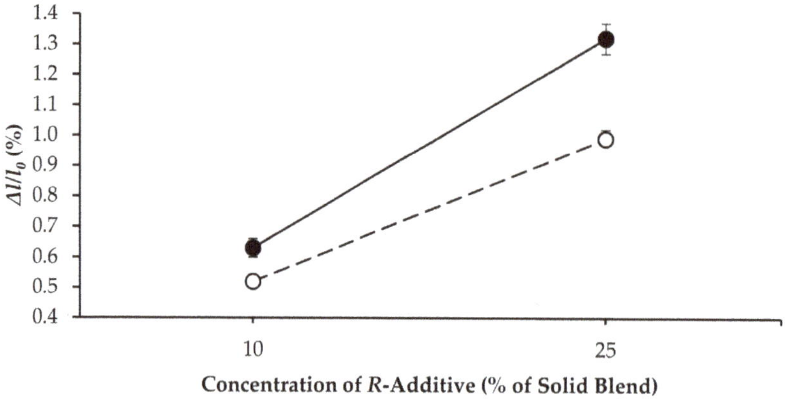

Figure 18. $\Delta l/l_0$, at 10% and 25% concentrations of *R*-additive, of GPC samples after 60 days of curing.

The formulations adopted in the present research, R10, R15, R20 and R25, fulfilled the requirements of the oil and gas industry and are aligned with the recommendations of Abbas et al. [63], Eric et al. [19], Mao et al. [64] and Richhariya et al. [21]. Multiple trial mixes were designed and tested prior to selecting the formulations to fulfill the specification requirements of API. Failure modes of the samples subjected to compressive strength tests are displayed in Figure 19, which shows that the samples were not completely crushed when the ultimate load was reached, and hence the samples were capable of resisting the load even after cracking. The compressive strengths of the formulations are acceptable for most well cementing operations, as they exceed 3.45 MPa [20].

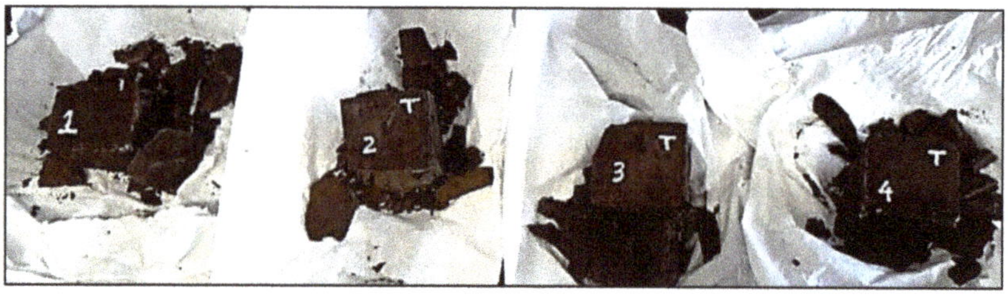

Figure 19. Failure mode of the samples subjected to compressive-strength tests.

As fly ash is a by-product of coal-fired power plants, its properties may vary according to, among others, age of the plant, ambient conditions and type and grade of coal employed. Accordingly, fly ash produced from different plants or batches of the same plant possesses inconsistent properties, hence presenting a limitation to the research. In the present study, screening was conducted on fly-ash samples collected from four different power plants in Malaysia that were selected based on the availability of samples, supply-chain management and approval from the Department of Environment of Malaysia for the supply prior to selecting the fly ash that was employed to produce the GPC samples. In addition to compliance to requirements for classification as a Class F fly ash as per ASTM C618-19 [49], the selection of the fly ash was determined with the aim of minimizing CaO content by virtue of its potential for resistance to CO_2 when exposed to the high-CO_2 environment of the oil and gas reservoir.

4. Conclusions

The present study aimed to investigate the effect of an expandable additive on the compressive strength and linear expansion of GPC, which is an alternative to OPC, for oil-well cementing. Fly-ash-based strength-enhanced GPC samples, with the addition of slag cement as the strength enhancer, were prepared by using different concentrations of *R*-additive and cured at 60 °C and atmospheric pressure. Mixability, amount of free water and slurry density were studied, and the effects of the concentration of *R*-additive on the compressive strength and linear expansion of the samples were analyzed. All formulations exhibited stability during mixing, mixability and homogeneity were adequate and the amount of free water collected was zero. An increase in the concentration of *R*-additive from 10% to 25% led to the reduction in the slurry density from 1.76 to 1.64 g/cm^3 and an increase in the rheological properties of the slurry, where *PV* increased from 48 to 104 cP and *YP* also increased from 3.8 to 12.3 N/m^2. An addition of *R*-additive at a concentration of 20% is recommended for optimization of the rheological properties. As the curing time was extended, *F* and $\Delta l/l_0$ increased with varying gradients. When cured at 60 °C and at atmospheric pressure, the highest *F* of 15.01 MPa was obtained when the concentration of *R*-additive was 10%, while the highest $\Delta l/l_0$ of 0.9985% was obtained when the concentration of *R*-additive was 25%. An increase in the curing temperature and pressure to 90 °C and 3000 psi (≈20.68 MPa) resulted in the reduction in *F* from 15.01 to 14.62 MPa and from 10.33 to 9.61 MPa, and the increase in $\Delta l/l_0$ from 0.52% to 0.63% and from 0.99% to 1.32%, when the concentrations of *R*-additive were 10% and 25%, respectively. The formulations adopted in the present research, which contain *R*-additive at concentrations ranging from 10% to 25%, fulfilled the requirements of the oil and gas industry.

Future research on the effect of adding elastomeric expandable additives to GPC to address volumetric shrinkage concerns in oil-well cementing can be augmented by performing an extensive study on the fluid flow for a wider and higher temperature range with cement qualification tests. In addition, the effect of downhole pressure can also be included in the investigations to assure that GPC can be adopted for a wide range of oil-well types.

Furthermore, in view of the inconsistent properties that fly ash produced from different plants or batches in the same plant, the development of pretreatment methods to standardize the quality of fly ash is recommended to facilitate the generation of findings that are consistent and reliable for future research that may employ fly ash from various sources.

Author Contributions: Conceptualization, S.H.A.R.; methodology, Y.A.S.; software, S.H.A.R.; validation, L.H.S.; formal analysis, N.N.Z.; investigation, N.H.; resources, M.F.H.; data curation, A.I.A.H.; writing—original draft preparation, S.A.F.; writing—review and editing, S.H.A.R.; visualization, L.H.S.; supervision, N.S.; project administration, Y.A.S.; funding acquisition, N.S. All authors have read and agreed to the published version of the manuscript.

Funding: This research was funded by PETRONAS Research Sdn. Bhd., Malaysia, via Project Cost Centre 015MD0-064, as administered by the Research Management Centre of Universiti Teknologi PETRONAS, Malaysia, and the Project Cost Centre E.025.GST.02019.003, as per the record of PETRONAS Research Sdn. Bhd.

Institutional Review Board Statement: Not applicable.

Informed Consent Statement: Not applicable.

Data Availability Statement: The data that support the findings of this study are available from PETRONAS Research Sdn. Bhd. Restrictions apply to the availability of these data, which were used under license for this study. Data are available from the corresponding author, S.H.A.R., with the permission of PETRONAS Research Sdn. Bhd.

Acknowledgments: The authors are thankful to the technologists of Universiti Teknologi PETRONAS and PETRONAS Research Sdn. Bhd. for the commitment and support to carry out various activities related to the experimental work in the laboratory.

Conflicts of Interest: The authors declare no conflict of interest.

Nomenclature

Abbreviations	
API	American Petroleum Institute
ASTM	American Society for Testing Materials
atm	atmospheric pressure
DPAM	dual-coated polyacrylamide
GPC	geopolymer cement
OPC	ordinary Portland cement
R-additive	elastomeric expandable additive used in the present study
SCP	sustained casing pressure
Chemical Formulae	
$Al_2Ca_6H_{12}O_{24}S_3$	ettringite
Al_2O_3	aluminum oxide
Al–O–Si	polysialate
Al–O–Si–Si	polysialate siloxo
Al–O–Si–Si–Si	polysialate disiloxo
$CaAl_2O_4$	calcium aluminate
$CaCO_3$	calcium carbonate
CaO	calcium oxide
$CaSO_4$	calcium sulfate
Ca^{2+}	calcium cation
Cl	chlorine
CO_2	carbon dioxide
C-S-H	calcium-silicate-hydrate
Fe_2O_3	iron (III) oxide
H_2CO_3	carbonic acid
K_2O	potassium oxide
K^+	potassium cation
MgO	magnesium oxide
NaOH	sodium hydroxide
Na_2O	sodium oxide
Na_2SiO_3	sodium silicate
Na^+	sodium cation
SiO_2	silicon dioxide
SO_3	sulfur trioxide
TiO_2	titanium (IV) Oxide
Notations	
A	cross-sectional surface area of the sample
ECD	equivalent circulating density
F	compressive strength
L_f	final length as measured using the expansion cell
L_i	initial length as measured using the expansion cell
M	alkali cation
n	degree of polycondensation
P	compressive load at the point of failure
$P_{annular}$	annular pressure loss
PV	plastic viscosity
TVD	true vertical depth
v	kinematic viscosity
v_{100}	kinematic viscosity at 100 rpm
v_{300}	kinematic viscosity at 300 rpm
YP	yield point
$\Delta l/l_0$	linear expansion
ρ	slurry density

References

1. Abd Rahman, S.H.; Zulkarnain, N.N.; Shafiq, N. Experimental study and design of experiment using statistical analysis for the development of geopolymer matrix for oil-well cementing for enhancing the integrity. *Crystals* **2021**, *11*, 139. [CrossRef]
2. Zulkarnain, N.N.; Farhan, S.A.; Sazali, Y.A.; Shafiq, N.; Abd Rahman, S.H.; Abd Hamid, A.I.; Habarudin, M.F. Reducing the waiting-on-cement time of geopolymer well cement using calcium chloride ($CaCl_2$) as the accelerator: Analysis of the compressive strength and acoustic impedance for well logging. *Sustainability* **2021**, *13*, 6128. [CrossRef]
3. Bu, Y.; Du, J.; Guo, S.; Liu, H.; Huang, C. Properties of oil well cement with high dosage of metakaolin. *Constr. Build. Mater.* **2016**, *112*, 39–48. [CrossRef]
4. Khalifeh, M.; Saasen, A.; Hodne, H.; Godøy, R.; Vrålstad, T. Geopolymers as an Alternative for Oil Well Cementing Applications: A Review of Advantages and Concerns. In Proceedings of the 36th International Conference on Ocean, Offshore and Arctic Engineering, Trondheim, Norway, 25–30 June 2017; The American Society of Mechanical Engineers: New York, NY, USA, 2017. [CrossRef]
5. *API SPEC 10A*; Cements and Materials for Well Cementing. American Petroleum Institute (API): Washington, DC, USA, 2019. Available online: https://standards.globalspec.com/std/14208303/api-spec-10a (accessed on 30 September 2021).
6. Nasvi, M.C.M.; Ranjith, P.G.; Sanjayan, J. Comparison of Mechanical Behaviors of Geopolymer and Class G Cement as Well Cement at Different Curing Temperatures for Geological Sequestration of Carbon Dioxide. In Proceedings of the 46th U.S. Rock Mechanics/Geomechanics Symposium, Chicago, IL, USA, 24–27 June 2012; American Rock Mechanics Association: Alexandria, VA, USA, 2012. ARMA-2012-232.
7. Barlet-Gouédard, V.; Rimmelé, G.; Goffé, B.; Porcherie, O. Well technologies for CO_2 geological storage: CO_2-resistant cement. *Oil Gas Sci. Technol.-Rev. IFP* **2007**, *62*, 325–334. [CrossRef]
8. Liteanu, E.; Spiers, C.J.; Peach, C.J. Failure behaviour wellbore cement in the presence of water and supercritical CO_2. *Energy Procedia* **2009**, *1*, 3553–3560. [CrossRef]
9. Salehi, S.; Khattak, M.J.; Ali, N.; Ezeakacha, C.; Saleh, F.K. Study and use of geopolymer mixtures for oil and gas well cementing applications. *J. Energy Resour. Technol.* **2018**, *140*, 012908. [CrossRef]
10. Salehi, S.; Ezeakacha, C.P.; Khattak, M.J. Geopolymer Cements: How Can You Plug and Abandon a Well with New Class of Cheap Efficient Sealing Materials. In Proceedings of the SPE Oklahoma City Oil and Gas Symposium, Oklahoma City, OK, USA, 27–31 March 2017; Society of Petroleum Engineers: Richardson, TX, USA, 2017. [CrossRef]
11. Vrålstad, T.; Saasen, A.; Fjær, E.; Øia, T.; Ytrehus, J.D.; Khalifeh, M. Plug & abandonment of offshore wells: Ensuring long-term well integrity and cost-efficiency. *J. Pet. Sci. Eng.* **2018**, *173*, 478–491. [CrossRef]
12. Bois, A.-P.; Garnier, A.; Galdiolo, G.; Laudet, J.-B. Use of a mechanistic model to forecast cement-sheath integrity. *SPE Drill. Complet.* **2012**, *27*, 303–314. [CrossRef]
13. Santra, A.K.; Reddy, B.R.; Liang, F.; Fitzgerald, R. Reaction of CO_2 with Portland Cement at Downhole Conditions and the Role of Pozzolanic Supplements. In Proceedings of the SPE International Symposium on Oilfield Chemistry, The Woodlands, TX, USA, 20–22 April 2009; Society of Petroleum Engineers: Richardson, TX, USA, 2009. [CrossRef]
14. Reddy, B.R.; Xu, Y.; Ravi, K.; Gray, D.W.; Pattillo, P. Cement shrinkage measurement in oilwell cementing—a comparative study of laboratory methods and procedures. *SPE Drill. Complet.* **2009**, *24*, 104–114. [CrossRef]
15. Kosmatka, S.H.; Wilson, M.L. *Design and Control of Concrete Mixtures: The Guide to Applications, Methods, and Materials*, 15th ed.; Portland Cement Association: Skokie, IL, USA, 2011.
16. Nasvi, M.C.M.; Ranjith, P.G.; Sanjayan, J.; Bui, H. Effect of temperature on permeability of geopolymer: A primary well sealant for carbon capture and storage wells. *Fuel* **2014**, *117*, 354–363. [CrossRef]
17. Abd Rahman, S.H.; Irawan, S.; Shafiq, N.; Suppiah, R.R. Investigating the expansion characteristics of geopolymer cement samples in a water bath and compared with the expansion of ASTM Class-G cement. *Heliyon* **2020**, *6*, e03478. [CrossRef]
18. Nagral, M.R.; Ostwal, T.; Chitawadagi, M.V. Effect of curing temperature and curing hours on the properties of geo-polymer concrete. *Int. J. Comput. Eng. Res.* **2014**, *4*, 1–11.
19. Eric, B.; Joel, F.; Grace, O. Oil well cement additives: A review of the common types. *Oil Gas Res.* **2016**, *2*, 112. [CrossRef]
20. Ridha, S.; Abd Hamid, A.I.; Abdul Halim, A.H.; Zamzuri, N.A. Elasticity and expansion test performance of geopolymer as oil well cement. *IOP Conf. Ser. Earth Environ. Sci.* **2018**, *140*, 012147. [CrossRef]
21. Richhariya, G.; Dora, D.T.K.; Parmar, K.R.; Pant, K.K.; Singhal, N.; Lal, K.; Kundu, P.P. Development of self-healing cement slurry through the incorporation of dual-encapsulated polyacrylamide for the prevention of water ingress in oil well. *Materials* **2020**, *13*, 2921. [CrossRef]
22. Duguid, A.; Scherer, G.W. Degradation of oilwell cement due to exposure to carbonated brine. *Int. J. Greenh. Gas Control* **2010**, *4*, 546–560. [CrossRef]
23. Bourgoyne, A.T.; Scott, S.L.; Regg, J.B. Sustained Casing Pressure in Offshore Producing Wells. In Proceedings of the Offshore Technology Conference, Houston, TX, USA, 3–6 May 1999. [CrossRef]
24. Farkas, R.F.; England, K.W.; Roy, M.L.; Dickinson, M.; Samuel, M.; Hart, R.E. New Cementing Technology Cures 40-Year-Old Squeeze Problems. In Proceedings of the SPE Annual Technical Conference and Exhibition, Houston, TX, USA, 3–6 October 1999; Society of Petroleum Engineers: Richardson, TX, USA, 1999. [CrossRef]
25. Merlini, M.; Artioli, G.; Cerulli, T.; Cella, F.; Bravo, A. Tricalcium aluminate hydration in additivated systems. a crystallographic study by SR-XRPD. *Cem. Concr. Res.* **2008**, *38*, 477–486. [CrossRef]

26. Chenevert, M.E.; Shrestha, B.K. Chemical shrinkage properties of oilfield cements. *SPE Drill. Eng.* **1991**, *6*, 37–43. [CrossRef]
27. Kiran, R.; Teodoriu, C.; Dadmohammadi, Y.; Nygaard, R.; Wood, D.; Mokhtari, M.; Salehi, S. Identification and evaluation of well integrity and causes of failure of well integrity barriers (a review). *J. Nat. Gas Sci. Eng.* **2017**, *45*, 511–526. [CrossRef]
28. Davidovits, J. *Geopolymer Chemistry & Applications*, 3rd ed.; Institut Géopolymère: Saint-Quentin, France, 2011.
29. Kong, D.L.Y.; Sanjayan, J.G.; Sagoe-Crentsil, K. Comparative performance of geopolymers made with metakaolin and fly ash after exposure to elevated temperatures. *Cem. Concr. Res.* **2007**, *37*, 1583–1589. [CrossRef]
30. Thokchom, S.; Ghosh, P.; Ghosh, S. Resistance of fly ash based geopolymer mortars in sulfuric acid. *ARPN J. Eng. Appl. Sci.* **2009**, *4*, 65–70.
31. Palomo, A.; Grutzeck, M.W.; Blanco, M.T. Alkali-activated fly ashes: A cement for the future. *Cem. Concr. Res.* **1999**, *29*, 1323–1329. [CrossRef]
32. Dimas, D.; Giannopoulou, I.; Panias, D. Polymerization in sodium silicate solutions: A fundamental process in geopolymerization technology. *J. Mater. Sci.* **2009**, *44*, 3719–3730. [CrossRef]
33. Khalifeh, M.; Saasen, A.; Vrålstad, T. Potential Utilization of Geopolymers in Plug and Abandonment Operations. In Proceedings of the SPE Bergen One Day Seminar, Bergen, Norway, 2 April 2014; Society of Petroleum Engineers: Richardson, TX, USA, 2014. [CrossRef]
34. Nasvi, M.M.C.; Gamage, R.P.; Jay, S. Geopolymer as well cement and the variation of its mechanical behavior with curing temperature. *Greenh. Gas Sci. Technol.* **2012**, *2*, 46–58. [CrossRef]
35. Uehara, M. New concrete with low environmental load using the geopolymer method. *Q. Rep. RTRI* **2010**, *51*, 1–7. [CrossRef]
36. Liu, X.; Aughenbaugh, K.; Nair, S.; Shuck, M.; van Oort, E. Solidification of Synthetic-Based Drilling Mud using Geopolymers. In Proceedings of the SPE Deepwater Drilling and Completions Conference, Galveston, TX, USA, 14–15 September 2016; Society of Petroleum Engineers: Richardson, TX, USA, 2016. [CrossRef]
37. Khalifeh, M.; Hodne, H.; Saasen, A.; Integrity, O.; Eduok, E.I. Usability of Geopolymers for Oil Well Cementing Applications: Reaction Mechanisms, Pumpability, and Properties. In Proceedings of the SPE Asia Pacific Oil & Gas Conference and Exhibition, Perth, Australia, 25–27 October 2016; Society of Petroleum Engineers: Richardson, TX, USA, 2016. [CrossRef]
38. Khalifeh, M.; Todorovic, J.; Vrålstad, T.; Saasen, A.; Hodne, H. Long-term durability of rock-based geopolymers aged at downhole conditions for oil well cementing operations. *J. Sustain. Cem.-Based Mater.* **2017**, *6*, 217–230. [CrossRef]
39. Van Jaarsveld, J.G.S.; Van Deventer, J.S.J.; Lorenzen, L. The potential use of geopolymeric materials to immobilise toxic metals: Part I. Theory and applications. *Miner. Eng.* **1997**, *10*, 659–669. [CrossRef]
40. Diaz, E.I.; Allouche, E.N. Recycling of Fly Ash into Geopolymer Concrete: Creation of a Database. In Proceedings of the 2010 IEEE Green Technologies Conference, Grapevine, TX, USA, 15–16 April 2010; Institute of Electrical and Electronics Engineers: Piscataway, NJ, USA, 2010; pp. 1–7. [CrossRef]
41. Yang, Z.X.; Ha, N.R.; Jang, M.S.; Hwang, K.H. Geopolymer concrete fabricated by waste concrete sludge with silica fume. *Mater. Sci. Forum.* **2009**, *620–622*, 791–794. [CrossRef]
42. Hewayde, E.; Nehdi, M.; Allouche, E.; Nakhla, G. Effect of geopolymer cement on microstructure, compressive strength and sulphuric acid resistance of concrete. *Mag. Concr. Res.* **2006**, *58*, 321–331. [CrossRef]
43. Lloyd, N.; Rangan, B. Geopolymer Concrete with Fly Ash. In Proceedings of the Second International Conference on Sustainable Construction Materials and Technologies, Università Politecnica delle Marche, Ancona, Italy, 28–30 June 2010; University of Wisconsin-Milwaukee: Milwaukee, WI, USA, 2010; pp. 1493–1504.
44. Majidi, B. Geopolymer technology, from fundamentals to advanced applications: A review. *Mater. Technol. Adv. Perform. Mater.* **2009**, *24*, 79–87. [CrossRef]
45. Davidovits, J. Environmentally driven geopolymer applications. In Proceedings of the Geopolymer Conference, Melbourne, Australia, 28–29 October 2002; Institut Géopolymère: Saint-Quentin, France, 2002.
46. Lee, N.K.; Kim, E.M.; Lee, H.K. Mechanical properties and setting characteristics of geopolymer mortar using styrene-butadiene (SB) latex. *Constr. Build. Mater.* **2016**, *113*, 264–272. [CrossRef]
47. Ekinci, E.; Türkmen, I.; Kantarci, F.; Karakoç, M.B. The improvement of mechanical, physical and durability characteristics of volcanic tuff based geopolymer concrete by using nano silica, micro silica and styrene-butadiene latex additives at different ratios. *Constr. Build. Mater.* **2019**, *201*, 257–267. [CrossRef]
48. Our Business: Power Plant and Water Desalination Plant Locations. Available online: https://www.malakoff.com.my/Our-Business/Power-Plant-and-Water-Desalination-Plant-Locations/ (accessed on 5 November 2021).
49. *ASTM C618–19*; Standard Specification for Coal Fly Ash and Raw or Calcined Natural Pozzolan for Use in Concrete. ASTM International: West Conshohocken, PA, USA, 2019.
50. *GB 175-2007/XG3-2018*; Common Portland Cement, Including Amendment 3. Code of China: Beijing, China, 2019.
51. Yuhuan, B.; Rui, M.; Jiapei, D.; Shenglai, G.; Huajie, L.; Letian, Z. Utilization of metakaolin-based geopolymer as a mud-cake solidification agent to enhance the bonding strength of oil well cement–formation interface. *R. Soc. Open Sci.* **2020**, *7*, 191230. [CrossRef]
52. Kallesten, B.; Kakay, S.; Gebremariam, K. Synthesis and characterization of fly ash and slag based geopolymer concrete. *IOP Conf. Ser. Mater. Sci. Eng.* **2019**, *700*, 012032. [CrossRef]
53. Kanesan, D.; Ridha, S.; Rao, P. Formulation of geopolymer cement using mixture of slag and Class F fly ash for oil well cementing. *IOP Conf. Ser. Mater. Sci. Eng.* **2017**, *201*, 012014. [CrossRef]

54. *API RP 10B-2*; Recommended Practice for Testing Well Cements; 2nd ed. American Petroleum Institute (API): Washington, DC, USA, 2019.
55. *API RP10B-5*; Recommended Practice on Determination of Shrinkage and Expansion of Well Cement Formulations at Atmospheric Pressure. American Petroleum Institute (API): Washington, DC, USA, 2005.
56. Igbani, S.; Appah, D.; Ogoni, H.A. The application of response surface methodology in Minitab 16, to identify the optimal, comfort, and adverse zones of compressive strength responses in ferrous oilwell cement sheath systems. *Int. J. Eng. Mod. Technol.* **2020**, *6*, 20–39.
57. Zahid, M.; Shafiq, N.; Isa, M.H.; Gil, L. Statistical modeling and mix design optimization of fly ash based engineered geopolymer composite using response surface methodology. *J. Clean. Prod.* **2018**, *194*, 483–498. [CrossRef]
58. Baumgarte, C.; Thiercelin, M.; Klaus, D. Case studies of expanding cement to prevent microannular formation. In Proceedings of the SPE Annual Technical Conference and Exhibition, Houston, TX, USA, 3–6 October 1999; Society of Petroleum Engineers: Richardson, TX, USA, 1999. [CrossRef]
59. Sofi, A. Effect of waste tyre rubber on mechanical and durability properties of concrete—A review. *Ain Shams Eng. J.* **2018**, *9*, 2691–2700. [CrossRef]
60. Barlet-Gouédard, V.; Rimmelé, G.; Porcherie, O.; Quisel, N.; Desroches, J. A solution against well cement degradation under CO_2 geological storage environment. *Int. J. Greenh. Gas Control* **2009**, *3*, 206–216. [CrossRef]
61. Powers, P.O.; Billmeyer, B.R. Swelling of synthetic rubbers in mineral oils. Effect of temperature and aniline point. *Rubber Chem. Technol.* **1945**, *18*, 452–459. [CrossRef]
62. Shan, G.-R.; Xu, P.-Y.; Weng, Z.-X.; Huang, Z.-M. Oil-absorption function of physical crosslinking in the high-oil-absorption resins. *J. Appl. Polym. Sci.* **2003**, *90*, 3945–3950. [CrossRef]
63. Abbas, G.; Irawan, S.; Kumar, S.; Elrayah, A.A.I. Improving oil well cement slurry performance using hydroxypropylmethylcellulose polymer. *Adv. Mater. Res.* **2013**, *787*, 222–227. [CrossRef]
64. Mao, W.; Litina, C.; Al-Tabbaa, A. Development and application of novel sodium silicate microcapsule-based self-healing oil well cement. *Materials* **2020**, *13*, 456. [CrossRef]

applied sciences

MDPI

Article

Alkali-Activated Copper Slag with Carbon Reinforcement: Effects of Metakaolinite, OPC and Surfactants

Patrick Ninla Lemougna [1,2,3], Guillermo Meza Hernandez [1,2], Nicole Dilissen [1,4], Felicite Kingne [1], Jun Gu [1] and Hubert Rahier [1,*]

1 Research Group Sustainable Materials Engineering (SUME), Lab of Physical Chemistry and Polymer Science (FYSC), Vrije Universiteit Brussel, Pleinlaan 2, 1050 Brussels, Belgium; patrick.ninla.lemougna@vub.be or lemougna@yahoo.fr (P.N.L.); guillermo.meza.hernandez@vub.be (G.M.H.); nicole.dilissen@vub.be (N.D.); felicite.kingne.kingne@vub.be (F.K.); jun.gu@vub.be (J.G.)

2 Strategic Initiative Materials in Flanders (SIM), 9052 Zwijnaarde, Belgium

3 Department of Minerals Engineering, School of Chemical Engineering and Mineral Industries (EGCIM), University of Ngaoundere, Ngaoundere P.O. Box 454, Cameroon

4 Buildwise, Avenue P Holoffe 21, 1342 Limelette, Belgium

* Correspondence: hubert.rahier@vub.be

Highlights:

What are the main findings?

- Copper slag can be upcycled in alkali-activated carbon fabric-based composites;
- A 20 wt.% metakaolinite substitution of copper slag increases reaction heat and mechanical properties;

What is the implication of the main finding?

- Low reactivity at 20 °C, but instantaneous at 80 °C with K-solution SiO_2/K_2O of 2.25;
- Composites with an elastic modulus of 19 GPa and flexural strength of 88 MPa.

Abstract: Copper slag is an industrial residue with a large unutilized fraction. This study presents the development of alkali-activated composites from a copper slag named Koranel®. The effects of metakaolinite, ordinary Portland cement (OPC) and surfactants were investigated. The reactivity of Koranel with potassium silicate solutions with molar ratio R = SiO_2/K_2O varying from 1 to 2.75, with 0.25 intervals, was investigated using isothermal calorimetry. The reactivity was relatively low at 20 °C; the reaction started after a few hours with a low silica modulus, to several weeks with the highest silica modulus. The substitution of Koranel by OPC (5 wt.%) or by metakaolinite (10–20 wt.%), both led to higher reaction heat and rate; meanwhile, the addition of 2 wt.% polyethylene glycol/2-methyl 2,4 pentanediol delayed the reaction time in the system containing metakaolinite. Raising the curing temperature from 20 °C to 80 °C shortened the setting time of the low reactive systems, from several days to almost instantaneous, opening perspectives for their application in the production of prepreg composite materials. The use of carbon fabric as reinforcement in the alkali-activated matrix led to composite materials with flexural strength reaching 88 MPa and elastic modulus of about 19 GPa—interesting for engineering applications such as high-strength lightweight panels.

Keywords: copper slag; alkali-activated materials; laminate composite; elastic modulus; flexural strength

1. Introduction

Steel-reinforced concrete is the most important composite material used for structural applications in civil engineering [1,2]. However, textile-reinforced concretes (TRCs) have attracted interest for many civil engineering applications, owing their advantages over steel reinforcement, including more design freedom, limited use of cementitious matrices, and improved resistance to corrosion [1,3–5]. The TRC is a kind of concrete composite mainly made of a woven fabric embedded as reinforcement in a finely grained cementitious matrix [1,3,6,7]. The fabric ("a manufactured planar textile structure made of fibers and/or yarns assembled by various means such as weaving, knitting, tufting, felting, braiding, or bonding of webs to give the structure sufficient strength and other properties required for its intended use") [1] could be made of different type of materials including carbon, basalt, aramid, polyethylene (PE), polypropylene (PP), poly-vinyl-alcohol (PVA), glass and also cellulose-based natural fiber [4,8,9].

Ordinary Portland cement (OPC) has been commonly used as binder for cementitious matrices for TRC [2,3,10]. Research activities on other binder types, including alkali-activated materials (AAMs), have been performed to address sustainability issues or improve specific TRC properties [3,11–13]. AAMs are materials enclosing binder systems resulting from the reaction of a solid or dissolved alkali metal source with a solid silicate powder, and they have the potential to reduce up to 70–80% CO_2 emissions compared with the cement industry if optimally formulated [13–15].

It is worth noting that the development of AAMs could also offer the possibility for upcycling industrial residues [16,17]. For instance, about 2.2 tons of copper slag (such as Koranel slag) are generated per ton of copper produced [18,19]. Hence, the production of TRC from a cementitious matrix containing Koranel slag will contribute to waste management.

Potential applications of TRC include facing and sandwich panels; thin-walled lightweight elements; lightweight constructions and complex civil engineering structures; repair of existing buildings and constructions; fire resistance; and high temperature performance materials [5,20–22]. However, at variance to textile/organic resin-based composites, no prepreg with mineral matrix exists. Prepreg can be prepared and stored for a long period at low temperature and sold before the final shaping. Mineral-based matrices such as AAMs are generally subjected to relatively rapid setting time at ambient temperatures, limiting their flexibility. The development of a mineral-based matrix for TRC with storage flexibility comparable to that of organic resin will widen their potential applications.

The aim of this study is the development of an alkali-activated matrix based on Koranel slag for laminated TRC composites with woven carbon fiber reinforcement. The reactivity of Koranel with potassium silicate solutions having a silica-to-potassium molar ratio ($R = SiO_2/K_2O$) varying from 1 to 2.75, with 0.25 intervals, was investigated to select the modulus, giving a reasonable open pot life of several hours at room temperature. This will allow the use of such a matrix in an impregnation bath for several hours without the need to refresh the mixture. The impregnated textile will also be stable for some hours allowing the handling, for instance, to drape in a mold. Isothermal calorimetry at 20 °C was used to select a suitable alkaline solution with low reactivity at ambient temperature. If such a matrix has high reactivity above ambient temperature (50–100 °C), the composite could be hardened in the mold in one hour, for instance, as is the case for organic resin textile composites. Differential scanning calorimetry was used to investigate the temperature at which the selected matrix started to react with a reasonable rate. The effect of 10–20 wt.% metakaolinite, 5 wt.% OPC and surfactant was also investigated on the reaction heat and rheology. Indeed, metakaolinite is a finer precursor which could facilitate fiber impregnation by the matrix. Some surfactants were already proposed in the literature about alkali activation, although for different reasons than making composites [23]. Finally, the bending strength and elastic modulus of the composites prepared with carbon fabric and formulated compositions were assessed, as well as the elastic modulus of both matrix and composite upon heating until 700 °C.

2. Materials and Methods

2.1. Materials

The copper slag used in this study was Koranel® from Aurubis Beerse nv in Beerse, Belgium, with 2600 cm^2/g Blaine surface. The Blaine was measured with an air permeability apparatus as described in EN 196-6:2018 [24]. The particle size distributions of the Koranel slag, metakaolinite and OPC were determined by a laser particle size analyzer (Beckman Coulter LS 13320, Pasadena, CA, USA) (Figure 1a). X-ray powder diffraction (XRD) spectra of the slag were taken by a Bruker D2 Phaser, Karlsruhe, Germany, in the 6–70° 2θ range using Cu Kα radiation (voltage 40 kV and current 40 mA), a step size of 0.02° and scan speed of 0.5 s/step. The XRD pattern shows that the material is mainly amorphous (Figure 1b), with few crystalline features ascribed to Fayalite, Fe_2SiO_4 (PDF 34-0178), Ankerite, Ca (Fe, Mg)$(CO_3)_2$ (PDF 33-0282), and Fe-rich spinel, $(Al,Fe)_2O_4$ (PDF 21-0540) (see also [25]).

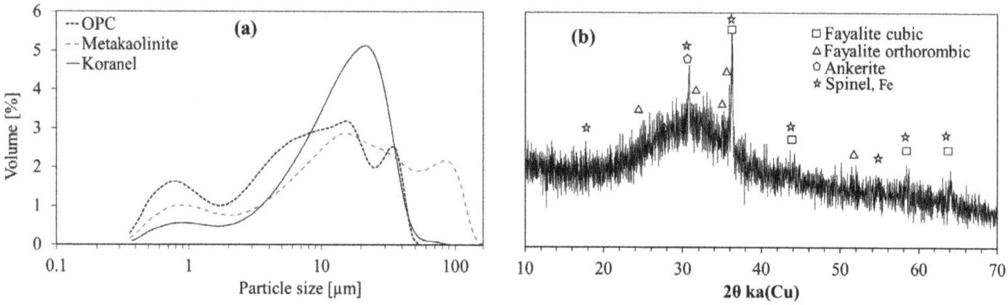

Figure 1. (**a**) Particle size distribution of Koranel® with Blaine 2600 cm^2/g, adapted from [25]; (**b**) XRD diffraction pattern of Koranel® with Blaine 2600 cm^2/g.

The metakaolinite (MK) used in this study is Argical™-M 1000 from IMERYS Fused Minerals, Paris, France. The ordinary Portland cement used was of CEM I 52.5 R from CBS Heidelberg cement group, Heidelberg, Germany. The chemical composition of Koranel, OPC and metakaolinite, as determined by X-ray fluorescence (PW 2400 Philips) on powder, is presented in Table 1. The d10, d50 and d90 of Koranel were respectively 2.51 µm, 13.22 µm and 27.87 µm. These values were, respectively, 0.87 µm, 7.88 µm and 30.40 µm for OPC and 1.31 µm, 16.01 µm and 80.57 µm for the metakaolinite. The 2-Methyl 2,4-Pentanediol and the Polyethylene Glycol 600 (PEG 600) used as surfactant were from Sigma Aldrich, Overijse, Belgium. The potassium hydroxide (Sigma Aldrich; 85 wt.% purity) and potassium silicate solution (Silmaco, Lanaken, Belgium) were used for the preparation of the activating solution. The potassium silicate solution was made of 27.22 wt.% SiO_2, 15.01 wt.% K_2O and 57.3 wt.% H_2O, with a molar ratio (R = SiO_2/K_2O) of 2.84 and density of 1.424 g/cm^3 at 20 °C. The potassium hydroxide was used to adjust the ratio in the solution. For instance, for 100 g of the potassium silicate solution, the amounts of KOH added to achieve the SiO_2/K_2O molar ratios of 1, 1.5, 2 and 2.25 were, respectively, about 38.7 g, 18.8 g, 8.9 g and 5.5 g. This corresponded to an addition of 32.9 g of pure KOH for the solution with R = 1, 15.9 g for R = 1.5, 7.5g for R = 2 and 4.7 g for R = 2.25; one hundred (100) g of these solutions contained a total mass of water (including water from dissolved KOH) of 44.4 g, 50.2 g, 53.9 g and 55.3 g, respectively. The properties of the 200 g/m^2 carbon fabric used (Fibermax Composite, London, UK) are presented in Table 2. Carbon fiber was chosen due to its high specific strength and its relatively chemical and thermal stability [26]. This allows the composite properties to be relatively stable over time, since no fiber degradation is expected due to the alkaline nature of the alkali-activated matrix.

Table 1. Normalized chemical composition in wt.% by XRF analyses for Koranel (K), metakaolinite (MK) and OPC. * Includes small amounts that are <0.1 wt.%. ** Loss on ignition at 950 °C.

Samples	SiO_2	Al_2O_3	Fe_2O_3	CaO	MgO	Cr_2O_3	K_2O	TiO_2	P_2O_5	SO_3	Others *	LOI **
K	26.6	9.5	57.4	2.8	0.9	0.7	0.2	0.2	0.8	0.9	0.0	-
MK	51.9	40.9	1.7	0.1	0.1	0	1	2.1	0	0	0.3	1.9
OPC	19.7	5.2	3.0	64.2	1.7	0	0	0.3	0	2.7	1.6	1.6

Table 2. Properties of individual carbon fiber from Fibermax Composite.

	Area Density (g/m^2)	Tensile Strength (MPa)	Tensile Modulus/E-Modulus (GPa)	Elongation at Break (%)
Carbon (Plain woven)	200	4410	235	1.9

2.2. Sample Preparation

To select the appropriate activating solution with low reactivity at ambient temperature and relatively high reactivity at 50–100 °C, a series of experiments with different potassium silicate solutions were performed. Different silica to potassium moduli (R = SiO_2/K_2O molar ratio), ranging from 1 to 2.75, with 0.25 intervals, were investigated with an L/S of 0.55, with L being the silicate solution. The prepared solutions were stored for 1 day in the laboratory environment for further equilibration. The dry components (Koranel (K) + metakaolinite (MK) + Portland cement (OPC)) and liquid components (prepared potassium silicate solution and 2-methyl 2,4-pentanediol ("m") or polyethylene glycol ("p") were first mixed separately; then, both liquid and solid components were mixed using an electric laboratory stirrer at a speed of 1000 rpm for 5 min to obtain a homogeneous slurry. The details on the formulated matrices for the preparation of the composite materials are presented in Table 3.

Table 3. Composition of the different matrices used for the preparation of the laminates with Koranel (K), metakaolinite (MK), Portland cement (OPC), 2-methyl 2,4-pentanediol (m) or polyethylene glycol (p). Sample names are composed of the letters mentioned between brackets and corresponding number of wt.% of the solid. * Discarded due to short setting time.

Sample Codes	Koranel (g)	Potassium Silicate Solution R = 2.25 (g)	Metakao-Linite (g)	OPC (g)	2-Methyl 2,4-Pentanediol (g)	PEG 600 (g)
K	100	55	-	-	-	-
Kp	100	55	-	-	-	2
Km	100	55	-	-	2	-
90K10MK	90	55	10	-	-	-
80K20MK	80	55	20	-	-	-
80K20MKm	80	55	20	-	2	-
80K20MKp	80	55	20	-	-	2
95K5OPC	95	55	-	5	-	-
90K10OPC *	90	55	-	10	-	-

For the composite preparation, a thin layer of about 0.5 mm of alkali-activated matrix was spread on a plastic PET film placed on a wooden mold in a pre-drawn square of 15 × 15 cm^2. The first layer of carbon-woven fabric (15 × 15 cm^2) was placed on top of the matrix layer. Thereafter, the carbon fabric was manually wet out (impregnated) with alkali-activated matrix using a roller (1; Figure 2). The process was repeated for 4 layers of the woven fabric and finally, a layer of geopolymer matrix was added; hence, we obtained a composite made of 5 matrix layers and 4 fiber layers (2; Figure 2). Caution was taken to use a minimum amount of matrix to ensure sufficient fiber impregnation with a high fiber volume fraction. Four weights of 5 kg each (3; Figure 2), equivalent of a total force of 196 N, were equally distributed on top of the laminate composites, and cured at 80 °C for 24 h in a conventional oven (4; Figure 2). After the curing period, the samples were cut in composite beams of 2.5 cm width and 14 cm length, then kept for 3 days in a laboratory environment before testing.

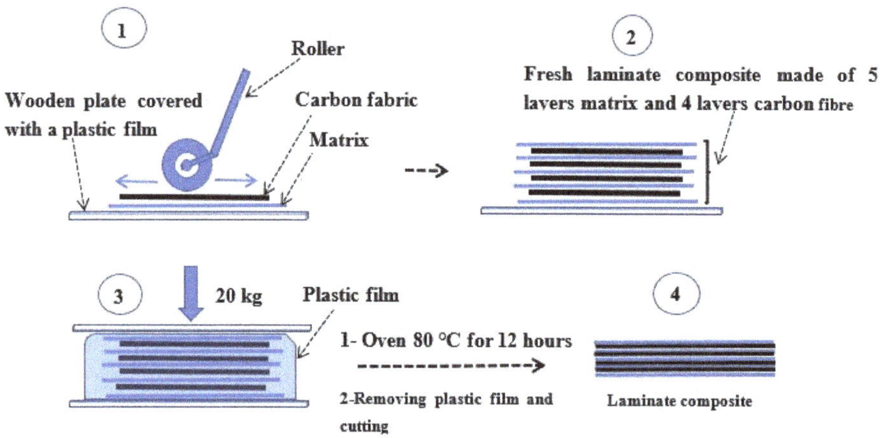

Figure 2. Sketch of laminate preparation. For details on the preparation, see text.

For comparative purposes, selected matrices were cast as paste in 4 × 4 × 16 cm^3 molds at once without vibrating and cured at 80 °C for 24 h.

2.3. Characterization Methods

2.3.1. Differential and Isothermal Calorimetry

Isothermal (20 °C) calorimetry (TAM Air, TA instruments, New Castle, DE, USA) was performed on 10 g of the solid part mixed with 7 g liquid. Mixing for 2 min at 1600 rpm occurred outside the calorimeter in the ampoule, which was placed within the calorimeter immediately after mixing. The data collection started 45 min after the start of mixing.

Non-Isothermal differential scanning calorimetry (DSC) was performed with a Mettler Toledo DSC822e, USA, using nitrogen (N_2, 100 mL/min) as a purge gas. Freshly mixed sample, 25 mg as seen in Table 3, was placed into a platinum crucible and heated from 0 to 250 °C at 5 °C/min to assess the temperature at which the matrix starts to react.

2.3.2. Scanning Electron Microscopy

Scanning electron microscopy was performed on an SEM device (Phenom Benelux Scientific X pro, Nazareth, Belgium) using an acceleration voltage of 5 kV and backscatter electron (BSE) for imaging. The samples were cut and put in the SEM sample holder; no conductive coating was applied. Possible remaining dust on the specimen was removed by compressed air before inserting it into the device.

2.3.3. Rheology

Viscosity analyses were performed at 25 °C with an AR-G2 rotational Rheometer from TA instruments, New Castle, DE, USA, using a cone plate and stainless steel geometry set up with a diameter of 15 cm.

2.3.4. Mechanical Properties

The mechanical testing included compressive strength and elastic modulus. The compressive tests were performed with an Instron 5885H test bench, Darmstadt, Germany. The instrument was equipped with a loadcell of 250 kN, and the tests were executed with a displacement rate of 1 mm/min. The elastic modulus was performed with a Resonalyser device [27], Bytec company, Merksplas, Belgium. This nondestructive method was performed on the samples prior to the flexural and compressive tests. The flexural strength was obtained with three-point bending tests performed with a Tinius Olsen 5ST, Surrey, UK.

2.3.5. Fiber Volume Fraction Estimation

The fiber volume fraction influences the mechanical properties and is determined as reported by Messiry [28] as:

$$V_f = \rho_f \times W_f/(\rho_f \times W_f + \rho_m \times W_m)$$

with V_f as the volume fraction of the fibers, W_f as the weight of the fibers, W_m as the weight of the matrix, ρ_f as the density of the fibers, and ρ_m as the density of the matrix.

The sketch of the experimental plan is presented in Figure 3.

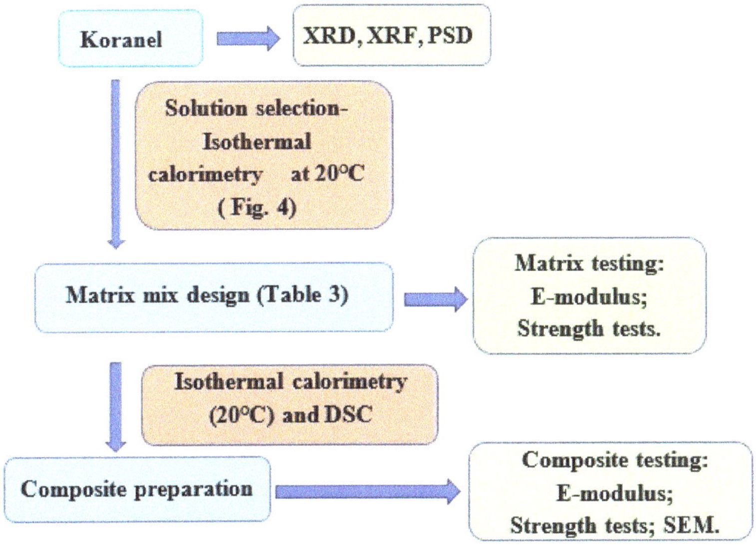

Figure 3. Sketch of the experimental design.

3. Results and Discussion

3.1. Reactivity of Matrices

3.1.1. Influence of Solution Silica Modulus on Reactivity at Room Temperature

The effect of the silica modulus of the potassium silicate solution on the reaction time was studied with isothermal calorimetry on Koranel (Figure 4). For potassium silicate solutions having a silica-to-potassium molar ratio (R) below 1.75, the reaction started earlier than the measurement (i.e., <45 min after mixing) and with a high rate. For R = 2.00, an induction period of 2 h was observed, and while increasing R, the induction time becomes 60 h (2.5 days) for R = 2.25 and 350 h (about 2 weeks) for R = 2.5 (Figure 4a). The peak of reaction also becomes broader and lower while increasing R from 2 to 2.5.

For R = 2.25, even a double exothermic peak is observed, i.e., one main peak at 3 days and the second smaller bump with maximum at 5.5 days. Such a smaller bump is not observed for R = 2, likely due to the fast reaction and overlap, and is neither observed for R = 2.5, likely because it is small and could occur at an even later time than the measurement of 24 days. R = 2.25 thus gives the unique insight into the complex reaction mechanism of potassium silicate-activated Koranel. Two exotherms for certain non-ferrous slags were also observed by others [17] who performed sodium silicate activation. The main peak observed is corresponding to the polymerization reactions [29,30], while the second, smaller and broader exotherm, seems to be depending on the silica content of the activator and is due to a secondary reaction or reorganization of the polymer network [17]. For none of the samples a split in decomposition, polymerization and stabilization was observed, as was described for metakaolinite [31]. The decomposition and polymerization seem to occur in the same exotherm. More recently, R. Caron [32] defined the steps in alkali activation more precisely as being first a nucleation and growth process, followed by volume contraction due to the polymerization, an induction period, a second nucleation and growth process corresponding to the second exotherm, new volume contraction process, and finally, slow diffusion controls further curing. According to our own insight, the first nucleation has not to be seen as a nucleation for crystals, but occurs as a result of polymerization and densification, comparable to the polymerization of an organic thermoset. An important difference is that no phase boundary, as between crystal and liquid, needs to be formed. The diffusion control occurs because of the formation of a dense layer around and in between the remaining slag particles, rendering the diffusion of ions more difficult than in the initial liquid activator. Although Koranel is an iron silicate (containing a high proportion of iron and silica), and thus chemically different from the slags and metakaolinite for which the mechanisms were described, one can assume that the general steps remain the same but that the products formed are different.

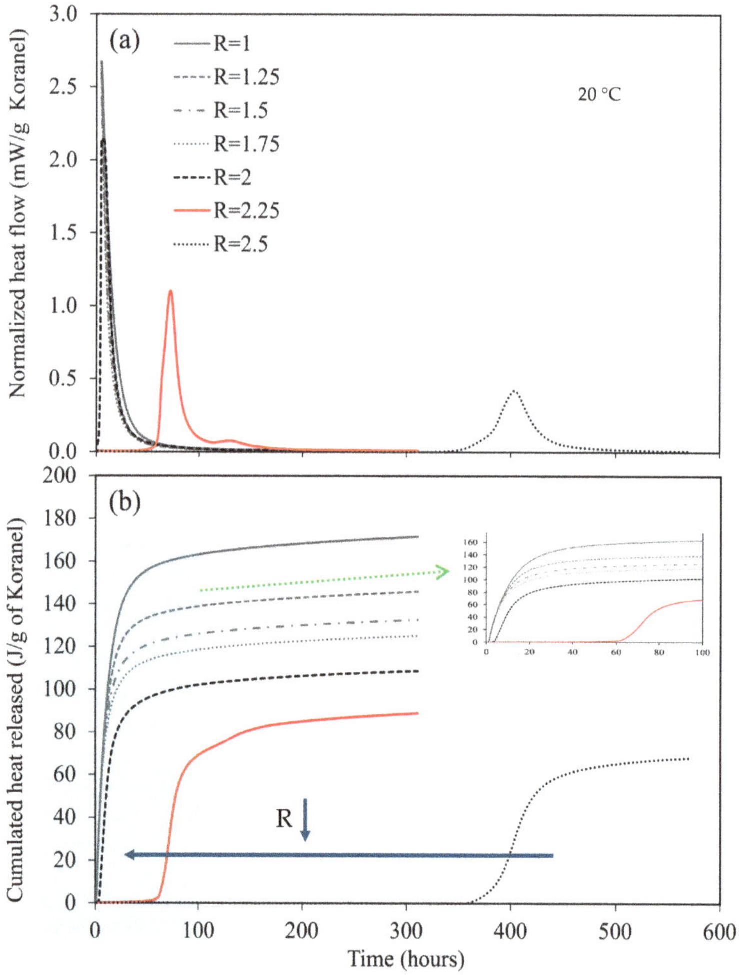

Figure 4. Effect of the silica-to-potassium molar ratio (R = SiO_2/K_2O) of the activating solution on (**a**) the normalized heat flow, and (**b**) the cumulative heat released per gram of solid of alkali-activated Koranel.

For a matrix to be used for textile, sufficient time before setting is needed if we consider a continuous production of several hours from prepared batches of matrix; the low reactivity will then prevent unwanted matrix setting during the production.

The normalized cumulative heat released from the different compositions (Figure 4b) shows that the reaction heat increases with the reduction of R due to the higher concentration of OH^- in the alkaline solution; this favors Koranel dissolution and polymerization reactions. However, a higher degree of dissolution is also associated with a faster reaction at room temperature. The matrix with R = 2.25 has therefore an ideal setting time (i.e., 60 h) for the purpose of laminate composite preparation in a production with batches of paste and is used in the following experiments. Its reactivity is low at room temperature but can be increased by increasing the curing temperature, as will be studied further on.

3.1.2. Influence of Additives and Koranel Replacement on Reactivity

With the selected activator solution, the influence of surfactants and small amounts of other precursors was investigated. The isothermal calorimetry at 20 °C (Figure 5) of the different compositions (Table 3) shows that the start of the reaction at room temperature differs depending on the composition of the solid phase. The addition of 5 wt.% OPC (brown curve; Figure 5a) decreased the open pot life with a factor of about 3; a sharp exotherm is observed. Meanwhile, the addition of metakaolinite (gray curves; Figure 5a) induced a delay in the appearance of the first exotherm. For the Koranel-based system (black curves; Figure 5a), the addition of surfactants slightly speeds up (2,4 methyl pentanediol) or delays (polyethylene glycol) the reaction; however, with the addition of metakaolinite, the effect of the surfactant was less obvious, a slight broadening of the peak might be observed.

The cumulative heat released (Figure 5b) of the different compositions clearly shows that the addition of OPC and metakaolinite leads to higher reaction heat, the latter being higher for the sample made of 20 wt.% metakaolinite in comparison to those made of 10 wt.% metakaolinite. The cumulative heat released per g of matrix varied from about 85 J/g for the composition made of Koranel to 140 J/g for the composition containing 20 wt.% metakaolinite. The increase from 85 J/g for Koranel to 115 J/g for 90K10MK is rather large, and the replacement of 10 wt.% more of Koranel by MK does not give rise to the same step in heat released, but only to an extra amount of 15 J/g. The cumulative heat released lowers upon the addition of both surfactants to the Koranel system. This was not expected but it shows that they interact chemically during the reaction and this interaction has a more important effect than being pure catalytic. Two evident ways the surfactants can play a role in the reaction rate are: (i) interaction with the surface of the precursor which could lead to a reduced reaction rate due to a reduced effective surface area; and (ii) interaction with the ions in solution, comparable to a chelating effect, although none of the surfactants are ionic, as is usually the case for chelating agents. This effect could also explain a reduction in reaction rate. No explanation was found for the shorter induction period observed for 2,4 methyl pentanediol. The maximum heat flow, thus maximum reaction rate, is for both pastes with surfactants substantially lower, as expected, according to the explanation given. These mechanisms can however not explain the slightly earlier start of the reaction and no other explanation could be found so far. Upon replacement of Koranel with metakaolinite, both surfactants have a different effect. While "p" does not influence the reaction too much, "m" decreases the total heat released substantially, though both had a slightly accelerating effect on the start of the reaction.

For an optimum dosage of alkali in alkali-activated materials from known precursors such as metakaolinite, the added alkali, M, should be enough to compensate the charge balancing of Al; hence, the M/Al molar ratio should be 1 [33]. Considering the chemical composition of Koranel and metakaolinite and the use of potassium silicate solution of R = 2.25 and a liquid-to-solid ratio of 0.7, the molar ratio K/Al in the alkali-activated mixture was 0.88 for matrix 80K20MK (Table 3) made of 20 wt.% metakaolinite and 80 wt.% Koranel. This means the alkaline reagent in matrix 80K20MK is theoretically not in excess for the alkali-activated reaction. Furthermore, for iron-rich slags, iron or iron-bearing phases in the slag also participate in the formation of the binding phase [14,34]. However, the degree of reaction of most of non-synthetic aluminosilicates is usually not 100% and some relics of the precursors are often evidenced by SEM [35]. Thus, for Fe silicates and Ca silicates, it is not only the K/Al ratio that is important, but other parameters also play a role as other reaction products such as calcium aluminosilicate hydrate (C-A-S-H) are formed. The exact stoichiometry is still not well known and the fact that reaching a 100% conversion with such systems is almost impossible hampers a more in-depth study.

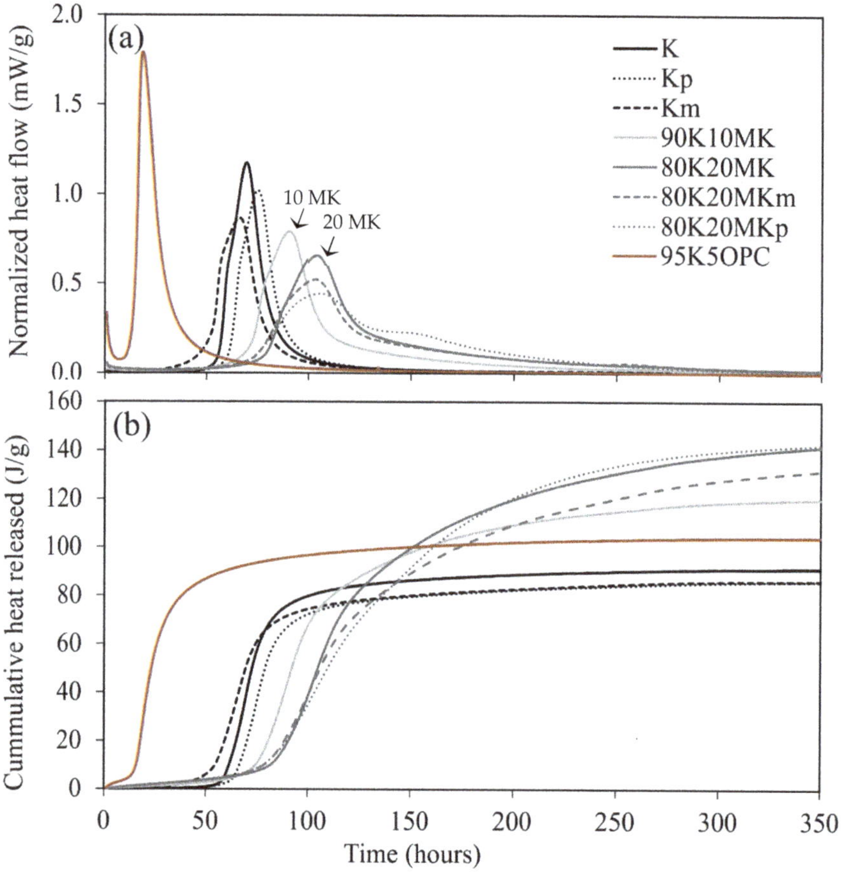

Figure 5. (**a**) Normalized heat flow and (**b**) cumulative heat released per gram of solid of alkali-activated materials with silicate modulus 2.25 and compositions defined in Table 3. Surfactant p is polyethylene glycol and m is 2-methyl 2,4-pentanediol.

In conclusion, one can state that the replacement of Koranel by OPC speeds up the reaction drastically, thus, the open pot life becomes too short for the application in mind. Replacement of Koranel with metakaolinite enhances the reaction. The addition of surfactants influences the reaction but the effect is limited. In the rest of this paper, the mixture of Koranel and metakaolinite as precursor will be further studied.

3.1.3. Reactivity with Increasing Temperature

The next step is to check if the reaction mixtures can set/harden within a time of about one hour at elevated temperature. Here again we have in mind a production method whereby the composite will be shaped and pressed at a higher temperature, as is normally performed for thermoset-based composites. The DSC trace of composition K with 100 wt.% Koranel as solid (reference composition) and matrix 80K20MK that presented the highest reaction heat due to metakaolinite addition is presented in Figure 6.

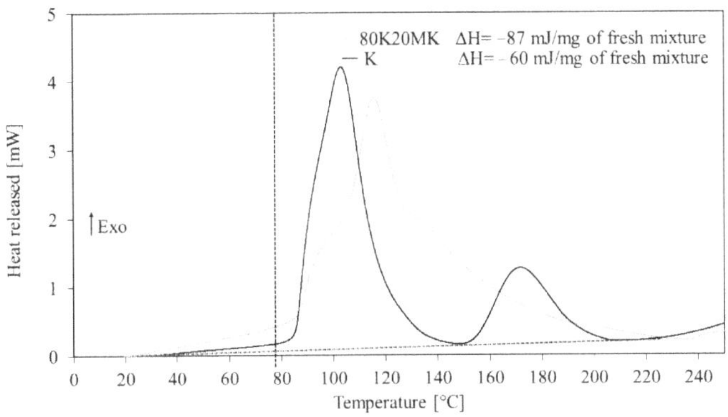

Figure 6. DSC curves of matrices K and 80K20MK (See Table 3). K shows two exothermic peaks at 100 °C and 150 °C with total enthalpy of ΔH = −60 J/g; 80K20MK shows one exothermic peak at 120 °C with two shoulders at 95 °C and 140 °C with ΔH = −87 J/g.

The reaction of Koranel with the alkaline solution is complex and occurs in two steps as indicated by two successively occurring exothermic peaks in the DSC pattern. The main exothermic reaction starts around 80–85 °C. The addition of 20 wt.% metakaolinite shifts the reaction start to a slightly lower temperature, 75 °C. This shift could be ascribed to the fineness and difference in activation energy of the added metakaolinite. The effect is however opposite to what was observed at 20 °C, where MK slowed down the reaction.

This observation from DSC indicates that 80 °C is a reasonable curing temperature for this matrix based on Koranel. The reaction enthalpies were −87 and −60 J/g for the 80K20MK and K compositions, respectively, or about −150 and −103 J/g of solid for the 80K20MK and K compositions, respectively; this is more or less comparable to the heat released after 2 weeks curing, measured using isothermal calorimetry at 20 °C (Figure 5).

3.2. *Effect of Surfactants on the Properties of Matrices*

3.2.1. Rheology of Fresh Matrices

Surfactants are often used to reduce the amount of water, thus the L/S ratio, given that they reduce the viscosity. They might also have an influence on the wetting of fibers in composites. The effect of two surfactants is discussed here. The effect of 2-methyl 2,4 pentanediol ("m") and PEG ("p") on the rheology of the fresh matrices K, 80K20MKm and 80K20MKp was compared each time with the selected matrix 80K20MK (Figure 7; see Table 3 for matrices).

The substitution of 20 wt.% copper slag by metakaolinite has a special effect on the viscosity (Figure 7a). At low shear rates, a decrease in the thixotropy is observed due to the addition of metakaolinite. This is interesting for the flow of the paste at almost steady conditions, for instance, when a textile is pulled through a bath of matrix. Under those conditions, the thixotropy of the pure Koranel-based matrix hinders a uniform deposition of the matrix on the textile. However, at intermediate shear rates there is no substantial difference between both pastes up to a rate of about 2/s, when the viscosity of the mixture with MK becomes slightly higher than the one without MK. The shear thinning observed at intermediate shear stresses can be related to the alignment of particles. The addition of 2 wt.% 2-methyl 2,4 pentanediol to 80K20MK reduced the viscosity and shear stress slightly up to 1/s (Figure 7b), opposite to polyethylene glycol, for which there is a slight increase in almost the complete studied range of shear rates (Figure 7c).

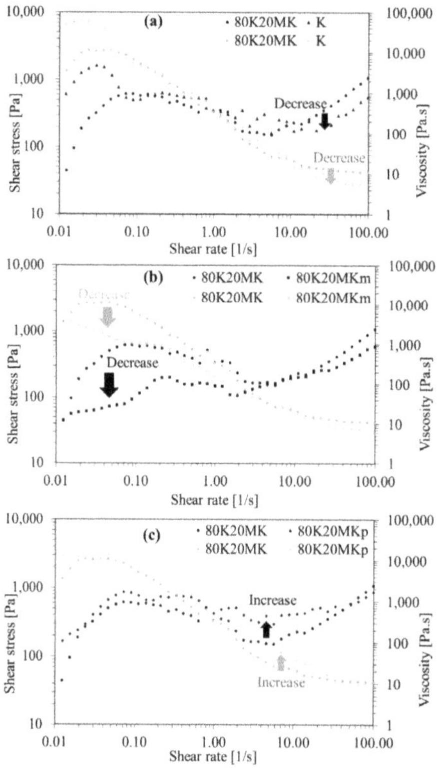

Figure 7. Rheology of matrices K (**a**), 80K20MKm (**b**) and 80K20MKp (**c**), all indicated with open dots in comparison with matrix 80K20MK, indicated with filled dots.

In our system, MK thus lowers the thixotropy; the surfactants did not change the rheology to a large extent. Only 2 wt.% 2-methyl 2,4 pentanediol lowered the viscosity in the intermediate range, with maximum a factor of 10. PEG slightly increased the viscosity in the studied composition range.

3.2.2. Mechanical Properties of Hardened Matrix Pastes

The effect of 2-methyl 2,4 pentanediol and polyethylene glycol on compressive strength, flexural strength and elastic modulus of 80K20MK is presented in Figure 8. Matrices 100K, 80K20MK, 80K20MKm and 80K20MKp were cured at 80 °C for 24 h with additional post-curing at room temperature for 3 days. Matrix 80K20MK resulted in a compressive strength of 30 MPa, somewhat better than the pure Koranel with the same activator. The addition of 2-methyl 2,4 pentanediol had no effect on the compressive strength; meanwhile, polyethylene glycol contributed to an increase of 35 MPa. The flexural strength for 80K20MK was 0.8 MPa, comparable to the one for pure Koranel. The addition of 2-methyl 2,4 pentanediol tripled the flexural strength to 2.5 MPa and that of polyethylene glycol helped even to reach 4 MPa. This positive effect is also noticeable in the elastic modulus, for which 80K20MK reaches 3 GPa, and 80K20MKm and 80K20MKp reach, respectively, 5.3 and 6.3 GPa in bending, whereas for axial, it reaches 2.5, 4.6 and 6.2. This large effect of surfactants is likely due to their influence on reducing microcrack formation [23], thus posing these surfactants forward as valuable additions for producing high strength lightweight panels. The fact that the compressive strength increases only slightly is because defects are closed upon compression, while for flexion, a crack means concentration of the force on the edges. Thus, the structure will become stronger if less cracks are formed.

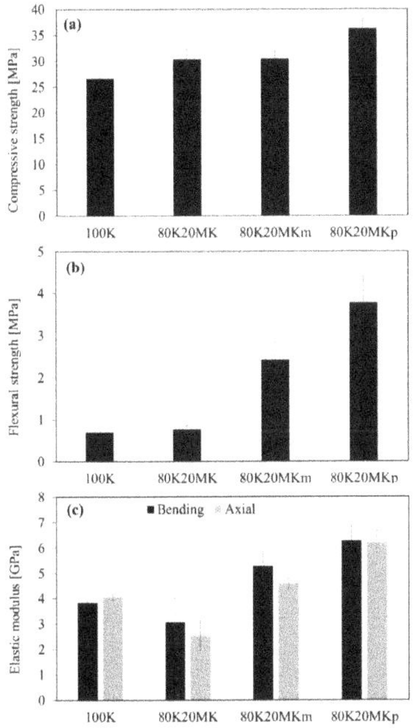

Figure 8. (**a**) Compressive strength, (**b**) flexural strength and (**c**) elastic modulus of matrices 100K, 80K20MK, 80K20MKm and 80K20MKp after curing for 1 day at 80 °C and 3 days post-curing at room temperature to compare the effect of 2-methyl 2,4 pentanediol ("m") and polyethylene glycol ("p") on these mechanical properties. Both surfactants increased the compressive strength, but the effect is most pronounced on the flexural strength that increases with a factor of about 4.

Compared to some of the literature data [17], a compressive strength of 35 MPa is rather low. The most likely reason is the lower OH^- concentration or higher modulus of the chosen activating solution, which was beneficial for the lower reaction rate but also leads to a higher fraction of undissolved slag in the hardened alkali-activated slag. The flexural strength of concrete is usually a factor of 8–10 lower than the compressive strength [36]. The flexural strength for the samples without surfactants is thus rather low, but the use of surfactants increases it to an average value.

3.3. Mechanical Properties of Composites

After matrix characterization, this section is presenting the mechanical properties of the prepared composites with the aforementioned matrices (Table 3) and plain woven carbon fabric (Table 2). Five samples per composite configuration were prepared according to Figure 2.

There is a rather large spread on the results of the stress–strain curves of samples with same composition (Figure 9). This is in part due to the manual production method that limits reproducibility. It is also worth pointing out that some ductile behavior was observed on the failure curves of the composites, as is the aim of making composites. In the first almost linear part, the textile and matrix work together. In the second part, the composite becomes damaged, and delamination or fiber failure can occur. This can be seen, for instance, from the sudden drop in stress. Hence, the prepared composites could be used for engineering applications where some ductility is required, including thin elements and lightweight panels.

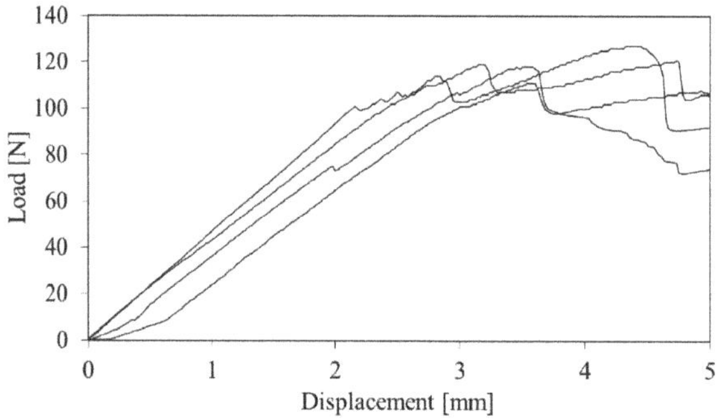

Figure 9. Sample of failure curves of composites made of four layers of carbon fiber 200 g/m^2 and five layers of matrix 80K20MK. Measurement on four different samples with the same composition are shown.

The volume fraction of fibers of the composites ranged between 25 to 40 vol% (Figure 10). This wide range is caused by the manual production and the differences in workability of the matrix. Due to the nature of the matrix, it was difficult to achieve composites with more than 40% fiber volume fraction. Considering the fact that flexural strength mainly depends on the fiber properties, because the matrix will crack long before the fibers, the higher the fiber volume fraction, the higher the flexural strength. It is somewhat unfortunate that the addition of MK renders the production of high volume fraction composites more difficult. As a result, these composites have the lowest fiber volume fractions of the series. The fiber volume fraction for the series 80K20MK increased only a bit with surfactants, but remained between 25–30 vol%.

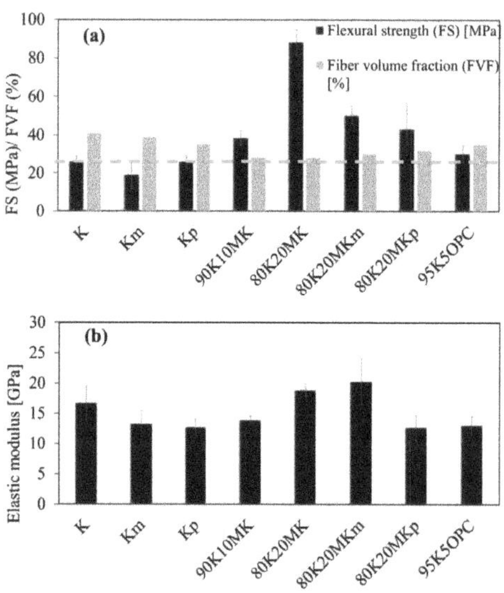

Figure 10. Fiber volume fraction, flexural strength (**a**) and elastic modulus (**b**) of the respective composites. Names indicate the matrices, as used in Table 3. For each composition, the average of five measurements is given.

Depending on the matrix composition, the flexural strength varied from about 19 to 83 MPa, while the Elastic modulus varied between 13 and 20 GPa (Figure 10). Both the addition of 2-methyl-2,4 pentanediol (Km) and polyethylene glycol (Kp) to Koranel-based matrices had no or little effect on the flexural strength, and even showed a decrease in the elastic modulus with respect to K (Figure 10). This is in contrast to the increase in the flexural strength and elastic modulus with the surfactant observed in the pure matrices (Figure 8b). The fact that the surfactants do not increase the strength in composites as they do for the pure matrix shows that carbon fiber reinforcement is dominant on the flexural strength and that the difference in matrix properties resulting from surfactant addition did not prevail.

The addition of 10 wt.% of metakaolinite (90K10MK) shows an increase to 40 MPa in flexural strength, but no increase in elastic modulus (5 GPa) with respect to K. Higher performance was observed for samples with 20 wt.% metakaolinite (80K20MK) in the matrix, up to 88 MPa in flexural strength and 20 GPa in elastic modulus. Samples prepared with metakaolinite presented lower fiber volume fraction (25 vol%), but higher mechanical properties than the reference (cf. green dashed line in Figure 10). Hence, the presence of metakaolinite was beneficial for strength development, in consistency with calorimetry results on cumulative heat released (Figure 5). Another reason might be the small particle size of metakaolinite that allows for better contact and maybe even penetration in the textile. The substitution of copper slag by 5 wt.% OPC (95K5OPC) only had a negligeable effect on the flexural strength and a decrease in elastic modulus despite the positive effect of OPC on reaction heat.

The flexural strength results are in some cases higher or comparable to the literature values of 40–45 MPa obtained with geopolymer laminate composites prepared with metakaolinite and basalt fabric [12]. These authors also observed that fiber reinforcement favored water evaporation from the composites at room temperature and improved the thermal stability in comparison to the pure matrix when thermally treated at 300–800 °C.

In another study, a bending strength of 18 MPa was observed with blast furnace slag composite reinforced with 1 wt.% of 6 mm long carbon fiber, representing an increase of 19% compared to the pure matrix [37]. Meanwhile, an elastic modulus of 37 GPa and flexural strength of 133 MPa were achieved on potassium-based metakaolin geopolymers with 20–25% volume of unidirectional carbon fiber [38]. The analysis of these results shows some divergences in the bending strength depending on the matrix, fiber type, size and weaving, but in all the cases, the use of fiber was beneficial for the bending strength.

In summary, for the mechanical properties, in comparison to composition K, the surfactant and OPC addition had no or little effect in the studied range of the prepared composites. However, metakaolinite addition had a beneficial effect on the mechanical properties, with about 100% increase in the composite made of matrix containing 20 wt.% MK (Figure 10).

3.4. Microstructural Observations

The microstructure of the surface of the composites from compositions K, 80K20MK, 80K20MKm and 80K20MKp is presented in Figure 11. While keeping the same liquid-to-solid ratio, the substitution of Koranel by metakaolinite led to a significant reduction in cracks and composites with smoother surfaces. The addition of both 2-methyl-2,4 pentanediol and polyethylene glycol, however, seemed to increase the void proportion at the surface. This is consistent with a previous study on the use of 2-methyl-2,4 pentanediol, which was observed to lead to an increased porosity by acting as air entrainer with drying shrinkage reduction [23], and may explain why the surfactants do not have a significant effect on the flexural strength and elastic modulus (Figure 10) of the composites. However, the appearance of voids in the samples with surfactants did not significantly influence the density, which was always in the range of (2.00 ± 0.05) g/cm^3.

The microstructure of the cross section of composite 80K20MK shows alternating layers of matrix and carbon fiber (Figure 12). The matrix is not perfectly adhering to the carbon fiber and is not really penetrating in the fiber bundles. This is ascribed to the fiber's hydrophobic properties, which makes them incompatible with water-based systems [39]. However, methods including plasma treatment, silica or calcite deposition on carbon fiber, were suggested to improve the wettability and enhance the adhesion between the fiber and cementitious matrices [39–42]. Fiber surface chemistry modification and nanomaterials' incorporation in the binder were also observed to enhance the fiber/matrix interface in alkali-activated composites [43–46]. For instance, the addition of 2 wt.% nano titanium dioxide in alkali-activated fly ash reinforced by micro carbon fiber was observed to increase the fracture toughness by 29% in comparison to the control sample, due to possible filling behavior and enhancement of the carbon fiber–matrix interfacial zone, which induced a hindrance to crack formation [46]. However, actions to improve fiber–matrix interactions are also costly, so need to be analyzed considering the possible improvement in properties and the composite application. Since in this paper the focus is on proving that the matrix can be used for textile-reinforced composites, no effort was exerted to improve the fiber–matrix interaction. Anyway, it is worth reminding that the durability of textile-reinforced alkali-activated composites do not only depend on their resistance to the external environment, but also on the chemical interaction between the fibers and binder within the composite. For instance, due to the alkaline environment of the binder, alkali-activated composites reinforced with glass and basalt fibers were found to have significant reductions in tensile strength their over time, while the strengths of those reinforced with carbon fibers were sufficiently stable [26,47].

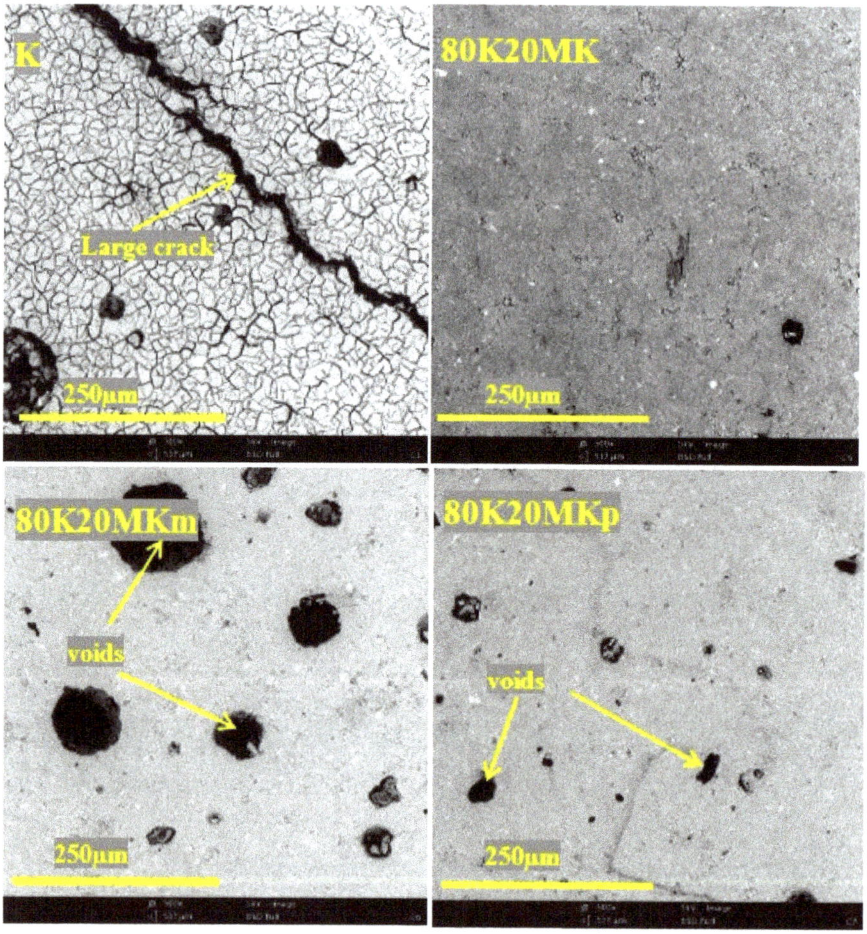

Figure 11. Microstructural surface of indicated composites (K), (80K20MK), (80K20MKm) and (80K20MKp) (cf. Table 3).

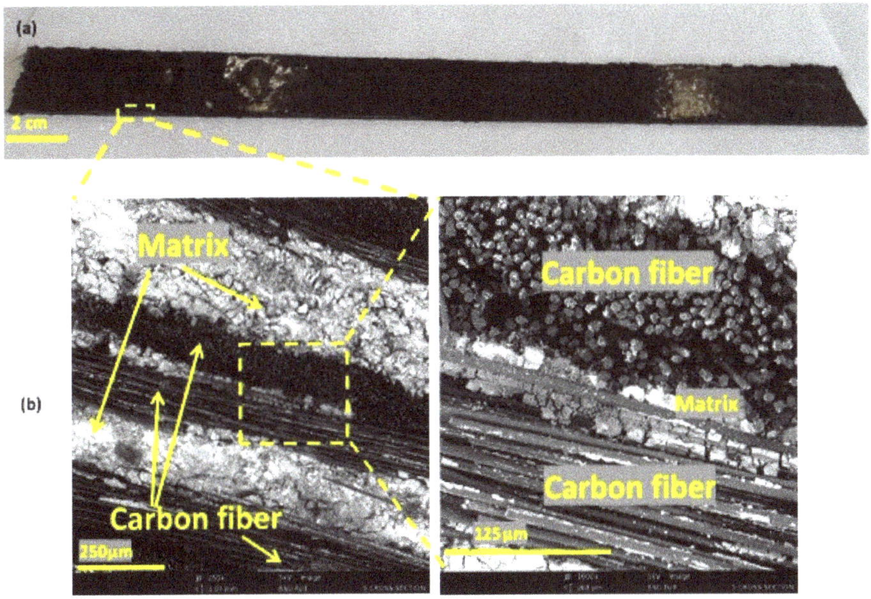

Figure 12. (**a**) photo of the composite; (**b**) microstructural cross section of laminate composite made of woven carbon fiber and matrix 80K20MK.

3.5. Thermal Stability of the Composites

The elastic modulus of the composite and matrix when heating until 700 °C, monitored via impulse excitation, is presented in Figure 13. The elastic modulus of the matrix first decreased from about 5 to 2 GPa, and from about 20 to 250 °C. Then, an increase was observed until around 620 °C, where it reached a value of 9.5 GPa before starting to decrease until it reached 7 GPa at 700 °C. The modulus continued to decrease during the dwell time to 5 GPa and remained almost constant on cooling. The decrease in the elastic modulus of the matrix from 20 to 300 °C occurs in two steps and may be ascribed to microcrack formation due to residual water loss on heating. The first step, around 100 °C, is broad while the second step is more pronounced. The E modulus of the composite has a comparable profile of reduction until about 200 °C. However, for the pure matrix, the decrease is more than 50% up to 250 °C, while for the composite, it is only about 25%. In part, it is due to the reduced amount of matrix because of the fibers in the composite. However, there is only about 25–30 vol% of fibers; thus, this cannot explain the large difference. The effect must thus be linked to reduced cracking, because fibers stop the crack propagation and the interaction between fibers and matrix, reducing the importance of the matrix. Before 300 °C, the elastic moduli of the matrix and composite start to increase, but above 300 °C, the one for the composite decreases again. The most probable explanation for the increase is that the system starts to heal as a result of mobility increase with temperature, although a glass transition could not be observed with calorimetry in this temperature range. The modulus keeps on increasing up to about 630 °C after which it decreases again. A similar behavior was observed in laterite-based alkali-activated materials, where a strength increase was observed on heating between 250 and 450 °C [48]. One should however be aware that the strength does not necessarily follow the same trend as the modulus, because the elongation at failure can be different. The composite, however, shows a reduction in modulus in two steps above 300 °C. The first is probably due to the thermal incompatibility between the matrix and the fiber that led to deterioration upon heating [49]. The second one is at least in part related to the decomposition of the carbon fibers in air. The elastic modulus signal could not be recorded above 640 °C due to sample deterioration upon heating. The reduction in the mechanical properties upon heating is in line with some reported studies on alkali-activated fiber-reinforced composites [12]. However, some studies also reported an increase in strength when metakaolin geopolymer with carbon fiber reinforcement was heated at 1100–1300 °C, due to matrix crystallization and densification that led to an enhancement of the fiber/matrix-bonding interface [38].

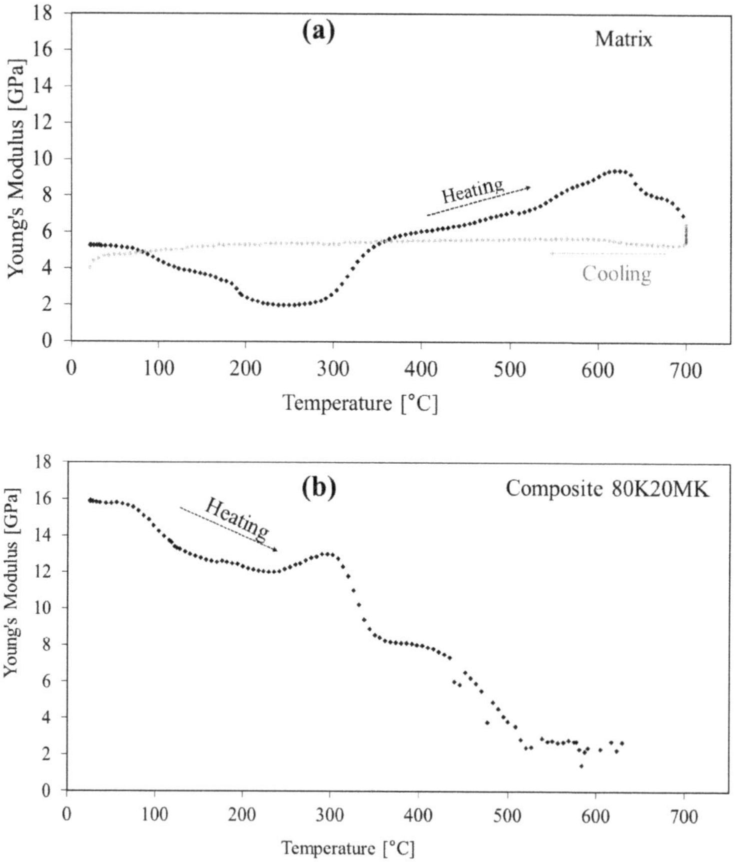

Figure 13. Elastic modulus vs. temperature of (**a**) the matrix and (**b**) composite 80K20MK.

4. Conclusions

Copper slag is an industrial residue with a large unutilized fraction. The present study investigated the development of alkali-activated composites from copper slag named Koranel and carbon textile-woven fabric. In the first step, the matrix for the composites was developed. The reactivity of Koranel with potassium silicate solutions, with molar ratio $R = SiO_2/K_2O$ varying from 1 to 2.75 at 0.25 intervals, was investigated using isothermal calorimetry. The results showed that the solution with R = 2.0 has an open pot life of only a few hours at room temperature, while the solution with R = 2.25 has an open pot life of several days at room temperature, but reacts quickly at 80 °C. Hence, the potassium silicate solution with the silica modulus of 2.25 is more suitable for a process in which the composite is impregnated, and in the following hours, would be used for making a laminated composite. The substitution of 5% and 10–20 wt.% Koranel, respectively, by OPC and metakaolinite, led to higher reaction heat. However, higher mechanical strength, especially in flexion, was only observed when 20 wt.% Koranel was substituted by metakaolinite. Meanwhile, OPC addition of only 5 wt.% induced a fast setting of the matrix, which was not deemed suitable for composite preparation. The effect of the 2 surfactants, 2-methyl 2,4 pentanediol and polyethylene glycol on the workability is limited, but PEG further increased the flexural strength.

Composites were made of four layers of carbon fabric and five layers of inorganic polymer matrix. Pure Koranel-based matrix presented a compressive and flexural strength of about 30–35 MPa and 1–5 MPa, respectively. An improvement in the bending strength was observed with a fiber addition in the prepared composites, with flexural strength values reaching 83 MPa. The addition of 2 wt.% polyethylene glycol/2-methyl 2,4 pentanediol was beneficial for the flexural strength of pure matrix, but the same was not observed in the mechanical properties of the composites.

The microstructure of the cross section showed alternating layers of matrix and carbon fiber, even though the matrix was not perfectly adhering on the carbon fiber and little to no impregnation of the fiber bundles was observed. The E modulus of pure matrix dropped by 50% above 200 °C. The pure matrix healed from 300 °C onwards, and its E modulus reached a maximum at 630 °C, after which it decreases again. At 700 °C, the modulus was about 5 GPa, and remained almost constant upon cooling. The composites have a modulus about 3 times higher at room temperature and lose only 25% of it up to 250 °C. Above 300 °C however, the modulus of the composite gradually decreases. The composite prepared with 20 wt.% metakaolinite still retained 50% of its elastic modulus on heating until 400 °C. However, upon further heating, the modulus decreases to almost 0 at 640 °C.

The results are interesting for the repurposing of Koranel copper slag in the development of alkali-activated composites for various engineering applications, including high strength lightweight panels and laminates. There are several aspects that need further investigation, such as the fiber–matrix interaction, the better impregnation of the fiber bundles, the thermal behavior of the matrix and composites, and the different processes responsible for deterioration and healing. All these aspects will be dealt with in future research.

Author Contributions: Conceptualization, P.N.L. and H.R.; Methodology, P.N.L., G.M.H., N.D., J.G. and H.R.; Formal analysis, G.M.H., F.K. and J.G.; Investigation, G.M.H. and F.K.; Data curation, G.M.H. and F.K.; Writing—original draft, P.N.L., G.M.H. and N.D.; Writing—review & editing, P.N.L., G.M.H., N.D., J.G. and H.R.; Supervision, H.R.; Project administration, H.R.; Funding acquisition, H.R. All authors have read and agreed to the published version of the manuscript.

Funding: This work was performed in the framework of STIF (ICON, supported by SIM, Flanders, N° HBC.2019.0120). The project has received funding from the European Union's Horizon 2020 research and innovation programme under N° 963603 Current Direct.

Institutional Review Board Statement: Not applicable.

Informed Consent Statement: Not applicable.

Data Availability Statement: The data presented in this study are available on request from the corresponding author.

Acknowledgments: This work was performed in the framework of STIF (ICON, supported by SIM, Flanders, N° HBC.2019.0120). The project has received funding from the European Union's Horizon 2020 research and innovation programme under N° 963603 Current Direct. The Authors also acknowledged the support from the Department Mechanics of Materials and Constructions (MeMC) of VUB for assistance in mechanical testing.

Conflicts of Interest: Author Nicole Dilissen is employed by Buildwise. The remaining authors declare that the research was conducted in the absence of any commercial or financial relationships that could be construed as a potential conflict of interest.

References

1. Brameshuber, W. *Report 36: Textile Reinforced Concrete-State-of-the-Art Report of RILEM TC 201-TRC*; RILEM Publications: Trondheim, Norway, 2006; Volume 36, ISBN 2-912143-99-3.
2. Li, Y.; Yin, S.; Lv, H. Combined effects of dry-wet cycles and sustained loads on the seismic behavior of TRC-strengthened RC columns. *Structures* **2021**, *33*, 2226–2237. [CrossRef]

3. Mechtcherine, V.; Schneider, K.; Brameshuber, W. 2—Mineral-based matrices for textile-reinforced concrete. In *Textile Fibre Composites in Civil Engineering*; Triantafillou, T., Ed.; Woodhead Publishing: Sawston, UK, 2016; pp. 25–43; ISBN 978-1-78242-446-8.
4. Ospitia, N.; Tsangouri, E.; Pourkazemi, A.; Stiens, J.H.; Aggelis, D.G. NDT inspection on TRC and precast concrete sandwich panels: A review. *Constr. Build. Mater.* **2021**, *296*, 123622. [CrossRef]
5. Van Driessche, A.; Aggelis, D.G.; Tsangouri, E. Complex fracture on thin-wall textile reinforced cement (TRC) shells monitored by acoustic emission. *Thin-Walled Struct.* **2021**, *167*, 108216. [CrossRef]
6. Mohan, A.; Madhavi, T.C. Development of binders for textile reinforced concrete. *Mater. Today Proc.* **2021**, *46*, 3297–3301. [CrossRef]
7. Tekle, B.H.; Messerer, D.; Holschemacher, K. Bond induced concrete splitting failure in textile-reinforced fine-grained concrete. *Constr. Build. Mater.* **2021**, *303*, 124503. [CrossRef]
8. Alzeer, M.I.M.; MacKenzie, K.J.D. Chapter 5—Fiber composites of inorganic polymers (geopolymers) reinforced with natural fibers. In *Composite Materials*; Low, I.-M., Dong, Y., Eds.; Elsevier: Amsterdam, The Netherlands, 2021; pp. 117–147; ISBN 978-0-12-820512-9.
9. Naaman, A.E. 18—Thin TRC products: Status, outlook, and future directions. In *Textile Fibre Composites in Civil Engineering*; Triantafillou, T., Ed.; Woodhead Publishing: Sawston, UK, 2016; pp. 413–439; ISBN 978-1-78242-446-8.
10. Valeri, P.; Fernàndez Ruiz, M.; Muttoni, A. Tensile response of textile reinforced concrete. *Constr. Build. Mater.* **2020**, *258*, 119517. [CrossRef]
11. Alomayri, T.; Shaikh, F.U.A.; Low, I.M. Synthesis and mechanical properties of cotton fabric reinforced geopolymer composites. *Compos. Part B Eng.* **2014**, *60*, 36–42. [CrossRef]
12. Ribero, D.; Kriven, W.M. Properties of Geopolymer Composites Reinforced with Basalt Chopped Strand Mat or Woven Fabric. *J. Am. Ceram. Soc.* **2016**, *99*, 1192–1199. [CrossRef]
13. Samal, S.; Marvalová, B.; Petríková, I.; Vallons, K.A.M.; Lomov, S.V.; Rahier, H. Impact and post impact behavior of fabric reinforced geopolymer composite. *Constr. Build. Mater.* **2016**, *127*, 111–124. [CrossRef]
14. Peys, A.; White, C.E.; Rahier, H.; Blanpain, B.; Pontikes, Y. Alkali-activation of CaO-FeOx-SiO2 slag: Formation mechanism from in-situ X-ray total scattering. *Cem. Concr. Res.* **2019**, *122*, 179–188. [CrossRef]
15. Provis, J.L.; van Deventer, J.S.J. *Alkali Activated Materials State-of-the-Art Report, RILEM TC 224-AAM*; Springer: Dordrecht, The Netherlands, 2014; ISBN 978-94-007-7672-2.
16. Flesoura, G.; Dilissen, N.; Dimitrakis, G.; Vleugels, J.; Pontikes, Y. A new approach for the vitrification of municipal solid waste incinerator bottom ash by microwave irradiation. *J. Clean. Prod.* **2021**, *284*, 124787. [CrossRef]
17. Van De Sande, J.; Peys, A.; Hertel, T.; Rahier, H.; Pontikes, Y. Upcycling of non-ferrous metallurgy slags: Identifying the most reactive slag for inorganic polymer construction materials. *Resour. Conserv. Recycl.* **2020**, *154*, 104627. [CrossRef]
18. Lemougna, P.N.; Yliniemi, J.; Adesanya, E.; Tanskanen, P.; Kinnunen, P.; Roning, J.; Illikainen, M. Reuse of copper slag in high-strength building ceramics containing spodumene tailings as fluxing agent. *Miner. Eng.* **2020**, *155*, 106448. [CrossRef]
19. Siddique, R.; Singh, M.; Jain, M. Recycling copper slag in steel fibre concrete for sustainable construction. *J. Clean. Prod.* **2020**, *271*, 122559. [CrossRef]
20. Saidi, M.; Reboul, N.; Gabor, A. Cyclic behaviour of textile-reinforced cementitious matrix composites (TRC) using distributed fibre optic sensors technology. *Compos. Part Appl. Sci. Manuf.* **2021**, *149*, 106531. [CrossRef]
21. Shen, L.; Wang, J.; Xu, S.; Zhao, X.; Peng, Y. Flexural behavior of TRC contained chopped fibers subjected to high temperature. *Constr. Build. Mater.* **2020**, *262*, 120562. [CrossRef]
22. Yin, S.; Cong, X.; Wang, C.; Wang, C. Research on flexural performance of composited RC beams with different forms of TRC permanent formwork. *Structures* **2021**, *29*, 1424–1434. [CrossRef]
23. Beersaerts, G.; Ascensão, G.; Pontikes, Y. Modifying the pore size distribution in Fe-rich inorganic polymer mortars: An effective shrinkage mitigation strategy. *Cem. Concr. Res.* **2021**, *141*, 106330. [CrossRef]
24. *EN 196-6:2018*; Methods of testing cement—Part 6: Determination of fineness. EN: Plzen, Czech Republic, 2018.
25. Lemougna, P.N.; Dilissen, N.; Hernandez, G.M.; Kingne, F.; Gu, J.; Rahier, H. Effect of Sodium Disilicate and Metasilicate on the Microstructure and Mechanical Properties of One-Part Alkali-Activated Copper Slag/Ground Granulated Blast Furnace Slag. *Materials* **2021**, *14*, 5505. [CrossRef]
26. Ranjbar, N.; Zhang, M. Fiber-reinforced geopolymer composites: A review. *Cem. Concr. Compos.* **2020**, *107*, 103498. [CrossRef]
27. De Baere, I.; Van Paepegem, W.; Degrieck, J.; Sol, H.; Van Hemelrijck, D.; Petreli, A. Comparison of different identification techniques for measurement of quasi-zero Poisson's ratio of fabric-reinforced laminates. *Compos. Part Appl. Sci. Manuf.* **2007**, *38*, 2047–2054. [CrossRef]
28. Messiry, M.E. Theoretical analysis of natural fiber volume fraction of reinforced composites. *Alex. Eng. J.* **2013**, *52*, 301–306. [CrossRef]
29. Kriskova, L.; Machiels, L.; Pontikes, Y. Inorganic Polymers from a Plasma Convertor Slag: Effect of Activating Solution on Microstructure and Properties. *J. Sustain. Metall.* **2015**, *1*, 240–251. [CrossRef]
30. Onisei, S.; Lesage, K.; Blanpain, B.; Pontikes, Y. Early Age Microstructural Transformations of an Inorganic Polymer Made of Fayalite Slag. *J. Am. Ceram. Soc.* **2015**, *98*, 2269–2277. [CrossRef]
31. Yao, X.; Zhang, Z.; Zhu, H.; Chen, Y. Geopolymerization process of alkali–metakaolinite characterized by isothermal calorimetry. *Thermochim. Acta* **2009**, *493*, 49–54. [CrossRef]

32. Caron, R.; Patel, R.A.; Dehn, F. Activation kinetic model and mechanisms for alkali-activated slag cements. *Constr. Build. Mater.* **2022**, *323*, 126577. [CrossRef]
33. Rahier, H.; Van Mele, B.; Biesemans, M.; Wastiels, J.; Wu, X. Low-temperature synthesized aluminosilicate glasses. *J. Mater. Sci.* **1996**, *31*, 71–79. [CrossRef]
34. Peys, A.; White, C.E.; Olds, D.; Rahier, H.; Blanpain, B.; Pontikes, Y. Molecular structure of CaO–FeOx–SiO2 glassy slags and resultant inorganic polymer binders. *J. Am. Ceram. Soc.* **2018**, *101*, 5846–5857. [CrossRef]
35. Luukkonen, T.; Abdollahnejad, Z.; Yliniemi, J.; Kinnunen, P.; Illikainen, M. One-part alkali-activated materials: A review. *Cem. Concr. Res.* **2018**, *103*, 21–34. [CrossRef]
36. Suda, V.B.R.; Priyatham Paul, S. Relationship between compressive, split tensile and flexural strengths of ternary blended concrete. *Int. Conf. Adv. Constr. Mater. Struct.* **2022**, *65*, 1112–1119. [CrossRef]
37. Vilaplana, J.L.; Baeza, F.J.; Galao, O.; Alcocel, E.G.; Zornoza, E.; Garcés, P. Mechanical properties of alkali activated blast furnace slag pastes reinforced with carbon fibers. *Constr. Build. Mater.* **2016**, *116*, 63–71. [CrossRef]
38. He, P.; Jia, D.; Lin, T.; Wang, M.; Zhou, Y. Effects of high-temperature heat treatment on the mechanical properties of unidirectional carbon fiber reinforced geopolymer composites. *Ceram. Int.* **2010**, *36*, 1447–1453. [CrossRef]
39. Mechtcherine, V.; Michel, A.; Liebscher, M.; Schneider, K.; Großmann, C. Mineral-impregnated carbon fiber composites as novel reinforcement for concrete construction: Material and automation perspectives. *Autom. Constr.* **2020**, *110*, 103002. [CrossRef]
40. Li, H.; Liebscher, M.; Ranjbarian, M.; Hempel, S.; Tzounis, L.; Schröfl, C.; Mechtcherine, V. Electrochemical modification of carbon fiber yarns in cementitious pore solution for an enhanced interaction towards concrete matrices. *Appl. Surf. Sci.* **2019**, *487*, 52–58. [CrossRef]
41. Lu, M.; Xiao, H.; Liu, M.; Li, X.; Li, H.; Sun, L. Improved interfacial strength of SiO2 coated carbon fiber in cement matrix. *Cem. Concr. Compos.* **2018**, *91*, 21–28. [CrossRef]
42. Raphael, N.; Namratha, K.; Chandrashekar, B.N.; Sadasivuni, K.K.; Ponnamma, D.; Smitha, A.S.; Krishnaveni, S.; Cheng, C.; Byrappa, K. Surface modification and grafting of carbon fibers: A route to better interface. *Prog. Cryst. Growth Charact. Mater.* **2018**, *64*, 75–101. [CrossRef]
43. Růžek, V.; Dostayeva, A.M.; Walter, J.; Grab, T.; Korniejenko, K. Carbon Fiber-Reinforced Geopolymer Composites: A Review. *Fibers* **2023**, *11*, 17. [CrossRef]
44. Rahman, A.S.; Jackson, P.; Radford, D.W. Improved toughness and delamination resistance in continuous fiber reinforced geopolymer composites via incorporation of nano-fillers. *Cem. Concr. Compos.* **2020**, *108*, 103496. [CrossRef]
45. Huang, L.; Tang, L.; Bachinger, A.; Li, Y.; Yang, Z. Improving the performance of alkali-activated slag mortar with electro/chemically treated carbon fiber textile. *J. Clean. Prod.* **2023**, *418*, 138214. [CrossRef]
46. Raza, A.; Azab, M.; Baki, Z.A.; El Hachem, C.; El Ouni, M.H.; Kahla, N.B. Experimental study on mechanical, toughness and microstructural characteristics of micro-carbon fibre-reinforced geopolymer having nano TiO_2. *Alex. Eng. J.* **2023**, *64*, 451–463. [CrossRef]
47. Funke, H.; Gelbrich, S.; Kroll, L. The Durability and Performance of Short Fibers for a Newly Developed Alkali-Activated Binder. *Fibers* **2016**, *4*, 11. [CrossRef]
48. Lemougna, P.N.; Madi, A.B.; Kamseu, E.; Melo, U.C.; Delplancke, M.-P.; Rahier, H. Influence of the processing temperature on the compressive strength of Na activated lateritic soil for building applications. *Constr. Build. Mater.* **2014**, *65*, 60–66. [CrossRef]
49. Behera, P.; Baheti, V.; Militky, J.; Louda, P. Elevated temperature properties of basalt microfibril filled geopolymer composites. *Constr. Build. Mater.* **2018**, *163*, 850–860. [CrossRef]

applied sciences

MDPI

Article

Physico-Mechanical Properties of Alkali-Activated Based Composites Using Recycled Tire Fibers

Edwin Gudiel [1,2], Jordi Payá [2,*], María Victoria Borrachero [2] and José María Monzó [2]

1 Centro de Investigación de la Arquitectura y la Ciudad CIAC, Pontificia Universidad Católica del Perú, Av. Universitaria 1801, San Miguel, 15088 Lima, Peru; egudiel@pucp.edu.pe

2 Instituto de Ciencia y Tecnología del Hormigón (ICITECH), Universitat Politècnica de València, Camino de Vera S/N, 46022 Valencia, Spain; vborrachero@cst.upv.es (M.V.B.); jmmonzo@cst.upv.es (J.M.M.)

* Correspondence: jjpaya@cst.upv.es

Abstract: Used tires (UTs) are a global problem, especially in developing countries due to inadequate management systems. During retreading, when the worn tread is replaced, waste is generated in the form of tire fibers (TFs) and particles, which can be reused as raw materials to produce economically and environmentally low-cost prefabricated elements. Using TFs as a lightweight aggregate in nonstructural geopolymer-based elements is a sustainable valorization option. This study aims to valorize used tires by incorporating them as TFs into lightweight geopolymer mixes and analyzing their physico-mechanical, thermal, and thermography properties for building and civil engineering applications. The geopolymer is produced from a precursor (spent catalyst residue from catalytic cracking, FCC) and an alkaline activator composed of rice husk ash (RHA), sodium hydroxide, and water. The control sample's (mortar with siliceous sand, CTRLSIL) compressive strength came close to 50 MPa, while the TF mixes ranged from 32 to 3 MPa, which meet the masonry standards. The thermal conductivity and thermography analyses showed that increasing the TF content reduced the heat transmission and achieved a similar performance to expanded-clay concrete and better performance than for conventional concrete.

Keywords: tire fibers; FCC precursor; rice husk ash; geopolymer; thermal conductivity; thermography

1. Introduction

In September 2015, the Summit for Sustainable Development approved Agenda 2030 for sustainable development (SD), which includes 17 Sustainable Development Goals (SDG) to end poverty, combat inequality and injustice, and address climate change [1,2]. SDGs 11, 12, and 13 emphasize that SD is not possible without drastically transforming the way we build. This study focused on using more sustainable construction materials, such as composites with recycled tire fiber (TFs), and reusing industrial waste like spent fluid catalytic cracking catalyst (FCC) from the petrol industry and rice husk ash (RHA) to reduce the ecological footprint by changes in production methods, waste reduction, and recycling. It is important to note that one of the primary greenhouse gases (GHGs) is carbon dioxide, and the Portland cement (PC) industry contributes 7% of the total carbon dioxide emitted to the atmosphere by human activity [3]. This makes it a very polluting activity. Complete PC substitution for alkali-activated systems represents a significant reduction in the carbon footprint [4].

The progressive increase in the amount of rubber waste generated from used tires (UTs) has resulted in an environmental problem. According to reports from major tire and rubber product manufacturing associations, global annual tire production is around 1.4 billion units, the equivalent to an estimated 17 million tons of UTs a year. The dynamic growth in the number of UTs is well-exemplified by the EU, where production increased from 2.1 million tons in 1994 to 3.3 million tons in 2010, and the annual cost of its disposal in EU countries has been estimated at around EUR 600 million [5]. In 2020, the production of 4.2 million tons of rubber tires was estimated [6]. Waste generation is inevitable and it is therefore essential to have rational environmental management systems that minimize waste generation and maximize reuse and recycling. Alternatively, recovering energy and materials from discarded tires, known as valorization, is crucial [7,8].

In the EU, producers operate according to the Extended Producer Responsibility (EPR) system [9], which covers about 15 countries. This appears to be the most suitable and robust system for addressing and resolving issues that arise from UTs in a sustainable manner by achieving a 100% recovery rate. In Spain, SIGNUS is responsible for ensuring environmentally sound UT management [10].

An example in Latin America is Brazil, where 18.3 million tires were produced in 2003. The demand for UTs (retreads) is very high due to low purchasing power and the high price of new tires. In recent years, the importation of UTs has increased, with up to 7 million tires in 2007 [11].

Retreading is an alternative in UT management and utilizes tire casings by replacing worn treads and shoulders to reconstruct the original structure [12]. During this valorization process, waste is generated in the form of TFs and particles in the scraping stage.

The reaction of an aluminosilicate mineral mixed with a highly concentrated alkaline activating solution (based on a mixture of sodium hydroxide and an alkaline silicate such as sodium or potassium silicate) produces a structure called a geopolymer [13,14]. This type of alkali-activated cement can provide a similar performance to conventional PC- based binders, and offers an additional environmental advantage by reducing GHG emissions [15].

This study analyzed the valorization of TFs as a lightweight aggregate in mortar based on alkali-activated cement (AAC). Previous studies conducted on PC concrete with TFs report decreased bulk density as the TF contents increase [16–18]. Additionally, the loss of mechanical strength in flexural and compressive strength has been identified, even with use of the fibers [19–25]. This consequence can be attributed to the poor or nonexistent adhesion between the cement matrix and rubber [24–26]. Other properties, such as thermal properties [27], have drawn the interest of some researchers.

In previously published research [28], composites with various lightweight aggregates (pumice, expanded clay, coconut shell, among others) and using alkaline activation matrices based on blast furnace slag and metakaolin (activated with sodium hydroxide and silicate) have been described. The novelty of the present research lies in the use of TFs and an alternative cementitious matrix. In this work, the mechanical properties and thermal conductivity of AAC matrix composites prepared from FCC (activated with NaOH/RHA suspension) with TF reinforcement were evaluated.

2. Materials and Methods

2.1. Materials

The chemical compositions of FCC and RHA are summarized in Table 1.

Table 1. The chemical compositions of FCC and RHA as weight percentages.

	SiO_2	Al_2O_3	Fe_2O_3	CaO	MgO	SO_3	K_2O	LOI	Na_2O	P_2O_5	TiO_2
FCC	47.76	49.26	0.6	0.11	0.17	0.02	0.02	0.51	0.31	0.01	1.22
RHA	85.58	0.25	0.21	1.83	0.5	0.26	3.39	6.99	-	0.67	-

Geopolymer mortars with aggregate were produced [29] at a water/binder (w/b) ratio of 0.6 and an aggregate/binder (a/b) ratio of 2.25, both by weight. The binding material was the spent catalyst residue from catalytic cracking (FCC), an aluminosilicate with high pozzolanic reactivity, a density of 2.44 g/cm^3, and a particle diameter of 17.1 μm. RHA was used as a source [29] to prepare the alkaline activator, with a density of 1.90 g/cm^3 and a particle diameter of 62.2 μm. The activator for mortars containing siliceous/limestone aggregate was prepared by mixing 270 g of tap water, 81 g of NaOH (pellets), and 78.75 g of RHA in a 1 L thermal bottle. This mixture was stored in the thermal bottle for 24 h. The suspension was used for mixing with 450 g of FCC and the corresponding quantity of the aggregates and TFs (see Table 2). The activator for mortars with only TFs was prepared by mixing tap water, NaOH, and RHA in the following proportions: water/FCC = 0.6, NaOH/RHA = 1.029, and water/NaOH = 3.33 (see Table 2) and prepared using the same procedure. The mechanical strength measurement of the mortars was then conducted following standard UNE-EN 196-1 [30].

Table 2. Doses of the studied mortars.

Mortar Type	FCC (g)	Activator * (Water+ NaOH + RHA, in g)	TFs (g)	Aggregate (g)
CTRLSIL	450	270.00 + 81.00 + 78.75	0	1350
CTRLCAL				
TF SIL15			99.70	1147.5
TF CAL15			97.53	
TF SIL25			166.16	1012.5
TF CAL25			162.55	
TF 2/1	675.00	405.00 + 121.50 + 118.13	337.50	0
TF 1.8/1	607.50	364.50 + 109.35 + 106.31		
TF 1.6/1	540.00	324.00 + 97.20 + 106.31		
TF 1.4/1	472.50	283.50 + 85.05 + 82.69		
TF1.2/1	405.00	243.00 + 72.90 + 70.88		
TF 1/1	337.50	202.50 + 60.75 + 59.06		

* The activating suspension was prepared in a thermal bottle for 24 h.

TFs were produced by scraping tire surfaces during retreading, where the majority of fibers and particles were generated. TFs were separated from particles, and the material retained by the #0.63 mm mesh was used (Figure 1).

The aggregates used to produce geopolymer mortars were siliceous and limestone aggregates, and the particle size distribution was analyzed by sieving according to the mesh size series of standard UNE-EN 196-1 [30]. The density of aggregates was determined by the pycnometer method for particle sizes between 0.063 mm and 4 mm (UNE-EN 1097-6, [31]). This yielded a density of the siliceous aggregate equaling 2.54 g/cm^3 and limestone aggregate equaling 2.60 g/cm^3. For TFs, the density was determined by the difference in displaced volume using acetone as the fluid. The density obtained for TFs was 1.25 g/cm^3. The particle size distributions of the aggregates used in this study are shown in Figure 2.

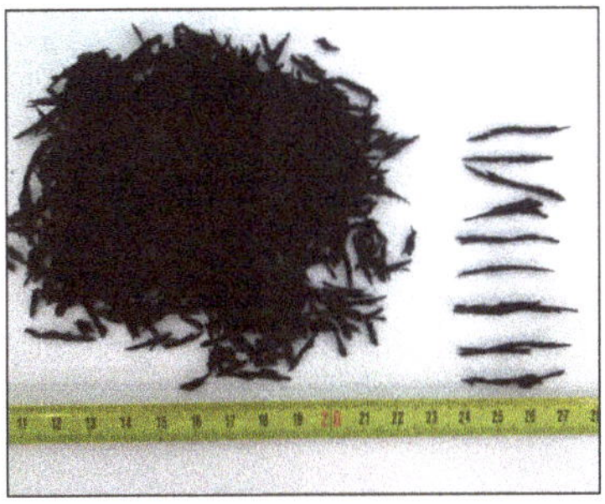

Figure 1. Tire fibers (TFs).

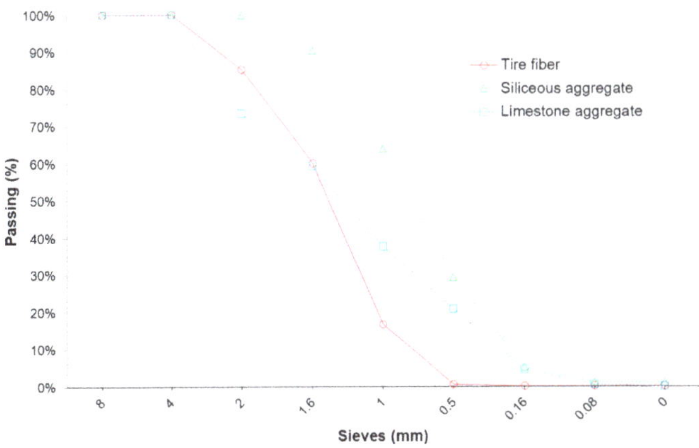

Figure 2. Particle size distributions for TFs and aggregates.

The geopolymer mortars with TFs were prepared based on the percentage content of TFs (TFaggregate15 and TFaggregate25), which corresponded to replacement with the volume of siliceous or limestone aggregate with TFs (15% and 25% replacement levels). To further highlight the properties of the mortars with TFs, two control mortars (CTRLSIL for mortars with the siliceous aggregate and CTRLCAL for those with the limestone aggregate) without TFs (100% siliceous or limestone aggregate) were prepared. Other geopolymer mortars were made with 100% TFs, where the FCC/TF volume ratios went from 2/1 to 1/1 (mortars TF2/1, TF1.8/1, TF1.6/1, TF1.4/1, TF1.2/1, and TF1/1). Table 2 summarizes the doses of the manufactured mortars.

2.2. Equipment and Procedures

Two batches were prepared per mortar type with sufficient quantity to fabricate three prismatic specimens measuring $40 \times 40 \times 160$ mm^3 each. Some specimens were left inside a humid chamber at 20 °C and 100% relative humidity (RH) for 7, 28, and 90 days, while others were exposed to 65 °C curing for 1, 3, and 7 days. After the curing period, each specimen's saturated and dry weight were determined: immersion in water for 24 h and subsequent drying in an oven at 60 °C until constant weight was achieved. Mechanical strengths (compressive strength Rc, and flexural strength Rf) were obtained under controlled displacement conditions at a speed of 1 mm/min using a universal testing machine (Instron model 3382). Rf was the average value of three values obtained from the prismatic samples, and the Rc (average value of six values) was measured with the six portions obtained. Thermal conductivity tests were performed with C-THERM

TCI equipment (C-Therm Technologies Ltd., Fredericton, New Brunswick, Canada). This method allowed us to make estimates of the materials' thermal conductivity within the 0.2–29 W/mK range.

Using the same mixtures as those above, plates of 150 × 150 × 20 mm^3 were prepared for thermographic analysis by employing a thermal imager, which allows for measurements of the thermal jumps. The utilized equipment was a TESTO 870-2 thermal imager (Testo Inc., Sparta, NJ, USA) with an infrared resolution of 160 × 120 pixels and emissivity of 0.93. Figure 3 shows the elements used for the thermography test, which consisted of a fully thermally insulated box of 400 × 600 × 600 mm^3 on all sides. The front part (centered position) housed the 150 × 150 × 20 mm^3 plate. On the central back part, the heat-emitting point was inside the box (a Philips HR-250 R 250 W—230–250 V incandescent and infrared luminaire was used to generate heat). The thermal imager was located on the front exterior.

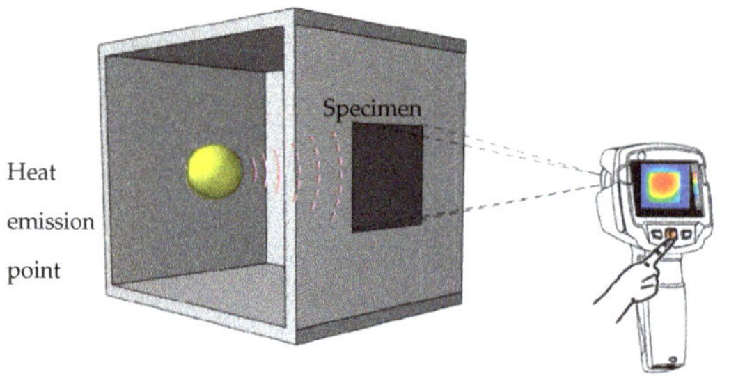

Figure 3. Box (thermally insulated sides) used for the thermography test.

Finally, fragments from the mechanical tests were selected and examined by field emission scanning electron microscopy (FESEM) with the ULTRA 55 model (Zeiss, Oberkochen, Germany), which allows for operations at very low potentials (0.02–5 kV). Sample preparation for visualization on this equipment required coating with a high vacuum system (LEICA EM MED020 model, Leica Microsystems, Wetzlar Germany).

3. Results

The physical (bulk density and water absorption) and mechanical (flexural and compressive strengths) properties of the AAC-based composites with TFs were characterized. Additionally, the thermal properties (thermal conductivity and thermography studies) were assessed. The following sections discuss the studied properties.

3.1. Dry Bulk Density and Water Absorption

The water absorption (Abs, as a percentage) and dry bulk density (ρ_d) for the control mortars and the mortars with a partial substitution of siliceous and limestone aggregates for TFs are depicted in Figure 4. It depicts how the ρ_d lowered when increasing the substitution of aggregates for TFs because the density of the siliceous and limestone aggregates was twice that of the TFs. In the mortars with 100% TFs without mineral aggregate, the ρ_d lowered due to the reduction in the amount of the cementing binder. In all cases, the ρ_d values were lower than those obtained for the control mortars [32].

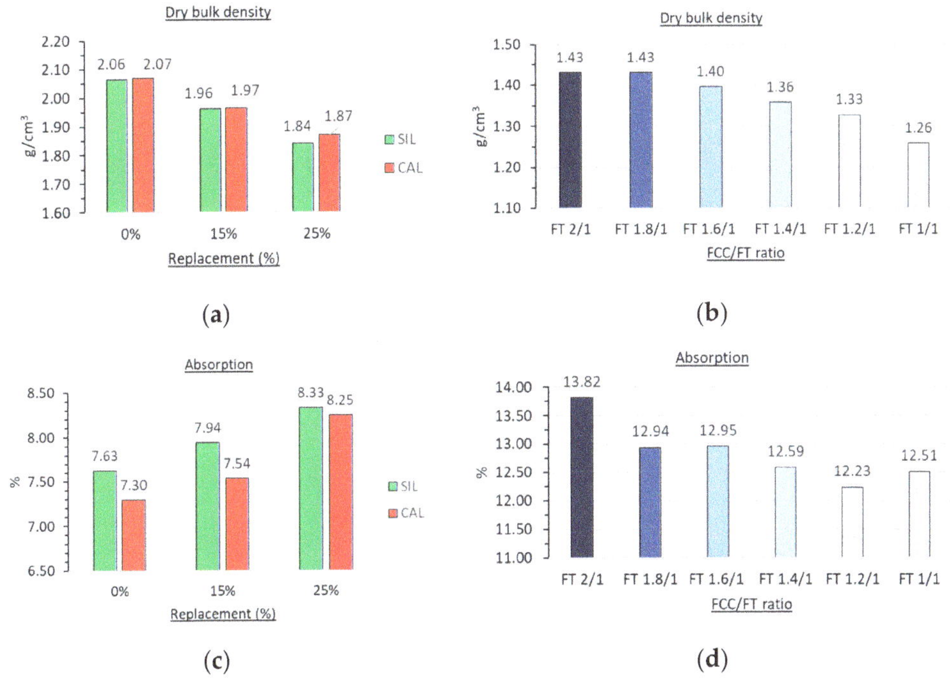

Figure 4. Physico-mechanical properties of mortars: (**a**) dry density for the control mortars and for those with 15% or 25% volume replacement with TFs; (**b**) dry density for the composite mortars without aggregate; (**c**) water absorption for the control mortars and the mortars with 15% or 25% volume replacement with TFs; (**d**) water absorption for the composite mortars without aggregate. In (**a**,**c**), SIL (green columns) corresponds to mortars with the siliceous aggregate, CAL (red columns) corresponds to mortars with the limestone aggregate.

The absorption value increased with the larger amount of TFs because TFs are capable of trapping air and causing an increased porosity of the matrix. These results align with previous research works, which indicate a higher air content in the rubber mortars compared with regular mortars [21,33]. This can be attributed to the rough surface of TFs and their ability to trap air [26].

The microstructural characterization (Figure 5) observed by FESEM allowed us to analyze the cementing matrix and the interfacial transition zone for mortars TF2/1, TF SIL25, and TF CAL25 after 7 curing days at 65 °C. Figure 5a,b shows smaller porosity in the cementing phase (geopolymer) and a thinner paste/TF gap compared with the 100% TF sample (Figure 5c), which showed that more pores were present in the matrix and had a thicker gap. This behavior would explain the greater absorption and lower density of these mortars compared with those made with 100% mineral aggregate. Obviously, this behavior would directly affect the mortar strength [19].

(**a**) (**b**) (**c**)

Figure 5. FESEM micrographs of the selected mortars cured for 7 days at 65 °C: (**a**) the mortar with siliceous sand and 25% TFs; (**b**) the mortar with limestone sand and 25% TFs; (**c**) the mortar with 100% TFs.

3.2. Mechanical Properties

Figures 6 and 7 present the mechanical strength (Rf, and Rc) values for the mortars with siliceous and limestone aggregates (with and without TFs).

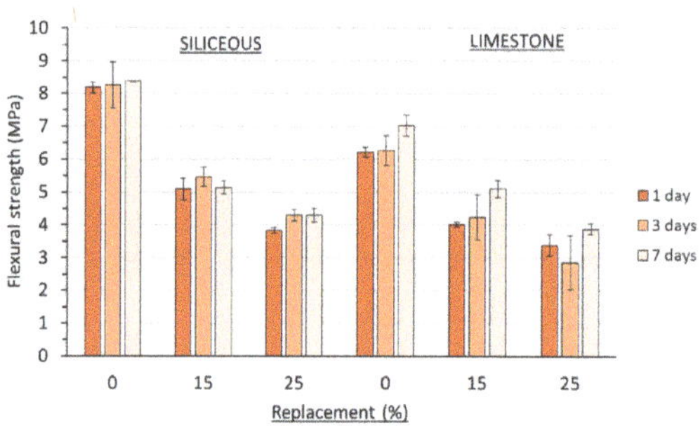

(**a**)

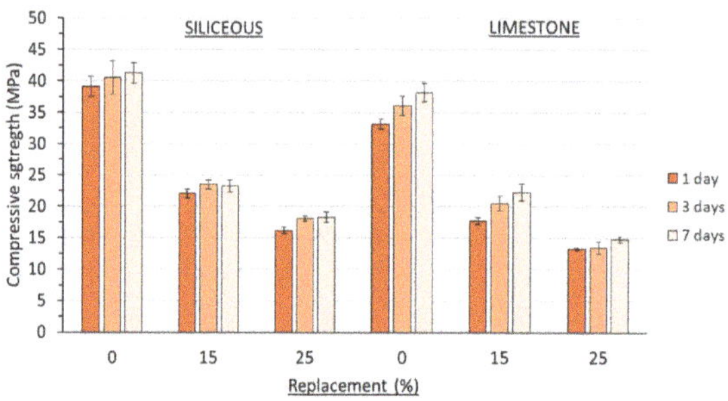

(**b**)

Figure 6. Mechanical properties of the mortars with siliceous and limestone aggregates incorporating TFs (0, 15, and 25% replacement by volume): (**a**) flexural strength (Rf) for the mortars cured at 65 °C; (**b**) compressive strength (Rc) for the mortars cured at 65 °C.

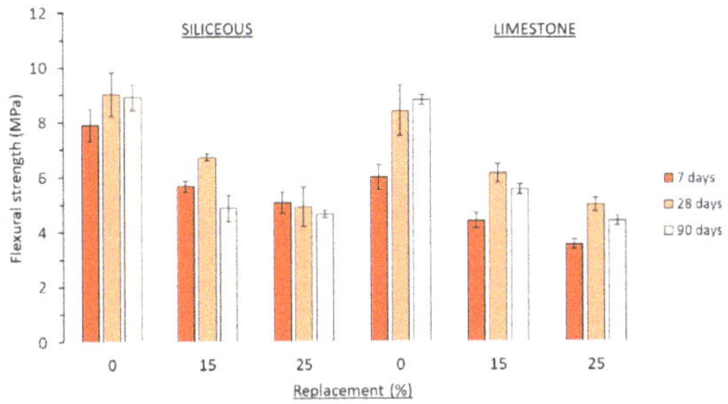

(**a**)

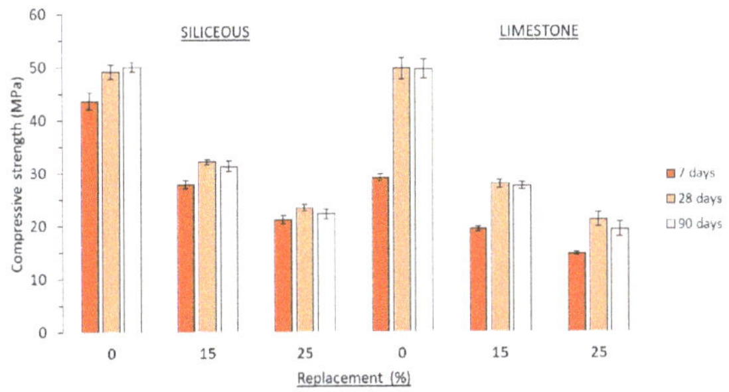

(**b**)

Figure 7. Mechanical properties of the mortars with siliceous and limestone aggregates incorporating TFs (0, 15 and 25% replacement by volume): (**a**) flexural strength (Rf) for the mortars cured at 20 °C; (**b**) compressive strength (Rc) for the mortars cured at 20 °C.

In Figure 6, the Rf and Rc of the mortars cured at 65 °C (1, 3 and 7 curing days) sharply decreased as the amount of TFs increased, and this trend was in accordance with the dry density values. No significant differences for the Rc values were noted between curing days 1 and 7, and their strength value gradually increased or remained until curing day 7. This means that the development of the cementing matrix was optimum after 1 curing day. This behavior was described by Rovnanik [34], who measured the Rc and Rf developments of metakaolin-based mortars cured at 60 °C and 80 °C, finding that the mechanical properties were similar for 1, 3, and 7 curing days. Bing-hui et al. [35] also found a similar behavior: these authors measured the Rc values for mortars cured for 24 and 72 h, which were very similar. The same authors demonstrated that the curing at 20 °C was insufficient to yield Rc values higher than 10 MPa after 24 h curing.

Figure 7 summarizes the results for the mortars cured at 20 °C (7, 28, and 90 curing days). The decrease in both fRf and Rc followed a similar pattern to those cured at 65 °C, which indicates a loss in mechanical strength as the amount of TFs increased.

Figure 8 shows that curing at 20 °C for 28 days generally yielded better results than curing at 65 °C for 7 days. This was also observed for the control mortar, which indicates that curing significantly affects the AAC matrix [36]. Aggregate use in the mortars was indifferent to the obtained values, which were similar among the mortars with limestone and siliceous aggregates. The Rc values obtained for the control mortars at both 65 °C and 20 °C agreed with the results obtained by other authors, but with similar alkali-activated mortars [29].

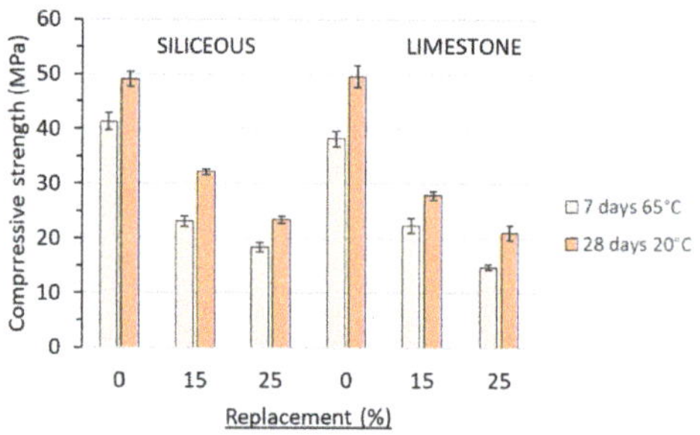

Figure 8. Comparison of the Rc values of the mortars with siliceous and limestone aggregates cured at 65 °C for 7 days vs. 20 °C for 28 days.

Figures 9–11 present the mechanical strength values (Rf and Rc) of the mortars with 100% TFs. Figure 9 shows a decrease in the Rf and Rc of the mortars cured at 65 °C when the binder quantity was reduced in relation to the TF volume. It generally depicts that the strengths reached at 1 curing day at 65 °C were similar to those for longer curing ages (3 and 7 days). This behavior means that geopolymeric binding matrix development is optimum for a short curing period, and it is not necessary to prolong the curing process, which reduces the energy use. Moreover, it highlights for the 1/1 composites that the mechanical properties are good for nonstructural purposes (e.g., masonry blocks for non-bearing walls [37]). Hence, the FCC and alkali activator uses are reduced, and consequently, so are the environmental and economic costs of composite production.

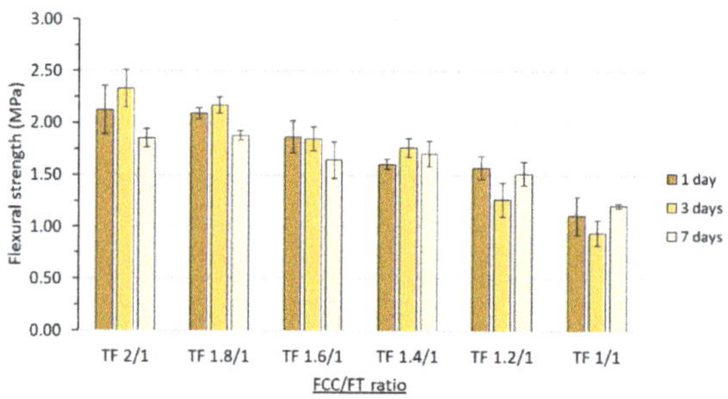

(**a**)

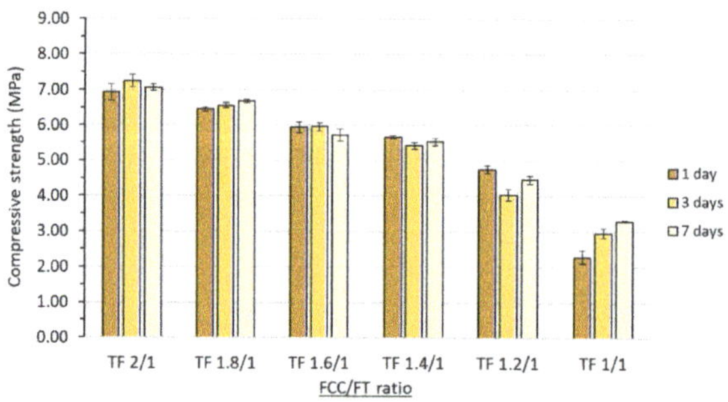

(**b**)

Figure 9. Mechanical properties of the composites with 100% TFs as an aggregate cured at 65 °C with different FCC/TF volume ratios: (**a**) flexural strength (Rf); (**b**) compressive strength (Rc).

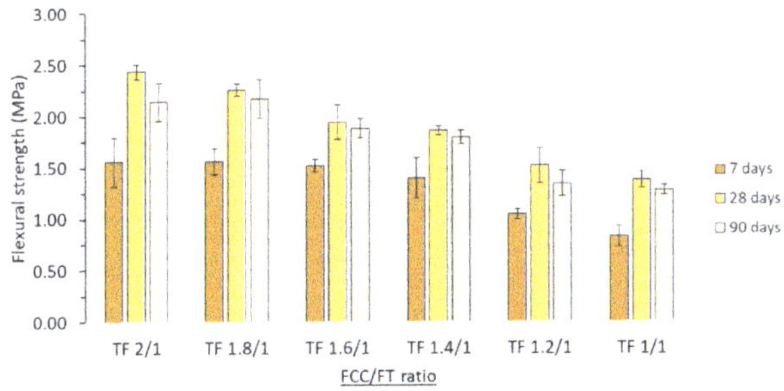

(a)

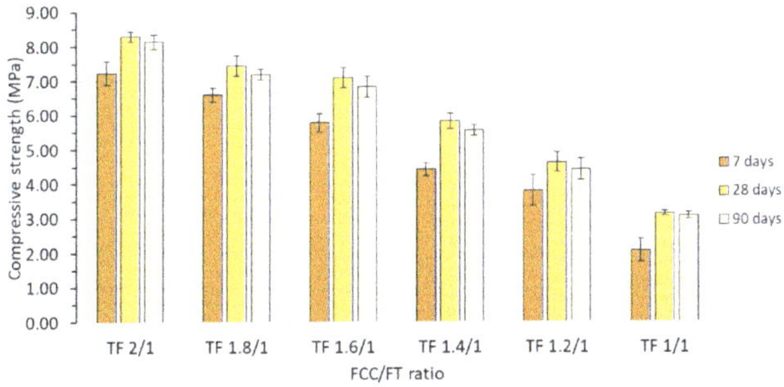

(b)

Figure 10. Mechanical properties of the composites with 100% TF aggregate cured at 20 °C, with different FCC/TF volume ratios: (**a**) flexural strength (Rf); (**b**) compressive strength (Rc).

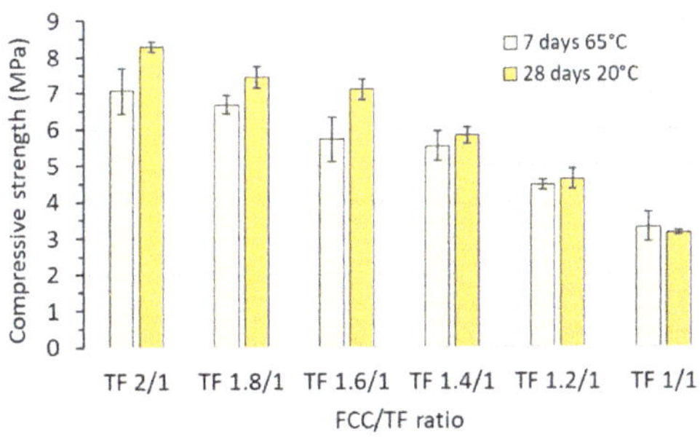

Figure 11. Comparison of the Rc of mortars with 100% TFs, cured at 65 °C for 7 days vs. 20 °C for 28 days.

In Figure 10, for the mortars cured at 20 °C, the Rf and Rc loss trend was similar to those cured at 65 °C, with a drop in mechanical strength as the binder quantity was lowered. In this case, the Rc gain from 7 to 28 curing days was notable, which suggests that appropriate geopolymeric matrix development requires 28 days, unlike that observed for the 65 °C curing process, for which only 1 day was required to achieve good mechanical properties.

Figure 11 highlights that curing at 20 °C for 28 days and for mortars with a low percentage of TFs (TF 2/1, TF 1.8/1 and TF 1.6/1) yielded better results than curing at 65 °C for 7 days, which was similar to that explained for the mortar with the partial mineral aggregate replacement. This outcome indicates that curing significantly affects the AAC matrix. Although curing at 65 °C uses more energy, it is advantageous to acquire good composite properties in a very short processing time. For composites with a high percentage of TFs (TF 1.4/1, TF 1.2/1, and TF 1/1), the effect of the AAC matrix was not significant in terms of the developed strength.

3.3. Thermal Conductivity

The thermal conductivity (k) results are summarized in Table 3. The incorporation of TFs into the mortar structure decreased the material's thermal conductivity as the amount of TFs in its composition increased. The mortars with partial siliceous and limestone aggregate substitution for TFs achieved lower thermal conductivity than conventional PC concrete (λ = 2.24 W/m·K) [38]. The mortars with the limestone aggregate obtained lower thermal conductivity values than those with the siliceous aggregate. This can be attributed to the conductivity difference between both aggregates, which is larger for quartz (7.69 W/mK) vs. calcite (3.59 W/mK) [39]. The network of pores generated by the nonpolar nature of TFs also influenced the results because of the fibers' capacity to trap air on their rough surface. As the TF content increased, the presence of pores also grew [26,27,40,41]. Figure 12 presents selected photographs of the specimen surfaces after testing, which confirmed the high pore content in the 100% TF composite. In the composites with 100% TF (2/1 to 1/1), the results had similar values among the six mixtures, with the ones containing a higher binder content achieving slightly better outcomes. This was because more compact and more uniform mortars were achieved on their testing surfaces compared with those with lower ratios. This resulted in less compact and less uniform specimens, given the lower binder content. The values of all six mixtures fell within the typical intermediate range for concrete with expanded clay [42]. ACI committee 213 R-03 proposed Equation (1) for correlating density (ρ) and the thermal conductivity value (k), while Asadi et al. [36] proposed Equation (2) to obtain them in concrete samples.

$$k = 0.0864\, e^{0.0125\rho} \tag{1}$$

$$k = 0.0625\, e^{0.0015\rho} \tag{2}$$

Table 3. Thermal conductivity (k) values.

Sample	k (W/mK)
CTRLSIL	0.969 ± 0.030
TF SIL15	0.948 ± 0.103
TF SIL25	0.926 ± 0.033
CTRLCAL	0.925 ± 0.043
TF CAL15	0.881 ± 0.024
TF CAL25	0.849 ± 0.049
TF 2/1	0.524 ± 0.097
TF 1.8/1	0.500 ± 0.068
TF 1.6/0	0.524 ± 0.033
TF 2.4/1	0.528 ± 0.027
TF 1.2/2	0.535 ± 0.006
TF 1/1	0.551 ± 0.006

In the TF 2/1 composite, when applying both equations, the corresponding values from Equations (1) and (2) were respectively 0.516 and 0.534 W/mK, which were similar to the experimental data (0.542 W/mK). However, the values for TF 1/1 obtained from the proposed Equations (1) and (2) were respectively 0.417 and 0.413 W/mK, and were significantly lower than for the experimental one (0.551 W/mK). The thermal conductivity of tire rubber fell within the 0.19–0.22 W/mK range [43]. Theoretically, the k value for TF 1/1 should be lower than that for TF 2/1, but the experimental results showed similar k values for both composites. This behavior could be attributed to the larger total volume related to the interfacial transition zone gap when increasing the TF volume fraction [44].

Figure 12. Surface of specimen breakage: left, the control mortar with the siliceous aggregate (CTRLSIL); center, the control mortar with the limestone aggregate (CTRLCAL); right, the composite with the 100% TF aggregate and the FCC-based geopolymer (TF2/1).

3.4. Thermography

The thermographic analysis was able to capture infrared thermal images every 5 min of the samples placed at the front of the thermally insulated box. Figure 13 shows the type of images recorded from the equipment. These images were compared and analyzed to quantitatively measure each specimen's surface temperature and to observe the temperature evolution [45].

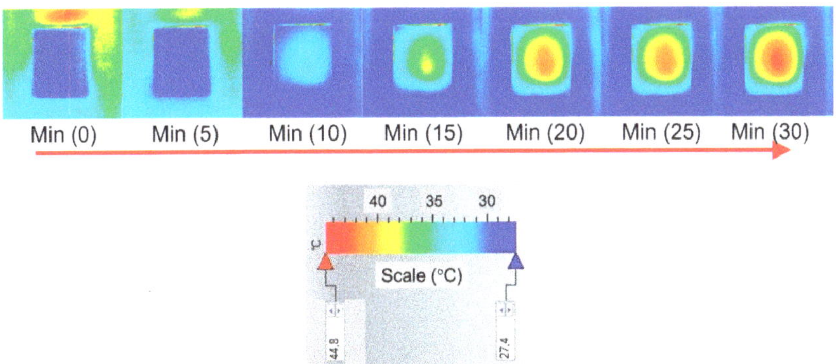

Figure 13. Infrared thermal images taken at different times (minutes) demonstrating the temperature increase during the test. Sample FT 2/1 subjected to 80 °C (heat source) from 0 to 30 min.

In Figure 14, a section was taken from each infrared thermal image to obtain a quantitative curve showing the temperature reached at each point on the image (pixel). Together with the other curves obtained in the other images taken every 5 min, it illustrates that the temperature increased until the maximum temperature was reached at 30 min. A temperature increase was only shown in the part in which the specimen was located, and not on the insulated box surface, where temperature increase was much lower.

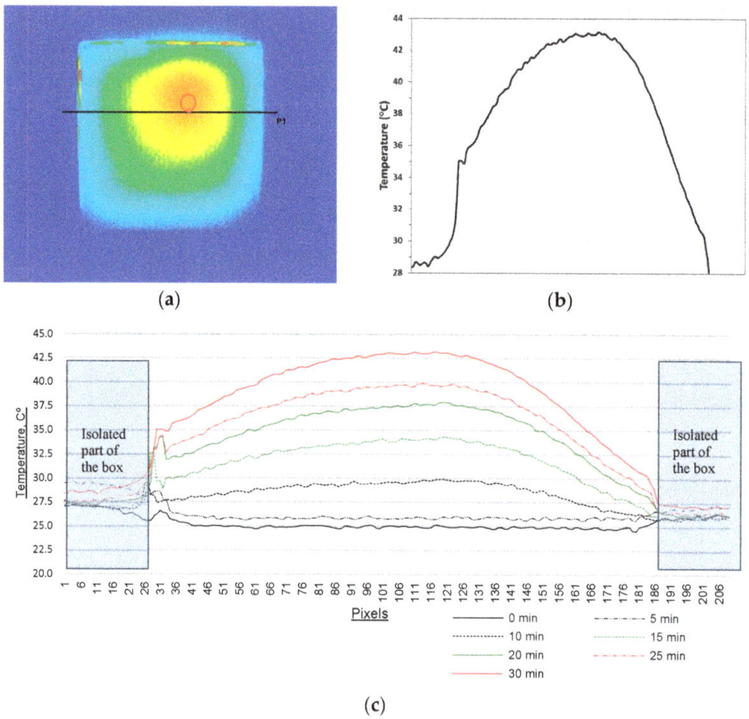

Figure 14. Thermography analysis: (**a**) infrared thermal image showing section P1, red circle represents the maximum temperature zone, and horizontal black line represent the points in which the temperature was calculated for representing in (**b**); (**b**) quantitative curve for section P1; (**c**) curves taken every 5 min for section P1.

The thermographic analysis results are summarized in Figure 15. For the mortars with the 25% TF replacing mineral aggregate, compared with the corresponding control mortars, no significant difference was observed in the temperature reached outside the box. However, a temperature differential of 4 °C was achieved for the composites with 100% TF.

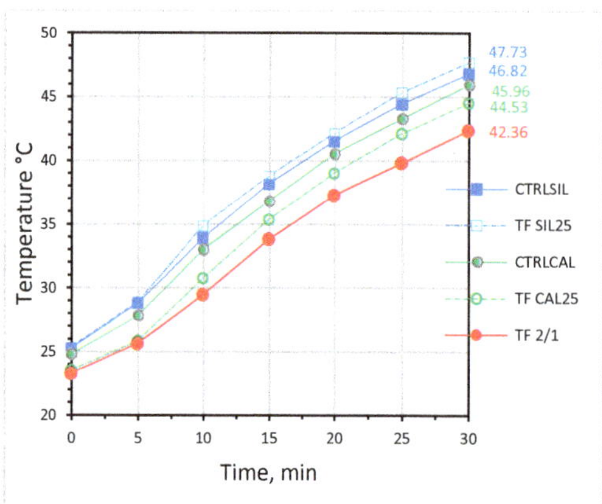

Figure 15. Thermography results: maximum temperature values measured on the external side of mortar plates taken every 5 min.

4. Conclusions

(1) TFs used as an aggregate in the AAC mortars reduced the dry bulk density and increased absorption compared with the values obtained for the reference mortars.

(2) The compressive strength of the mortars containing TFs ranged from 3 to 32 MPa. Compressive strength lowered as the TF content increased.

(3) The thermal conductivity values in the mortars in which the aggregate was replaced with TFs lowered as the TF content increased. The thermal conductivity of the mortars with 100% TFs came very close to 0.50 W/(m·K).

(4) Incorporating TFs into the FCC-based AAC cement mortars resulted in a lower temperature and thermal jump by the thermography technique (5 °C) compared with the values obtained with the reference mortars.

(5) In general, TFs used (derived from scrapping tires) in AAC-based materials can improve the thermal properties of these materials, and can lead to energy savings due to their low thermal conductivity. This is particularly beneficial when used in construction typologies that require thermal insulation properties. This material becomes an environmental benefit, especially as a result of working with waste materials and secondary raw materials, and confers them with added value, which cushions the impact of them accumulating in landfills. The obtained results highlight the potential of using TFs for preparing low-thermal conductivity composites with an AAC-based cementing phase, especially for systems with a high percentage of TFs (e.g., TF 2/1, which presented the lowest thermal conductivity and the best behavior in the thermography analysis).

Author Contributions: Conceptualization, E.G. and J.M.M.; Methodology, E.G., J.P. and M.V.B.; Software, E.G.; Validation, J.P. and J.M.M.; Formal analysis, M.V.B. and J.M.M.; Investigation, E.G., J.P., M.V.B. and J.M.M.; Resources, E.G., J.P., M.V.B. and J.M.M.; Data curation, E.G. and J.P.; Writing—original draft preparation, E.G.; Writing—review and editing, E.G., J.P., M.V.B. and J.M.M.; Visualization, E.G. and J.P.; Supervision, J.M.M.; Project administration, J.M.M.; Funding acquisition, J.P., M.V.B. and J.M.M. All authors have read and agreed to the published version of the manuscript.

Funding: This research received no external funding.

Institutional Review Board Statement: Not applicable.

Informed Consent Statement: Not applicable.

Data Availability Statement: The original contributions presented in the study are included in the article; further inquiries can be directed to the corresponding author.

Acknowledgments: We wish to thank the Microscopy Service of the Universitat Politècnica de València.

Conflicts of Interest: The authors declare no conflicts of interest.

References

1. Colglazier, W. Sustainable development agenda: 2030. *Science* **2015**, *349*, 1048–1050. [CrossRef] [PubMed]
2. Shulla, K.; Filho, W.L.; Lardjane, S.; Sommer, J.H.; Borgemeister, C. Sustainable development education in the context of the 2030 Agenda for sustainable development. *Int. J. Sust. Dev. World Ecol.* **2020**, *27*, 458–468. [CrossRef]
3. Supriya; Chaudhury, R.; Sharma, U.; Thapliyal, P.C.; Singh, L.P. Low-CO_2 emission strategies to achieve net zero target in cement sector. *J. Clean. Prod.* **2023**, *417*, 137466. [CrossRef]
4. Alsalman, A.; Assi, L.N.; Kareem, R.S.; Carter, K.; Ziehl, P. Energy and CO_2 emission assessments of alkali-activated concrete and ordinary Portland cement concrete: A comparative analysis of different grades of concrete. *Clean. Environ. Sys.* **2021**, *3*, 100047. [CrossRef]
5. Sienkiewicz, M.; Kucinska-Lipka, J.; Janik, H.; Balas, A. Progress in used tyres management in the European Union: A review. *Waste Manag.* **2012**, *32*, 1742–1751. [CrossRef]

6. Statista. Available online: https://www.statista.com/statistics/411255/eu-tire-production/ (accessed on 21 November 2024).
7. Sathiskumar, C.; Karthikeyan, S. Recycling of waste tires and its energy storage application of by-products: A review. *Sust. Mat. Technol.* **2019**, *22*, e00125. [CrossRef]
8. Xiao, Z.; Pramanik, A.; Basak, A.K.; Prakash, C.; Shankar, S. Material recovery and recycling of waste tyres: A review. *Clean. Mat.* **2022**, *5*, 100115. [CrossRef]
9. Campbell-Johnston, K.; Friant, M.C.; Thapa, K.; Lakerveld, D.; Vermeulen, W.J.V. How circular is your tyre: Experiences with extended producer responsibility from a circular economy perspective. *J. Clean. Prod.* **2020**, *270*, 122042. [CrossRef]
10. Signus. Available online: https://www.signus.es/ (accessed on 21 November 2024).
11. Milanez, B.; Bührs, T. Extended producer responsibility in Brazil: The case of tyre waste. *J. Clean. Prod.* **2009**, *17*, 608–615. [CrossRef]
12. Gaidhane, J.; Ullah, I.; Khalatkar, A. Tyre remanufacturing: A brief review. *Mater. Today Proc.* **2022**, *60*, 2257–2261. [CrossRef]
13. Davidovits, J.; Quentin, S. Geopolymers: Inorganic polymeric new materials. *J. Therm. Anal.* **1991**, *37*, 1633–1656. [CrossRef]
14. Gartner, E. Industrially interesting approaches to 'low-CO_2' cements. *Cem. Concr. Res.* **2004**, *34*, 1489–1498. [CrossRef]
15. Das, S.; Saha, P.; Jena, S.P.; Panda, P. Geopolymer concrete: Sustainable green concrete for reduced greenhouse gas emission: A review. *Mater. Today Proc.* **2022**, *60*, 62–71. [CrossRef]
16. Jones, K.; Durham, S.A. Beneficial uses of buffed rubber as fiber mesh in concrete mixtures. *Int. J. Construc. Environ.* **2013**, *2*, 113. [CrossRef]
17. Eiras, J.N.; Segovia, F.; Borrachero, M.V.; Monzó, J.; Bonilla, M.; Payá, J. Physical and mechanical properties of foamed Portland cement composite containing crumb rubber from worn tires. *Mater. Des.* **2014**, *59*, 550–557. [CrossRef]
18. Onuaguluchi, O.; Panesar, D.K. Hardened properties of concrete mixtures containing pre-coated crumb rubber and silica fume. *J. Clean. Prod.* **2014**, *82*, 125–131. [CrossRef]
19. Huang, X.; Ranade, R.; Ni, W.; Li, V.C. On the use of recycled tire rubber to develop low E-modulus ECC for durable concrete repairs. *Constr. Build. Mater.* **2013**, *46*, 134–141. [CrossRef]
20. Khaloo, A.R.; Dehestani, M.; Rahmatabadi, P. Mechanical properties of concrete containing a high volume of tire-rubber particles. *Waste Manag.* **2008**, *28*, 2472–2482. [CrossRef]
21. Ataei, H. Experimental study of rubber tire aggregates effect on compressive and dynamic load-bearing properties of cylindrical concrete specimens. *J. Mater. Cycles Waste Manag.* **2016**, *18*, 665–676. [CrossRef]
22. Dhonde, H.B.; Mo, Y.L.; Hsu, T.T.C.; Vogel, J. Fresh and hardened properties of self-consolidating fiber-reinforced concrete. *ACI Mater. J.* **2007**, *104*, 491–500. [CrossRef]
23. Xue, J.; Shinozuka, M. Rubberized concrete: A green structural material with enhanced energy-dissipation capability. *Constr. Build. Mater.* **2013**, *42*, 196–204. [CrossRef]
24. Bu, C.; Zhu, D.; Lu, X.; Liu, L.; Sun, Y.; Yu, L.; Xiao, T.; Zhang, W. Modification of rubberized concrete: A review. *Buildings* **2022**, *12*, 999. [CrossRef]
25. Dou, Y.; Feng, G.; Xu, L.; Yang, Y.; Zhang, Z.; You, L.; Zhong, S.; Gao, Y.; Cui, X. Modification of rubber particles and its application in rubberized concrete. *J. Build. Eng.* **2022**, *51*, 104346. [CrossRef]
26. Siddique, R.; Naik, T.R. Properties of concrete containing scrap-tire rubber: An overview. *Waste Manag.* **2004**, *24*, 563–569. [CrossRef] [PubMed]
27. Herrero, S.; Mayor, P.; Hernández-Olivares, F. Influence of proportion and particle size gradation of rubber from end-of-life tires on mechanical, thermal and acoustic properties of plaster-rubber mortars. *Mater. Des.* **2013**, *47*, 633–642. [CrossRef]
28. Tan, S.Q.; Lim, N.H.A.S.; Saleh, A.T.; Wei, K.E.L.; Samadi, M.; Huseien, G.F. A Bibliometric Review of Lightweight Aggregate Geopolymer Concrete. *CivilEng* **2024**, *5*, 892–927. [CrossRef]
29. Bouzón, N.; Payá, J.; Borrachero, M.V.; Soriano, L.; Tashima, M.M.; Monzó, J. Refluxed rice husk ash/NaOH suspension for preparing alkali activated binders. *Mater. Lett.* **2014**, *115*, 72–74. [CrossRef]
30. *UNE-EN 196-1:2018*; Methods of Testing Cement—Part 1: Determination of Strength. AENOR: Madrid, Spain, 2018.
31. *UNE-EN 1097-6:2014*; Tests for Mechanical and Physical Properties of Aggregates—Part 6: Determination of Particle Density and Water Absorption. AENOR: Madrid, Spain, 2014.
32. Aziz, F.N.A.; Bidas, S.M.; Nasir, N.A.M.; Jaafar, M.S. Mechanical properties of lightweight mortar modified with oil palm fruit fibre and tire crumb. *Constr. Build. Mater.* **2014**, *73*, 544–550. [CrossRef]
33. Fattuhi, N.I.; Clark, L.A. Cement-based materials containing shredded scrap truck tyre rubber. *Constr. Build. Mater.* **1996**, *10*, 229–236. [CrossRef]
34. Rovnaník, P. Effect of curing temperature on the development of hard structure of metakaolin-based geopolymer. *Cons. Build. Mat.* **2010**, *24*, 1176–1183. [CrossRef]
35. Bing-hui, M.; Zhu, H.; Cui, X.; He, Y.; Gong, S. Effect of curing temperature on geopolymerization of metakaolin-based geopolymers. *App. Clay Sci.* **2014**, *99*, 144–148. [CrossRef]

36. Mohamed, O.A. Effects of the curing regime, acid exposure, alkaline activator dosage, and precursor content on the strength development of mortar with alkali-activated slag and fly ash binder: A critical review. *Polymers* **2023**, *15*, 1248. [CrossRef] [PubMed]
37. Blash, A.A.A.; Bakar, B.H.A.; Udi, U.J.; Dabbour, B.S.A.; Jaafar, A.A.; Yanhao, L.; Abu Bakar, I.A.; Rashed, M. Performance of unreinforced masonry walls in compression: A review of design provisions, experimental research, and future needs. *Appl. Sci.* **2023**, *13*, 12306. [CrossRef]
38. Asadi, I.; Shafigh, P.; Hassan, Z.F.B.A.; Mahyuddin, N.B. Thermal conductivity of concrete: A review. *J. Build. Eng.* **2018**, *20*, 81–93. [CrossRef]
39. Schön, J.H. Chapter 9—Thermal properties. In *Developments in Petroleum Science*; Schön, J.H., Ed.; Elsevier: Amsterdam, The Netherlands, 2015; Volume 65, pp. 369–414. ISBN 9780081004043. [CrossRef]
40. Khatib, Z.K.; Bayomy, F.M. Rubberized Portland cement concrete. *J. Mat. Civ. Eng.* **1999**, *11*, 206–213. [CrossRef]
41. Sukontasukkul, P.; Chaikaew, C. Properties of concrete pedestrian block mixed with crumb rubber. *Constr. Build. Mater.* **2006**, *20*, 450–457. [CrossRef]
42. Rashad, A.M. Lightweight expanded clay aggregate as a building material: An overview. *Constr. Build. Mater.* **2018**, *170*, 757–775. [CrossRef]
43. Yang, Y.; Zhang, T.; Reddy, K.R.; Li, J.; Liu, S. Thermal conductivity of scrap tire rubber-sand composite as insulating material: Experimental investigation and predictive modelling. *Constr. Build. Mater.* **2022**, *332*, 127387. [CrossRef]
44. Agustini, N.K.A.; Triwiyono, A.; Sulistyo, D.; Suyitno, S. Effects of water to solid ratio on thermal conductivity of fly ash-based geopolymer paste. *IOP Conf. Ser. Earth Environ. Sci.* **2020**, *426*, 012010. [CrossRef]
45. Möllmann, K.P.; Vollmer, M. Infrared thermal imaging as a tool in university physics education. *Eur. J. Phys.* **2007**, *28*, S37. [CrossRef]

applied sciences

MDPI

Article

Performance of Eco-Friendly Zero-Cement Particle Board under Harsh Environment

Arman Hatami Shirkouh [1], Farshad Meftahi [1], Ahmed Soliman [1,*], Stéphane Godbout [2] and Joahnn Palacios [2]

[1] Building Civil and Environmental Engineering, Gina Cody School of Engineering and Computer Science, Concordia University, Montreal, QC H3G 1M8, Canada; arman.hatamishirkouh@mail.concordia.ca (A.H.S.); farshad.meftahi@concordia.ca (F.M.)

[2] Research and Development Institute for the Agri-Environment (IRDA), Quebec, QC G1P 3W8, Canada

* Correspondence: ahmed.soliman@concordia.ca

Abstract: The increasing scarcity of virgin natural resources and the need for sustainable waste management in densely populated urban areas have heightened the importance of developing new recycling technologies. One promising approach involves recycling agricultural waste in construction applications and transforming it into secondary products. This is anticipated to reduce the demand for new resources and lower the environmental impact, aligning with industrial ecology principles. Combined with a low carbon emission binder (i.e., alkali-activated), utilizing agro-waste to produce zero-cement particle boards is a promising method for green construction. Traditionally, particle boards are engineered from wood or agricultural waste products that are pressed and bonded with a binder, such as cement or synthetic resins. However, alternative binders replace cement in zero-cement particle boards to address environmental concerns, such as the carbon dioxide emissions associated with cement production. This study investigated the effects of accelerated aging on the performance of alkali-activated agro-waste particle boards. Accelerated aging conditions simulate natural aging phenomena. Repeated wetting–drying and freezing–thawing cycles increased water absorption and thickness swelling and reduced flexural strength. The thermal performance of the alkali-activated particle boards did not exhibit significant changes. Hence, it was confirmed that agro-waste has a high potential for utilization in producing particle boards provided that the working environment is carefully selected to optimize performance.

Keywords: alkali-activated materials; agro-waste; particle board; swelling; accelerated aging conditions

1. Introduction

Alkali-activated concrete provides a sustainable and innovative alternative to traditional Portland cement-based concrete [1]. Alkali-activated materials offer several advantages. They reduce carbon dioxide emissions associated with cement production by utilizing industrial by-products like fly ash and slag, which would otherwise be disposed of as waste and sent to landfills [2]. This makes them an environmentally friendly alternative. Additionally, alkali-activated materials demonstrate excellent durability, resistance to chemical attack, and reduced shrinkage compared to traditional concrete [3]. As a result, alkali-activated materials are distinguished from cement-based materials as they are considered sustainable, durable, and high-performance materials for various construction applications, contributing to the development of more environmentally conscious and long-lasting structures [4]. In this regard, alkali-activated materials are used for different applications. One of these applications is an alkali-activated particle board.

An alkali-activated particle board is a new product in the construction industry, bringing high durability and resistance to traditional particle boards [5]. This type of board is treated with alkali substances that enhance its performance characteristics, most notably its resilience against moisture and its potential for higher mechanical strength [6]. The process involves activating aluminosilicate materials using alkaline solutions, creating chemical reactions that form durable bonds within the structure [7]. Mixing such binder with wood-based materials forms alkali-activated particle boards with improved environmental stability over cement-bonded particle boards, showing higher resistance to water absorption and less susceptibility to swelling or warping [5]. In addition, it presents an eco-friendly alternative due to its low-energy processes and potential for incorporating recycled materials [8].

Aras et al. explored the impact of incorporating olive mill solid waste (OSW) residue and different cement types on the properties of cement-bonded particle boards (CBPBs) [9]. In another study, Hou et al. investigated the fabrication of a value-added cement-bonded particle board (CBPB) using Masson pine processing residues and Portland cement. The results demonstrated good compatibility between Masson pine residues and cement, with higher cement/wood ratios leading to improved physical and mechanical properties in the CBPB [10]. Also, Odeyemi et al. investigated the utilization of agricultural wastes for producing cement-bonded particle boards, focusing on their physical and mechanical properties [11]. Ohijeagbon et al. developed composite ceiling boards bonded with cement using locally sourced wood residue from teak and African locust bean tree. The physico-mechanical properties were evaluated, including the moisture content, density, water absorption, drying shrinkage, and strength characteristics [12]. Bufalino et al. investigated the influence of the wood's chemical composition, anatomical traits, and density on the performance of cement-bonded particle boards (CBPBs); the research elucidates their impact on cement curing, the matrix–reinforcement interface, and the final CBPB's performance [13]. Another study by Faria et al. investigated the compressive behavior of cement-bonded particle board (CBPB) elements, aiming to expand its application to structural elements [14]. Using alkali-activated binders in particle board production as an alternative to cement has garnered attention due to their potential for enhancing performance and sustainability [5,15]. Researchers have explored different alkali-activated binders derived from industrial by-products and waste materials, evaluating their effects on alkali-activated particle boards' mechanical and physical properties [5]. Furthermore, efforts have been made to optimize manufacturing techniques, such as the particle size distribution and mixing parameters, to enhance the overall quality and performance of cement-bonded particle boards and alkali-activated particle boards [5,9]. These findings in the literature highlight the potential of cement-bonded particle boards and alkali-activated particle boards to be used as viable alternatives to conventional particle boards, paving the way for further advancements and practical applications in the construction industry.

Despite significant research progress regarding particle boards, several notable gaps remain in understanding their long-term durability and performance, particularly under harsh environmental conditions. Further exploration is needed to assess how particle boards withstand moisture, temperature fluctuations, and external stresses over time [16]. Moreover, while their mechanical properties have been studied, comprehensive research has not considered the effects of parameters like the particle size, binder composition, and manufacturing processes on the overall performance [11]. Additionally, there is limited investigation into the practical implementation of particle boards in real-world structural applications and their compatibility with other building materials. Also, a particle board is a cornerstone material in furniture manufacturing, renowned for its cost-effectiveness, versatility, and adaptability to a wide range of finishes and treatments. Predominantly utilized in producing flat-pack furniture, cabinets, shelving units, and countertops, a particle board's widespread application is a testament to its suitability for lightweight yet durable furniture designs. Its capacity to be engineered from recycled wood fibers aligns with contemporary sustainability trends, offering an eco-friendly alternative to solid wood. However, research on a particle board's long-term durability, particularly in environmental stressors such as moisture and temperature fluctuations, remains insufficient [17,18]. A deeper investigation into how various factors—such as the particle size, binder composition, and manufacturing techniques—affect its performance could reveal potential enhancements for its application in not only furniture, but also in broader structural contexts, thereby bridging critical knowledge gaps in its use within the construction and furniture industries [19].

On the other hand, forecasts indicate the growth of agricultural production in the coming years. The purpose of these products is not only to provide food for the human population but also to meet industrial needs. Furthermore, agricultural commodities play a pivotal role in the textile industry by providing natural fibers such as cotton, flax, and hemp for manufacturing textiles and apparel [20]. The increasing growth of bioenergy from biofuel has been an example of the diversification of agricultural products in recent years. This is more evident in products with starch and cellulose [21]. In addition, agricultural plant residues have a special place as biomass and are considered to have good potential for green energy production [22]. In 2006, bioenergy accounted for about 10% of the world's energy [20]. The demand for crops experienced a significant upward trend from 2000 to 2015 due to the topic of biofuel production [23]. This demand is anticipated to produce a massive amount of agro-waste. Hence, the present study evaluates the possibility of utilizing agro-waste in producing alkali-activated particle boards and its effects on mechanical properties and durability under different curing conditions.

This research endeavors to revolutionize sustainable practices in the construction and agricultural sectors by utilizing agro-waste, waste reduction, and renewable resources. The primary aim is to develop a viable product, namely a zero-cement agro-waste particle board, that meets the rigorous technical standards of construction applications and delivers substantial economic and environmental benefits to the agricultural industry. By focusing on the production of a zero-cement particle board, this study aims to advance the understanding and application of alkali-activated particle board technology within engineered agro-waste composites. Through comprehensive experimentation and analysis, the findings of this research are expected to significantly contribute to the field, paving the way for the widespread adoption of sustainable alternatives in construction and agriculture.

2. Materials and Methods

Granulated blast furnace slag (hereafter called slag) was used as a binder for all zero-cement particle board mixtures with an average particle size of 14.5 μm. The physical and chemical properties of the slag are presented in Table 1. The dry-powder activator, sodium meta-silicate (Na_2SiO_3), with a 10% ratio, was mutually utilized to activate the zero-cement particle board mixtures. Na_2SiO_3 had a density of 1.09 g/cm^3 and a molar ratio of 1.0. The chemical and physical properties of the anhydrous sodium meta-silicate are shown in Table 2 in the manufacture datasheet. A water-to-binder (w/b) ratio of 0.40 was determined based on the mass of the binder. An agro-waste commercially known as Topimamerboor was used as an aggregate to produce zero-cement-bonded particle boards using 1:2, 1:3, and 1:4 ratios with respect to the binder. Figure 1 shows the agro-waste aggregate used in this study.

Table 1. Chemical compositions of used slag.

Items	SiO_2	Al_2O_3	CaO	Fe_2O_3	SO_3	MgO	K_2O	Na_2O	TiO_2	M_nO_2
(%)	36.5	10.2	37.6	0.5	3.1	11.8	0.4	0.3	1.0	0.4

Table 2. Chemical and physical properties of anhydrous sodium meta-silicate.

Property	W_t % Na_2O	W_t % SiO_2	W_t % H_2O	Density (kg/m^3)	Particle Size	Melting Point (°C)	Heat of Solution (kJ/mol)
Typical Data	50.5	46.2	<3	1090	93% in 20 to 65 mesh	1088	−31.7

Figure 1. Agro-waste used in this study (commercial name is Topimamerboor).

2.1. Mixture Procedure and Properties

In this study, three mixed designs were cast, as shown in Table 3. The agro-waste was submerged in water for 24 h to reduce its high-water absorption (saturation level) and then compressed with a hydraulic jack with 4 MPa of pressure before mixing. Afterward, the agro-waste was blended with other ingredients according to ASTM C305 [24] for 5 min. Due to the wet condition of the agro-waste, the water-to-binder ratios were set at approximately 0.4. The specimens were cast into molds and rested for 24 h at a temperature and relative humidity of 23 ± 2 °C and 50 ± 3%, respectively. The molds were clipped on the top and bottom from two sides to prevent the elastic expansion of the agro-waste for the first 24 h.

Table 3. Mix properties of zero-cement particle boards.

Mix Code	W/B	Meta-Silicate (%)	Agro-Waste/Binder Ratio	Slag (g)	Water (g)
C200	0.4	10	1:2	200	80
C300	0.4	10	1:3	300	120
C400	0.4	10	1:4	400	160

C reflects the exposure and curing conditions. For ambient conditions: NC for 25 freeze–thaw cycles is 25FT; for 25 drying–wetting cycles: 25DW.

2.2. Exposure Conditions

After pressing, the compressed material was released, and the boards were removed from the mold. Three exposure conditions were applied after the specimen was cured for 28 days: ambient conditions, wetting and drying cycles, and freezing and thawing cycles.

2.2.1. Repeated Wetting and Drying

Repeated curing in wetting and drying conditions were performed with the aim of evaluating material degradation, including risks like cracking and spalling, and to compare the durability of different materials under standardized cyclic conditions of wetting and drying. First, to maintain stable environmental conditions and prevent water loss, the boards were covered within a closed box and kept in the laboratory for 28 days. For the wetting and drying curing condition, 25 wetting and drying cycles were conducted on specimens by alternating the temperature and relative humidity after 28 days of curing, as per ASTM C1185 [25]. Repeating the wetting and drying cycles could simulate the rain–heat cycles in natural weathering. Due to this aging condition, some fundamental chemical and physical deterioration mechanisms are stimulated. Moreover, these conditions increase the alkaline pore water, which is expected to attack the agro particles and the migration (through dissolution and re-precipitation) of some alkali-activated hydration products into the interfaces.

2.2.2. Modified Freezing and Thawing

This investigation examines the impact of repeated freezing–thawing cycles on product integrity, focusing on the potential for internal pressures and cracking caused by the expansion of water as it freezes within the capillary pores of cement paste. This study applied 25 freezing–thawing cycles to specimens of alkali-activated agro-waste particle boards following the same temperature and cycle mentioned in the ASTM C666 standards [26]. First, to maintain stable environmental conditions and prevent water loss, the boards were covered within a closed box and kept in the laboratory for 28 days. Before cycling, each specimen was submerged in water at less than 5 °C (41 °F) for 48 h. The specimens were then exposed to alternating cycles of freezing and thawing, fluctuating between 20 ± 3 °C (68 ± 5 °F), with each cycle comprising 16 h of freezing followed by 8 h of thawing.

2.3. *Tests*

2.3.1. Mass Loss

The change in the mass of tested specimens was used to indicate the particle boards' degradation after exposure to different conditions. The mass of each specimen was measured immediately after finishing curing and before exposure to assigned conditions, representing the initial mass. The mass was measured again after exposure to various numbers of cycles.

2.3.2. Density

Following BS EN 12390–7:2019 [27], the density of the samples was determined. Two specimens were cut out, and their weights were recorded based on their volumes. The bulk densities of the samples were calculated using Equation (1).

$$\text{Density } (\text{g}/\text{cm}^3) = \text{W}_\text{a}/\text{V}_\text{a} \tag{1}$$

where W_a = air-dried weight, and V_a = air-dried volume.

2.3.3. Water Absorption

Water absorption was determined according to BS 1881–122:2011 [28]. The initial weight was recorded using a weighing balance of two specimen samples cut from the board. The specimens were weighed after being immersed in water for 2 and 24 h (at room temperature), and the weight of each specimen was recorded as the final weight. Equation (2) was used to determine the percentage of water absorbed by each specimen.

$$WaterAbsorption\ (\%) = \frac{Wf - Wi}{Wi} \times 100 \tag{2}$$

where W_f = the final weight of the particle board; W_i = the initial weight of the particle board.

2.3.4. Thickness Swelling

The thickness swelling assessment was conducted per the specifications outlined in the ASTM standard D1037-03 [29]. Initial measurements of the samples' thicknesses were precisely recorded utilizing a Vernier caliper. Subsequently, the samples underwent immersion in water for 2 and 24 h, after which their final thickness measurements were taken. The quantification of water absorption for each sample, expressed as a percentage, was calculated using Equation (3), thereby facilitating the analysis of water-induced dimensional changes.

$$ThicknessSwelling\ (\%) = \frac{Tf - Ti}{Ti} \times 100 \tag{3}$$

where T_f = the final thickness of the particle board; W_i = the initial thickness of the particle board.

2.3.5. Thermal Performance

This test focused on assessing the thermal performance of concrete when exposed to radiant heat from a heat and cold cycle. Specimens with dimensions of 150 × 150 × 10 mm from alkali-activated agro-waste particle boards were tested. The goal was to simulate the heat exposure that particle board surfaces might experience, with real-world conditions indicating a surface temperature of approximately 30 °C. We aimed to better understand how heat penetrates particle boards, as it can significantly impact a building's indoor temperatures. The variations in the sample surface's temperature were monitored using a high-resolution camera forward-looking infrared (FLIR—160 × 120 IR resolution) every 10 min. A thermocouple attached to a data logger was used to monitor the temperature inside the chamber (Figure 2). The heating and cooling cycle details are depicted in Figure 3.

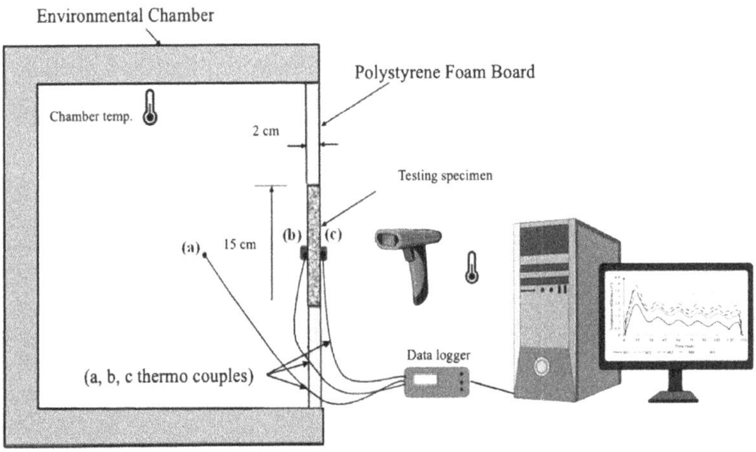

Figure 2. Schematic of heat transfer rate setup.

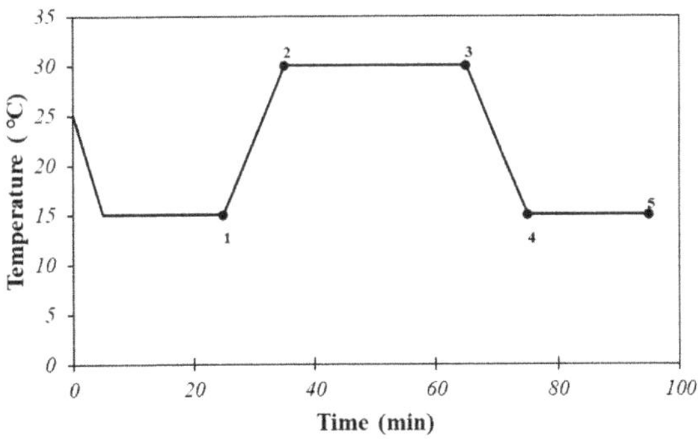

Figure 3. Thermal performance heating and cooling cycle and 5 monitoring points.

2.3.6. Flexural Strength

The flexural strength of the zero-cement particle board was determined based on ASTM D 1037 [29]. The particle boards were placed flat onto a flexural beam apparatus and subjected to a three-point loading setup, with the maximum load being recorded. The load was gradually applied until the particle boards failed, and a load versus deflection curve was plotted. The force that caused the particle board to fail was identified.

3. Results

3.1. Density

Figure 4 shows the result of the alkali-activated particle board density. The results show that specimen 25WD200 had the lowest density of 1052 kg/m^3, and specimen NC400 had the highest density of 1691 kg/m^3. According to IS 14276 [30], the suggested lower limit for density is 1250 kg/m^3, while JIS.A.5908 [31] advocates for a minimum density value of 800 kg/m^3. All manufactured particle boards adhered to these minimum standards as recommended. Figure 4 shows that the particle board density increases when the slag content increases. However, the density of the particle boards decreased when the specimens were exposed to the freezing and thawing cycles as well as the drying and wetting cycles. Also, the results show that the density of the particle boards decreased when the number of cycles increased. For instance, the density mixtures with 300 kg/m^3 of slag exposed to 10 FT cycles was 1486 kg/m^3, while after 24 FT cycles, it was 1310 kg/m^3. Also, the specimens cured under wetting and drying cycles had a lower density than those cured under freezing and thawing cycles. When comparing 25FT400 with 25WD400, it can be seen that it exhibited 8.5% less density. This is attributed to the change in the thickness of the particle board specimens when cured in the freezing and thawing cycles or wetting and drying cycles, as illustrated later.

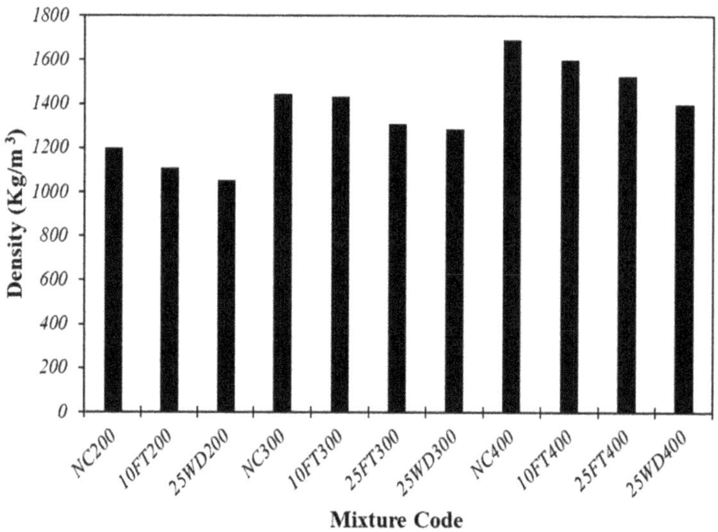

Figure 4. The density of the particle board with different slag contents and curing conditions.

3.2. Thickness Swelling

The thickness swelling of the alkali-activated particle board submerged in water for 2 and 24 h is illustrated in Figure 5. According to the guidelines set by the ASTM standard D1037-03 [29], the maximum allowable thickness swelling is 8%. As a result, all of the tested specimens met the requirement for thickness swelling. The results showed that 25FT200 and NC200 had the most significant changes in thickness, which were 5.12% and 4.65% after 24 h of submersion in water, respectively. However, the lowest change in thickness was exhibited by NC400, which was 2.17% after 2 h of submersion in water. This can be attributed to the ratio of agro-waste to slag, indicating that a higher slag content decreases thickness swelling. The widely accepted fact is that when the ratio of agro-waste to slag in particle boards rises, it leads to a notable increase in the thickness swelling, resulting in a decline in the boards' dimensional stability [11].

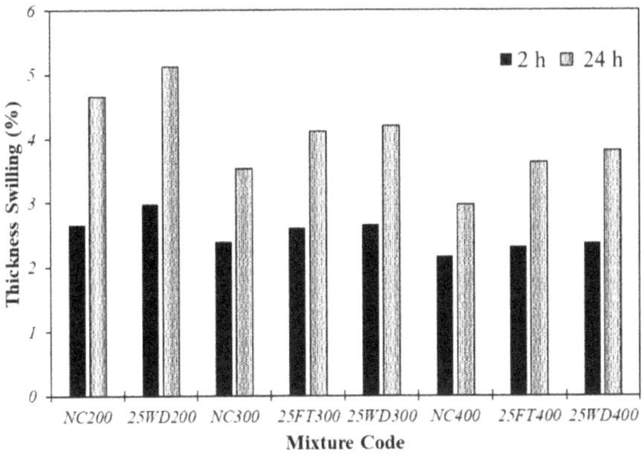

Figure 5. The result of the thickness swelling of the particle board with different slag contents and curing conditions.

On the other hand, the thickness swelling was increased when the particle boards were cured in freezing and thawing or drying and wetting conditions. For instance, it can be shown that specimens 25FT300 and 25WD300 had more changes in comparison with NC300, which were 4.11% and 4.20%, respectively. Also, it can be shown that curing in drying and wetting conditions has a greater effect than curing in freezing and thawing conditions. For instance, specimen 25WD400 had more thickness swelling changes than 25FT400. This can be attributed to the increase in the amount of pores when the specimens are cured in these conditions.

3.3. Water Absorption

The result of the water absorption test of the alkali-activated agro-waste particle boards is shown in Figure 6. The specimens were tested after being immersed in water for 2 and 24 h. The results show that the highest water absorption was achieved for 25WD200, which was about 46.88% after 24 h of water immersion. However, the lowest water absorption was achieved for NC400, at about 28.8%. Also, when the slag content increased in the particle board specimens, it decreased water absorption—for instance, NC200 achieved about 29.44% in 2 h of water immersion. However, the lowest water absorption was achieved for NC400, at about 24.35%. This can be attributed to the ratio of agro-waste to slag. After increasing the ratio of slag to agro-waste, the water absorption of the specimen of particle boards decreased due to a higher degree of hydration and the formation of hydration produced that seal-coated the agro-waste particles. As shown in Figure 6, the particle board specimens had higher water absorption when immersed for 24 h than those immersed for 2 h. In addition, the specimens that were cured in drying–wetting and freezing–thawing conditions had higher water absorption than those in ambient curing conditions. For instance, 25WD400 had more water absorption than NC400. Also, the water absorption of the particle board specimens follows the order from low to high: ambient curing condition, wetting and drying cycle, and freezing and thawing cycle. This can be attributed to the reduction in the polymerization process and the formation of hydration products to seal coat the agro particles.

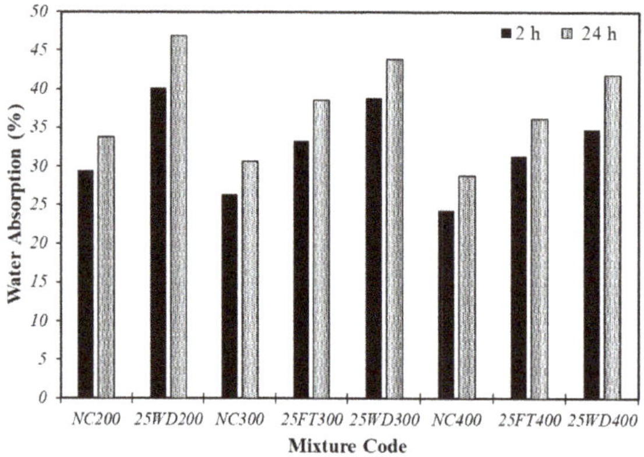

Figure 6. The result of the water absorption of the particle board with different slag contents and curing conditions.

3.4. Failure Mode

Figure 7 illustrates the failure modes of the alkali-activated particle boards and various degrees of damage. Obviously, the surfaces of all specimens exhibited rough textures, and there was an increase in the extent of corner loss in the specimens subjected to freezing–thawing cycles. For instance, considering the FT300 specimen as a representative, after 10 cycles, its surface underwent minimal alteration compared to the NC300 specimen. However, after 25 freezing–thawing cycles, the specimen's surface became notably rougher. This was accompanied by the detachment of small mortars that fell off from the corners and the emergence of micro-cracks on the surface. Additionally, noticeable swelling was observed on the specimens' surfaces due to agro-waste in the material composition. This swelling became more pronounced after 25 cycles of freezing and thawing conditions.

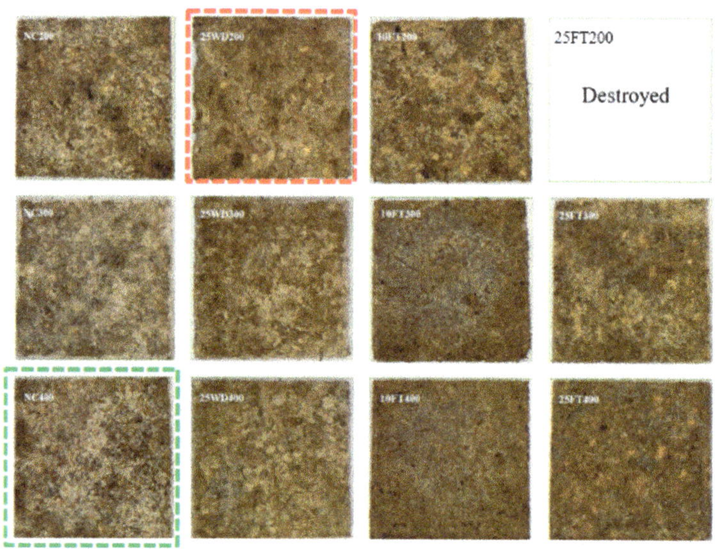

Figure 7. Failure rate of alkali-activated particle board specimens. (Destroyed: significant damage induced by exposure conditions, fail to parts).

Furthermore, the amount of slag content exerts a notable influence on the deterioration process of the specimens. At the same freezing–thawing cycles, it is evident that lower amounts of slag within the particle board specimens lead to more pronounced degradation. This trend is illustrated in Figure 7, where the NC400 specimen displayed negligible changes (Green Dashed). The specimen exhibits only slight swelling with no noticeable alterations. In contrast, the 25FT200 specimen experienced significant deterioration, while the 25FT300 and 25FT400 specimens demonstrated even less pronounced degradation when compared to the specimens subjected to standard curing conditions.

Upon the completion of 25 freezing–thawing cycles, discernible damage became apparent on the edges of the 25FT300 prismatic specimens. Furthermore, the edges of the 10FT300 specimens began to peel off following 10 freezing–thawing cycles. Remarkably severe surface damage was observed in the case of the 10FT200 and 25FT300 specimens. Notably, the mortar layer on the surface of the 25FT300 specimens had entirely dislodged, leaving the agro-waste aggregate exposed. The degradation of the 10FT200 specimen was even more pronounced, with substantial peeling of the concrete edges and wider cracks on the surface.

For the wetting and drying cycles, a higher content of slag was found to contribute to a reduced level of degradation in the particle board specimens, and this was similar to the trend observed during the freezing and thawing cycles. Figure 6 illustrates a discernible contrast in degradation severity between the 25WD200 (Red Dashed) and 25WD400 specimens. The amount of degradation witnessed during the wetting and drying cycles substantially exceeded that observed during the freezing and thawing cycles. For instance, the 25WD300 specimen showed more degradation than its freezing and thawing cycle counterpart, 25FT300. Upon closer examination, it became evident that after undergoing 25 cycles of curing under wetting and drying conditions, the surfaces of the specimens exhibited significant changes. The mortar layer underwent complete detachment, forming small pits, thereby exposing the agro-waste on the surface. Furthermore, at the edges of the specimens, severe peeling of the alkali-activated material occurred, resulting in wider cracks on the surface.

3.5. Mass Loss

Figures 8 and 9 illustrate the effects of freezing–thawing and wetting–drying cycles on mass loss. Generally, the particle board specimens under freezing–thawing or wetting–drying conditions exhibited mass loss. A consistent increase in mass loss was observed as the number of freezing–thawing and wetting–drying cycles increased. For instance, the FT300 specimen exhibited a more substantial mass loss after 25 cycles (i.e., 5.31%) than after exposure to 10 cycles (i.e., 3.94%), as shown in Figure 8.

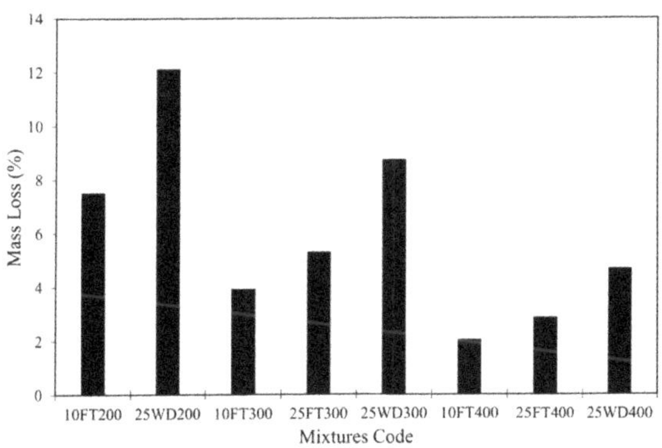

Figure 8. Effect of freezing–thawing and wetting–drying cycles on mass loss of agro-waste alkali-activated particle board.

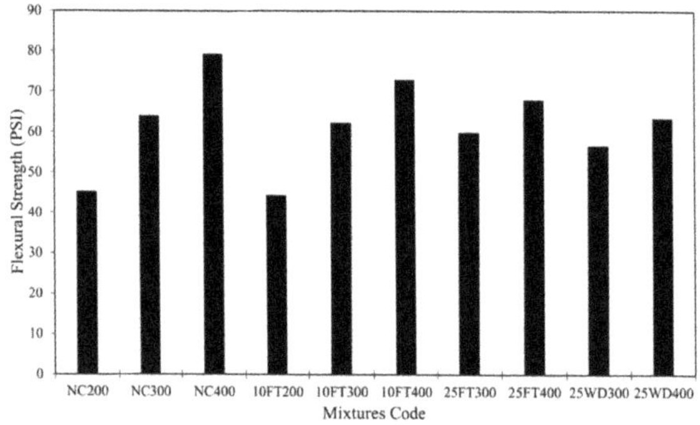

Figure 9. Effect of freezing–thawing and wetting–drying cycles on flexural strength of agro-waste alkali-activated particle board (1 Psi = 0.0069 MPa).

This observation can be attributed to the following rationales: In the initial phase, the expansion stress induced by freezing triggered specimen cracking. Consequently, a transfer of external water permeated internal voids and cracks, contributing to an augmented particle board specimen mass. However, as the freezing–thawing progressed, escalating freeze–thawing damage resulted in more degradation fragments from the specimen. This led to a reduction in the specimens' mass. Subsequently, when the mass loss attributed to spalling surpassed the mass gain stemming from absorbed water, an overall decrease in the particle board specimen mass ensued, accentuating the magnitude of mass loss.

On the other hand, the 25WD200 specimen displayed the highest change in mass variations, amounting to approximately 12.11%. Additionally, the higher amount of slag in the specimen led to a lower percentage of mass loss. For example, 25WD200 had more mass loss than 25WD300 and 25WD400, which had 12.11%, 8.74%, and 4.68% mass losses, respectively. Moreover, it is notable that wetting–drying cycles exhibit a more significant degradation potential when compared to freezing–thawing curing conditions, even when considering an equivalent number of cycles, even at higher binder mixtures. The mass loss was 4.68% for 25WD400, while it was 2.86% for the 25FT400 specimen.

3.6. Flexural Strength

3.6.1. Flexural Strength in Repeated Wetting and Drying Cycles

Wetting–drying curing conditions were used to analyze the aging characteristics of the alkali-activated particle board. Flexural strength tests were carried out on an alkali-activated particle board under 25 cycles of repeated wetting–drying curing conditions. The impact of cyclic wetting–drying on the flexural behavior and results of the flexural tests, encompassing parameters such as strength and flexural load deflection, is graphically shown in Figure 10 and Figure 11, respectively.

Generally, the repeated wetting–drying cycles resulted in decreased flexural strength. The flexural performance of the alkali-activated particle board under wetting–drying curing conditions typically declined compared to the specimens that underwent standard curing conditions. For instance, the 25WD200 particle board specimen experienced more pronounced degradation, resulting in a complete loss of flexural strength. In contrast, the NC 200 variant exhibited a flexural strength of 42.24 Psi (0.29 MPa). The results were better with an agro-waste/slag ratio of 0.25 compared to 0.5. For instance, the flexural strengths of 25WD300 and 25WD400 are about 56 Psi (0.38 MPa) and 63 Psi (0.43 MPa), respectively. Moreover, for the same deflection value, mixtures exposed to freezing and thawing cycles sustained lower loads. This finding reflects the internal degradation and assassinated damage induced by wetting and drying cycles.

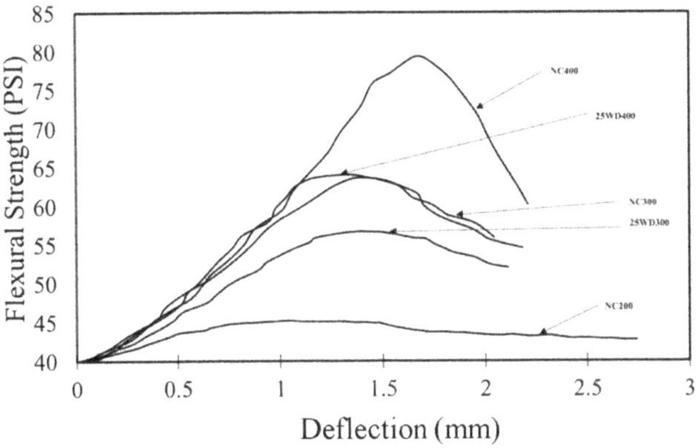

Figure 10. Effects of wetting–drying cycles on flexural curve characteristics of alkali-activated particle board (1 Psi = 0.0069 MPa).

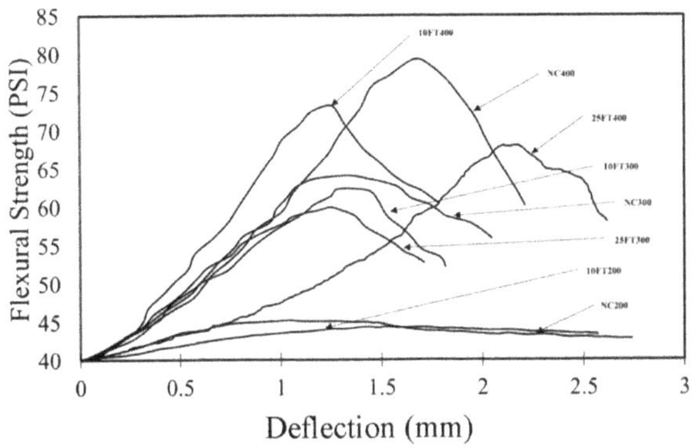

Figure 11. Effects of freezing–thawing cycles on flexural curve characteristics of alkali-activated particle board (1 Psi = 0.0069 MPa).

3.6.2. Flexural Strength in Freezing and Thawing Cycle

Freezing–thawing curing conditions were used to analyze the aging characteristics of the alkali-activated particle board. Flexural strength tests were performed on an alkali-activated particle board under 10 and 25 cycles of freezing and thawing curing conditions. The impact of cyclic freezing–thawing on the flexural behavior and results of the flexural tests, encompassing parameters such as strength, flexural load, and deflection, is graphically shown in Figure 10 and Figure 11, respectively.

Generally, the freezing and thawing curing conditions resulted in a decreasing trend for flexural strength. The flexural performance of the alkali-activated particle board under freezing–thawing curing conditions typically declined compared to the specimens that underwent standard curing condition. For instance, the 25FT200 particle board specimen experienced a complete loss of flexural strength. In contrast, the NC 200 variant exhibited a flexural strength of 42.24 Psi (0.29 MPa). Moreover, the results showed that increasing the number of cycles from 10 to 25 reduced the flexural strength more. For instance, the FT300 particle board specimen had a flexural strength of 62.3 Psi (0.43 MPa) after 10 cycles of freezing and thawing, while it exhibited a flexural strength of 59.9 Psi (0.41 MPa) after 25 cycles. The results were better at a lower agro-waste content. For instance, the flexural strengths of FT300 and FT400 are about 59.9 Psi (0.41 MPa) and 70 Psi (0.48 MPa), respectively.

3.7. Thermal Performance

All particle board specimens' surfaces were exposed to various temperatures as the environmental chamber's temperature was changed. The specimens were exposed to a 15 °C temperature for 25 min from an ambient temperature. Then, the specimens were exposed to a temperature of 30 °C for 30 min. Then, the temperature was lowered to 15 °C over 10 min and kept for 20 min. The result of the thermal performance of the alkali-activated agro-waste particle board is shown in Figure 12. It shows the temperature changes over time from the surface of the alkali-activated particle board specimens. The results show that using higher amounts of slag in the particle board specimens led to higher temperatures on the surfaces of the specimens. For example, the surface temperature of NC400 in all monitoring periods was much higher than that of NC200. Also, wetting–drying and freezing–thawing curing conditions followed the same trend as the ambient curing conditions. However, as shown in Figure 12, the wetting–drying and freezing–thawing curing conditions negatively affected the thermal performance of the alkali-activated particle board. The thermal performance when the specimens were cured under the wetting–drying or freezing and thawing curing conditions did not significantly change. For instance, 25FT300 and 25WD300 showed slightly lower temperatures than NC300. Also, the wetting–drying curing condition had a slightly lower temperature than the freezing and thawing curing condition. This can be attributed to two compensation effects: damage creating air gaps and a high water content. The thermal insulation induced by air gaps developed due to cracking and damage is compensated by the high water content and, consequently, higher thermal conductivity for the specimens.

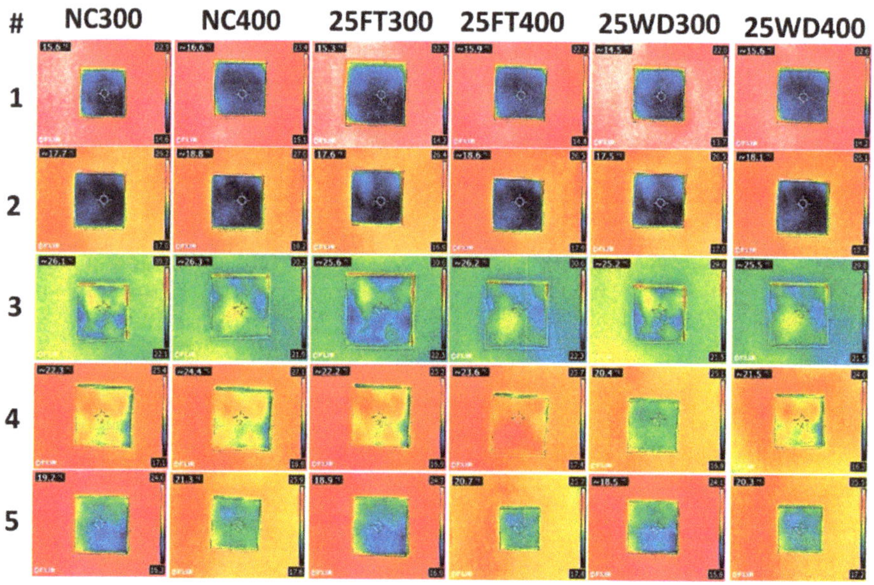

Figure 12. Result of thermal performance of alkali-activated agro-waste particle boards.

4. Conclusions

This study examined the effects of aging on the characteristics of an alkali-activated agro-waste particle board composite. Based on the findings and discussion presented, it can be firmly concluded that utilizing agricultural waste to produce alkali-activated particle boards is feasible but requires more caution when selecting the exposure conditions. Additionally, the influence of varying slag amounts was assessed, with the general trend showing that increasing the slag content resulted in a higher flexural strength. The key observations from the test results include the following:

- A high biner content compensates for the high water absorption of agro-waste.
- Repeated wetting–drying curing conditions and freezing–thawing cycles led to a decreased flexural strength over time.
- Repeated wetting–drying curing conditions had a higher degradation impact than that induced by freezing–thawing cycles.
- Agricultural waste exhibited a lower specific gravity, resulting in a more porous structure and lower product density during curing.
- From a practical perspective, the ratio of agricultural waste to slag and its effect on flexural strength was relatively high.
- The exposure conditions greatly affected the thermal performance based on the degree of damage and water content induced.

Author Contributions: Conceptualization, A.S.; methodology, F.M. and A.S.; validation, A.H.S. and A.S.; formal analysis, A.H.S.; investigation, A.H.S.; resources, A.S., S.G. and J.P.; data curation, A.H.S.; writing—original draft preparation, A.H.S.; writing—review and editing, A.S.; visualization, A.H.S.; supervision, A.S.; project administration, A.S., S.G. and J.P.; funding acquisition, A.S. All authors have read and agreed to the published version of the manuscript.

Funding: This research was funded by the Natural Sciences and Engineering Research Council of Canada, Discovery Funding program, grant number RGPIN-2018-05094.

Institutional Review Board Statement: Not applicable.

Informed Consent Statement: Not applicable.

Data Availability Statement: All data are available in Arman Hatami Shirkouh Master thesis aviable at Concordia Lib. Website.

Acknowledgments: The authors would like to thank Lafarge for donating the slag and Institut de recherche et de développement en agroenvironnement (IRDA) for donating the agriculture waste.

Conflicts of Interest: The authors declare no conflicts of interest.

References

1. Valente, M.; Sambucci, M.; Chougan, M.; Ghaffar, S.H. Reducing the emission of climate-altering substances in cementitious materials: A comparison between alkali-activated materials and Portland cement-based composites incorporating recycled tire rubber. *J. Clean. Prod.* **2022**, *333*, 130013. [CrossRef]
2. Huseien, G.F.; Hamzah, H.K.; Sam, A.R.; Khalid, N.H.; Shah, K.W.; Deogrescu, D.P.; Mirza, J. Alkali-activated mortars blended with glass bottle waste nanopowder: Environmental benefit and sustainability. *J. Clean. Prod.* **2020**, *243*, 118636. [CrossRef]
3. Mohamed, O.A. A review of durability and strength characteristics of alkali-activated slag concrete. *Materials* **2019**, *12*, 1198. [CrossRef] [PubMed]
4. Provis, J.L. Alkali-activated materials. *Cem. Concr. Res.* **2018**, *114*, 40–48. [CrossRef]
5. Kuqo, A.; Mayer, A.K.; Amiandamhen, S.O.; Adamopoulos, S.; Mai, C. Enhancement of physico-mechanical properties of geopolymer particle-boards through seagrass fibres. *Constr. Build. Mater.* **2023**, *374*, 130889. [CrossRef]
6. Thi, K.-D.T.; Liao, M.-C.; Vo, D.-H. The characteristics of alkali-activated slag-fly ash incorporating the high volume wood bottom ash: Mechanical properties and microstructures. *Constr. Build. Mater.* **2023**, *394*, 132240.
7. Athira, V.S.; Charitha, V.; Athira, G.; Bahurudeen, A. Agro-waste ash based alkali-activated binder: Cleaner production of zero cement concrete for construction. *J. Clean. Prod.* **2021**, *286*, 125429. [CrossRef]
8. Ghanad, D.A.; Soliman, A. Bio-based alkali-activated controlled low strength material: Engineering properties. *Constr. Build. Mater.* **2021**, *279*, 122445. [CrossRef]
9. Aras, U.; Kalaycıoğlu, H.; Yel, H.; Kuştaş, S. Utilization of olive mill solid waste in the manufacturing of cement-bonded particle-board. *J. Build. Eng.* **2022**, *49*, 104055. [CrossRef]
10. Hou, J.; Jin, Y.; Che, W.; Yu, Y. Value-added utilization of wood processing residues into cement-bonded particle-boards with admirable integrated performance. *Constr. Build. Mater.* **2022**, *344*, 128144. [CrossRef]
11. Odeyemi, S.O.; Abdulwahab, R.; Adeniyi, A.G.; Atoyebi, O.D. Physical and mechanical properties of cement-bonded particle board produced from African balsam tree (*Populous balsamifera*) and periwinkle shell residues. *Results Eng.* **2020**, *6*, 100126. [CrossRef]

12. Ohijeagbon, I.O.; Bello-Ochende, M.U.; Adeleke, A.A.; Ikubanni, P.P.; Samuel, A.A.; Lasode, O.A.; Atoyebi, O.D. Physico-mechanical properties of cement bonded ceiling board developed from teak and African locust bean tree wood residue. *Mater. Today Proc.* **2021**, *44*, 2865–2873. [CrossRef]
13. Bufalino, L.; de Souza, T.M.; Lima, N.N.; de Sá, V.A.; Tonoli, G.H.; Ferreira, C.A.; Junior, H.S.; de Sousa, R.B.; Zidanes, U.L.; de Paula Protásio, T.; et al. Contrasting the significant characteristics of pinewood and Amazon hardwoods to provide high-quality cement-bonded particle-boards. *Constr. Build. Mater.* **2023**, *394*, 132219. [CrossRef]
14. Faria, G.; Chastre, C.; Lúcio, V.; Nunes, Â. Compression behaviour of short columns made from cement-bonded particle board. *Constr. Build. Mater.* **2013**, *40*, 60–69. [CrossRef]
15. Olayiwola, H.O. *Development of Geopolymer Bonded Wood Composites*; Stellenbosch University: Stellenbosch, South Africa, 2021.
16. Soroushian, P.; Won, J.-P.; Hassan, M. Durability and microstructure analysis of CO_2-cured cement-bonded wood particle-board. *Cem. Concr. Compos.* **2013**, *41*, 34–44. [CrossRef]
17. El-Sayed, G.; Atallah, M.M.; Ahmad, M. Production of Fire-Resistant Particle Boards from some Agricultural Residues. *J. Soil Sci. Agric. Eng.* **2021**, *12*, 145–151. [CrossRef]
18. Ferrandez-Garcia, C.C.; Garcia-Ortuño, T.; Ferrandez-Garcia, M.T.; Ferrandez-Villena, M.; García, C.E. Fire-resistance, physical, and mechanical characterization of binderless rice straw particle-boards. *BioResources* **2017**, *12*, 8539–8549. [CrossRef]
19. Emmanuel, O.U.; Kuqo, A.; Mai, C. Non-conventional mineral binder-bonded lignocellulosic composite materials: A review. *BioResources* **2021**, *16*, 4606. [CrossRef]
20. Hazell, P.; Pachauri, R.K. *Bioenergy and Agriculture: Promises and Challenges*; IFPRI: Washington, DC, USA, 2006.
21. Food and Agriculture Organization of the United Nations. *The State of Food and Agriculture 2008*; Food and Agriculture Organization of the United Nations: Rome, Italy, 2008; 128p.
22. UN Energy. *A Decision Support Tool for Sustainable Bioenergy*; Prepared by FAO and UNEP for UN Energy; UN Energy: New York, NY, USA, 2010.
23. ERCD-FAO Agricultural Outlook. *OECD-FAO Agricultural Outlook 2017–2026*; ERCD-FAO Agricultural Outlook: Rome, Italy, 2017; Volume 2026.
24. *ASTM C305*; Standard Practice for Mechanical Mixing of Hydraulic Cement Pastes and Mortars of Plastic Consistency. ASTM International: West Conshohocken, PA, USA, 2006.
25. *ASTM C1185*; Standard Test Methods for Sampling and Testing Non-Asbestos Fiber-Cement Flat Sheet, Roofing and Siding Shingles, and Clapboards. ASTM International: West Conshohocken, PA, USA, 2016.
26. *ASTM C666*; Standard Test Method for Resistance of Concrete to Rapid Freezing and Thawing. ASTM International: West Conshohocken, PA, USA, 2017.
27. *BS EN 12390-7*; Testing Hardened Concrete-Density of Hardened Concrete. British-Adopted European Standard; BSI: London, UK, 2019.
28. *BS 1881-122:2011*; Testing Concrete. Method for Determination of Water Absorption. British-Adopted European Standard; BSI: London, UK, 2011.
29. *ASTM D1037-12*; Standard Test Methods for Evaluating Properties of Wood-Base Fiber and Particle Panel Materials. ASTM International: West Conshohocken, PA, USA, 2012.
30. *IS 14276*; Cement Bonded Particle Boards—Specification. Bureau of Indian Standards: Old Delhi, India, 1995.
31. *JIS A 5908*; Particle-Board. Japanese Industrial Standard/Japanese Standards Association: Tokyo, Japan, 2022.

applied sciences

MDPI

Article

High-Temperature Behavior of CaO-FeO_x-Al_2O_3-SiO_2-Rich Alkali Activated Materials

Guilherme Ascensão [1,2,3,4,*], Flora Faleschini [2], Maurizio Marchi [3], Monica Segata [3], Jorn Van De Sande [4,5], Hubert Rahier [5], Enrico Bernardo [6] and Yiannis Pontikes [4]

1 RISCO, Department of Civil Engineering, University of Aveiro, 3810-193 Aveiro, Portugal
2 Department of Civil, Environmental and Architectural Engineering, University of Padova, Via Marzolo, 9, 35131 Padova, Italy; flora.faleschini@dicea.unipd.it
3 HeidelbergCement Group, Italcementi S.p.A, Via Stezzano, 87, 24126 Bergamo, Italy; m.marchi@italcementi.it (M.M.); m.segata@italcementi.it (M.S.)
4 Department of Materials Engineering, KU Leuven, Kasteelpark Arenberg, 44, 3001 Leuven, Belgium; jorn.van.de.sande@vub.be (J.V.D.S.); yiannis.pontikes@kuleuven.be (Y.P.)
5 Department of Physical Chemistry and Polymer Science, Vrije Universiteit Brussel, Pleinlaan 2, 1050 Brussels, Belgium; hubert.rahier@vub.be
6 Department of Industrial Engineering, University of Padova, 35131 Padova, Italy; enrico.bernardo@unipd.it
* Correspondence: guilhermeascensao@ua.pt; Tel.: +35-19-6694-4248

Abstract: Alkali-activated materials (AAMs) provide an opportunity to up-cycle several residues into added-value materials. Although generally praised for their performance under thermal loads, the thermal behavior of AAMs is dictated by a multitude of factors and the performance of CaO-FeO_x-rich systems may differ from geopolymers. Therefore, this work ascertains the high-temperature resistance of CaO-FeO_x-Al_2O_3-SiO_2-rich AAMs. Mortars were exposed to different heating rates ($\leq$10 °C/min) and temperatures ($\leq$1100 °C), and volume and mass loss, apparent density, compressive strength (CS), mineralogical composition, and morphology were evaluated. At low heating rates, the main effects noted were densification and a gradual lightening of color as the temperature rose. CS underwent an abrupt decline at 750 °C and recovered at higher temperatures, reaching a maximum value of 184 $\pm$ 13 MPa at 1100 °C. With an increased heating rate to 10 °C/min, the strength loss at 750 °C persisted, but maximum CS was halved when firing at 900 °C. At 1100 °C, a significant reduction of CS was observed, but all samples maintained their integrity. Except for 1100 °C at 10 °C/min, all sintered-AAMs presented residual CS above 40 MPa. These results demonstrate that CaO-FeO_x-Al_2O_3-SiO_2-rich AAMs present interesting thermal behavior and can be potentially used to produce glass-ceramics or refractory materials from secondary resources.

Keywords: alkali activated materials; high-temperature behavior; secondary resources

1. Introduction

Increasing the use of alternative feedstock materials is crucial to reduce the current pressure on natural resources and to diminish the environmental impact of our industrial sectors. Moreover, using low-grade materials and residues as raw materials also represents an important and long identified economic opportunity for manufacturers. In the field of secondary resources, thermochemical conversion processes belong to a group of emerging technologies that aim to reclaim large volumes of low-value materials—such as industrial, municipal, and even mined solid wastes from historical landfills—and reintroduce them into materials' cycles in the form of added-value materials. The primary outputs of these processes are valuable energy carriers and metals, but an inorganic solid material is also generated. This inorganic fraction is molten and often vitrified by water quenching, afterwards being either disposed of or used in low-value applications (e.g., road paving). As thermochemical conversion processes continue to thrive, and their worldwide implementation grows, the volumes of these vitrified residues are expected to increase. The development of added-value products using such secondary outputs can have a remarkably positive effect on the sustainability of thermochemical technologies themselves, either directly, by providing a supplementary value stream, or indirectly, by abolishing associated environmental impacts and costs related to landfill. The use of such residues as precursors in the production of alkali-activated materials (AAMs) has been identified as a promising valorization route and suggested applications include, among others, pavers [1], mortars [2], and insulating building materials [3]. Another attractive property of AAMs is their performance at high temperatures. Unlike concrete, which suffers considerable degradation and spalling even below 400 °C [4,5], the physico-chemical characteristics of AAMs allow them to act as fire spreading and thermal energy barriers that prevent failure of construction elements. Yet, the thermal behavior of AAMs depend on several synthesis conditions, such as the chemical composition of the parent precursors, the type (silicate or hydroxide solutions), concentration, and charge balancing cation (Na^+ or K^+) in the activating solution, and Si/Al molar ratio [6,7]. The synthesis conditions dictate the pore structure and liquidus temperature of AAMs, as well as the amount and type of crystalline phases formed. Depending on their characteristics, the use of additives can also have a significant thermodynamic impact when firing alkali-activated materials.

The typical chemical composition of the slag produced during the thermochemical conversion of municipal and mined wastes can be described as a CaO-FeO_x-Al_2O_3-SiO_2 quaternary system, which sometimes also contains a significant MgO content [8]. This chemical composition could be adjusted to promote the formation of more refractory phases upon thermal exposure, i.e., by adding alumina-rich materials [7,9,10], but changing the composition of such low-value raw material may entail undesirable environmental and economic costs. Considering the above, the present work did not aim to enhance the thermal properties of CaO-FeO_x-Al_2O_3-SiO_2-rich AAMs by altering their chemical composition with beneficial admixtures but rather to characterize the thermal performance of potassium-based AAMs previously optimized for general construction purposes. Thus, the thermal performance of conventional construction materials such as OPC-concrete, and not fire-resistant or refractory ones, should be used as a benchmark throughout this work.

Pure (Na_2O/K_2O)-Al_2O_3-SiO_2 ternary systems are known to present excellent thermal resistance due to the crystallization of feldspar and feldspathoid phases, i.e., framework alumino-silicates, featuring various proportions of both SiO_4 and AlO_4 units, the latter being stabilized by alkali ions. These phases comprise albite (Na feldspar, $NaAlSi_3O_8$ or $Na_2O \cdot Al_2O_3 \cdot 6SiO_2$, having a liquidus temperature, 1118 °C), nepheline (Na-K feldspathoid, $Na_3KAl_4Si_4O_{16}$, stable up to 1256 °C), and carnegieite (Na feldspathoid, $NaAlSiO_4$ or $Na_2O \cdot Al_2O_3 \cdot 2SiO_2$, having a liquidus temperature, 1526 °C) in the case of sodium-based activators. In the case of potassium-based activators, the developed alumino-silicate compounds exhibit even higher melting points, as in the case of orthoclase (K feldspar, $KAlSi_3O_8$ or $K_2O \cdot Al_2O_3 \cdot 6SiO_2$, having a liquidus temperature of 1250 °C), leucite (K feldspathoid, $KAlSi_2O_6$ or $Na_2O \cdot Al_2O_3 \cdot 4SiO_2$, with a congruent melting point of 1693 °C), and kalsilite (K feldspathoid, $KAlSiO_4$ or $K_2O \cdot Al_2O_3 \cdot 2SiO_2$, having a liquidus temperature, 1750 °C) [10].

The refractory performance of potassium-based ternary systems further benefits from the higher viscosity of potassium melts. This reduces deformation due to flow and reinforces the advantageous character of potassium-based AAMs for high-temperature applications. Nevertheless, it should be mentioned that Na_2O-SiO_2-Al_2O_3 systems can also present excellent resistance to high temperature if a specific set of synthesis conditions that promote carnegieite formation (melting point, 1526 °C) are provided. On the other hand, K_2O-SiO_2-Al_2O_3 systems present a broader set of synthesis conditions where refractory phases are formed upon sintering [10], and this is particularly relevant in industrial contexts where versatile and resilient mixture designs are of utmost importance. From this point of view, the use of a potassium-based activating solution is advantageous both in terms of thermal performance and industrial processing.

The chemical composition of the slag produced in thermochemical conversion processes is characterized by the presence of significant calcium and iron content, which, however, have a direct influence on the phase composition and morphology of the material after heating. Low calcium systems (i.e., geopolymers) are generally preferable for producing high-temperature resistance AAMs as calcium reduces the liquidus temperature of K_2O-SiO_2-Al_2O_3 systems. Furthermore, the near absence of hydrated and chemically bonded water in those low-calcium structures [11,12] reduces the deleterious microstructural effects of dehydration, which results in more thermally stable AAMs. The presence of iron can, however, be deleterious to K_2O-Al_2O_3-SiO_2 systems, depending on the amount of alumina present in the systems [10]. If insufficient alumina is available, the iron compounds formed are characterized by much lower melting temperatures relative to their iron-free analogs. Hence, the thermal performance of AAMs made from FeO_x-Al_2O_3-SiO_2 precursors with low-aluminum content is expected to be significantly impaired relative to their iron-free equivalents. On the other hand, in alumina-rich K_2O-Al_2O_3-SiO_2 systems, the introduction of iron is not expected to preclude the crystallization of highly stable and refractory phases such as mullite, leucite, and kalsilite, and can even promote the crystallization of refractory hercynite [10]. The presence of these phases enhances the thermal resistance of AAMs made from alumina-rich FeO_x-Al_2O_3-SiO_2 precursors and are expected to deliver excellent fire-resistant properties [9,13].

Several authors have investigated the phase transitions and thermal characteristics of AAMs produced from precursors such as metakaolin and fly-ash, which can be mainly described as Al_2O_3-SiO_2 binary systems [14,15], and calcium or iron-rich ones, such as ground granulated blast furnace slag [15], ferronickel slag [9], and copper slags [7,16]. The thermal characteristics of more complex CaO-FeO_x-Al_2O_3-SiO_2 quaternary systems have not yet been thoroughly investigated. As discussed, calcium- and iron-containing phases will play vital roles in determining the thermal performance of those AAMs. Thus, examining the properties of CaO-FeO_x-Al_2O_3-SiO_2-rich AAMs after thermal exposure (the sintered counterparts) is crucial to establish these AAMs as reliable temperature resistant materials, with a place in the market, and deliver a long-term, large-scale, upcycling solution for CaO-FeO_x-Al_2O_3-SiO_2-rich residues generated during thermochemical conversion processes.

2. Materials and Methods

2.1. Raw Material Characterization and Mortar Preparation

A synthetic granular slag was produced with a chemical composition that mimics the expected inert vitrified material produced during the thermochemical conversion of combustible components of municipal and mined solid wastes. The detailed description of its production process is provided by Machiels et al. [1,2]. X-ray fluorescence (Bruker AXS S8 Tiger spectrometer, Billerica, MA, USA) was used to determine the bulk chemical composition of the slag, which can be generally described as a CaO-FeO_x-Al_2O_3-SiO_2 quaternary system. The complete chemical composition of the slag is given in Table 1. The Fe oxidation state was quantified as described by Close et al. [17] and the iron phases were found to be predominantly in the bivalent oxidation state (92%).

Table 1. Chemical composition of the synthetic slag used as main raw material (% by mass).

	LOI	SiO_2	CaO	FeO_x	Al_2O_3	MgO	TiO_2	K_2O	Na_2O	MnO	Fe^{3+}/Fe_{total}
Synthetic slag	1.9	34.9	23.1	22.8	16.2	1.3	0.6	0.5	0.3	0.1	0.08

Aiming to deliver sustainable materials, the use of ordinary Portland cement and natural aggregates was completely avoided in the mortars' mix design; the vitreous slag was used both as the main precursor and aggregate. For this purpose, the granular slag was subjected to distinct pre-processing procedures to achieve different particle size distributions according to the envisioned usage (Figure 1). Particle size distribution was determined using a laser particle analyzer (Malvern Mastersizer 3000, Malvern Panalytical, Malvern, UK). Finely powdered slag with a specific surface area of 1120 m^2/kg (determined by the Brunauer–Emmett–Teller method, BET) and density of 2.97 g/cm^3 was used as the binders' main precursor (94.5 wt.% of binders' solid load). Silica fume (SF, Elkem, >95% SiO_2) with a BET specific surface area of approximately 22,200 m^2/kg was used as an admixture to increase the availability of Si species and support strength development. Increasing the Si/Al ratio was also expected to promote the formation of highly thermally stable phases [14] and produce compact microstructures with low porosity levels after thermal exposure [18], thus improving the high-temperature resistance and residual strength of the produced materials. A potassium-based activating solution was used following a mix design previously optimized. Theoretical predictions [10] and empirical experimentations [6] in the available literature confirm the superior thermal performance of potassium-based AAMs relative to their sodium-based counterparts. Potassium hydroxide (14 M) and potassium silicate solutions (23.8 wt.% SiO_2, 9.5 wt.% K_2O, and 66.7 wt.% H_2O) were prepared by dissolving potassium hydroxide beads and anhydrous potassium silicate in demineralized water. The water content in the activating solution was adjusted (0.68 SiO_2/K_2O molar ratio and 83.4 wt.% H_2O) to achieve the pre-defined solid over liquid ratio (S/L) of 2.85 and molar ratios in the binder composition.

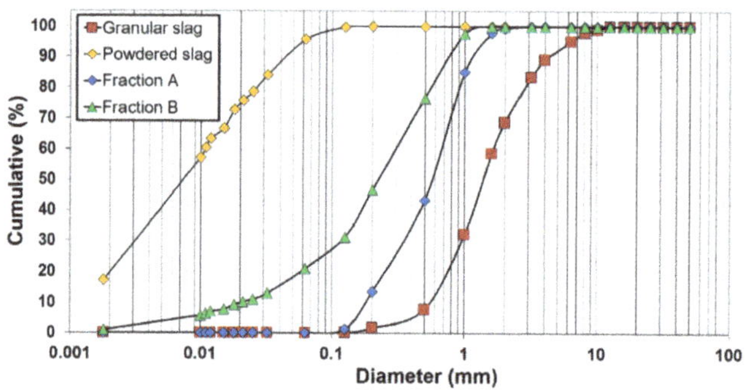

Figure 1. Particle size distribution of as-produced granular slag and after different pre-processing procedures. Powdered slag refers to finely milled slag used as binders' main precursor while fraction A and B designate roughly milled slags used as small size aggregates.

Based on our previous laboratory experimentation, the mortar mixture was designed for a binder-to-aggregate mass ratio of 0.88. Fuller distribution was applied to optimize aggregate particle size distribution (PSD). The PSD of the two fractions of the slag used as aggregates is shown in Figure 1. The density of the aggregate fractions was determined to be approximately 3.0 $\pm$ 0.03 g/cm^3, and their mass ratio in the mortars' design was set to be 1.4:1. The characteristics of the reference mortar mix design are provided in Table 2.

Table 2. Reference mortar composition (wt.%).

Powdered Slag (PS)	Silica Fume (SF)	Solution	Aggregates	
			Fraction A	Fraction B
32.7	**1.9**	12.2	31.2	22.0

The mortars were produced following the procedure described by Ascensão et al. [19]. In short, it involved: (a) manually mixing the powdered slag and silica fume in a plastic container; (b) mixing the binders' solid components (PS+SF) and the alkaline solution at 140 rpm for 600 s in a mortar mixer (Testing Bluhm & Beuerherdt Gmbh, Berlin, Germany);

and (c) mixing the pre-mixed aggregate fractions with the binder for 120 s at the same speed. The mortars were cast into 5 × 5 × 5 cm^3 metallic molds and cured for 24 h in controlled conditions (20 ± 0.5 °C and 95 ± 1.0% relative humidity). After demolding, the mortar specimens were individually sealed with a plastic film and thermally treated for 72 h at 50 °C and then kept at room conditions (20 ± 1.0 °C and 65 ± 5.0% relative humidity) until testing.

2.2. Methods

2.2.1. High-Temperature Processing

After 28 d of curing, mortar specimens were heated from room temperature up to 1100 °C in a chamber furnace (Carbolite RHF). Five different target temperatures were defined: 105, 600, 750, 900, and 1100 °C. These target temperatures were chosen on the grounds of previous studies [5,7] that demonstrated the existence of relevant chemical transformations occurring in these temperature ranges. For each target temperature, a thermal profile with a heating rate of 1 °C/min [5,20,21] and an isothermal step of 1 h at the maximum temperature was followed. The heated specimens were left to cool down naturally inside the closed furnace. This procedure was replicated with a heating rate of 10 °C/min to assess the impact of heating rate on AAMs mineralogy, morphology, and post-heating properties. A minimum of two specimens per heating rate and maximum temperature were tested.

2.2.2. Physical-Mechanical Characterization

The weight (0.1-g accuracy) and dimensions (0.01-mm accuracy) of each specimen were measured before and after thermal exposure to evaluate the mass loss and the volumetric changes that occurred during thermal treatment (and eventual anisotropic features), respectively. The apparent density of the mortars was determined by the relation between the weight and volume of each sample. Compressive strength was determined using a Universal Testing Machine (Controls) according to EN196-1:2016 [22].

2.2.3. Crystallinity and Morphological Characterization

The crystallinity of the starting slag, silica fume, and mortars produced was assessed by X-ray diffraction (XRD), carried out on a conventional Bragg-Brentano Bruker D8 Advance diffractometer equipped with Lynxeye detector (Cu Kα radiation λ = 1.54059 Å, divergence slit 0.5°, Soller slit set 2.5° + 2.5°, 5–70° 2θ, step/size 0.02°, and t/step 0.04 s.) and phase identification by EVA software (database ICDD-PDF-2). Samples from green and thermally treated mortars were collected from mechanical tests, then ground, sieved ($\leq$45 μm), and kept sealed prior to testing. Scanning electron microscopy (SEM—EVO® MA 15) equipped with energy dispersive X-ray spectrometry (EDS, AZtecEnergy, Karnataka, India) was used to further evaluate the differences in each sample's morphology before and after thermal exposure. All backscattered electron images (BSE) were acquired using a 20-kV acceleration voltage and a working distance of 10.0 mm.

3. Results and Discussion

3.1. Physical-Mechanical Characterization

The AAMs' residual properties were characterized as a function of temperature (up to 1100 °C) and heating rate (1 and 10 °C/min). Figure 2 illustrates the color change, cracking, and deformation of the mortars after high-temperature exposure. Below 750 °C, there were no visible effects on the surface of the specimens independent of the heating rate used. Similar observations have been made by Khan et al. [23] in fly ash/GGBFS mortars. Only a slight lightening of the color could be observed from 20 °C (Figure 2a) to 600 °C (Figure 2b), which can be attributed to gradual dehydration and some Fe oxidation as shown by Rickard et al. [24]. In fact, Figure 3 shows that mass loss mainly occurred in this temperature range. The mass loss at 105 °C can be attributed to the evaporation of free water located in existing pores. Previous works [7] have reported a higher mass loss in this temperature range, which can be explained—along with the differences in binder chemistry—by the curing conditions (50 °C for 3 d) and curing duration (28 d) prior to thermal testing employed in this work. Thus, part of the water has already evaporated before the experiment when following this procedure. The further mass loss observed at temperatures up to 600 °C can be attributed to the egress of water present in smaller sized pores, interstitial water, and chemically bonded water in hydroxyl groups [11,14]. No anisotropic features were detected, and a uniform volumetric contraction that did not exceed 20 vol% was observed. Rickard et al. [24] reported significant lower shrinkage values in Fe-rich AAMS (3% at 900 °C) while He et al. [25] refer to shrinkage values of 5% in FA geopolymers fired at temperature of 800 °C. The influence of the heating rate started to become evident from 750 °C, as shown in Figure 2c–h. The specimens heated at lower rates (1 °C/min) changed color from a light grey to a gradually more brownish and yellow color (see Figure 2c,e,g). No cracks were visible on the surfaces of these samples, and the absence of color gradients indicates uniform mineralogical transformations [24]. The mineralogical transformations that occurred at these temperatures explain the changes in color and are discussed later in this work.

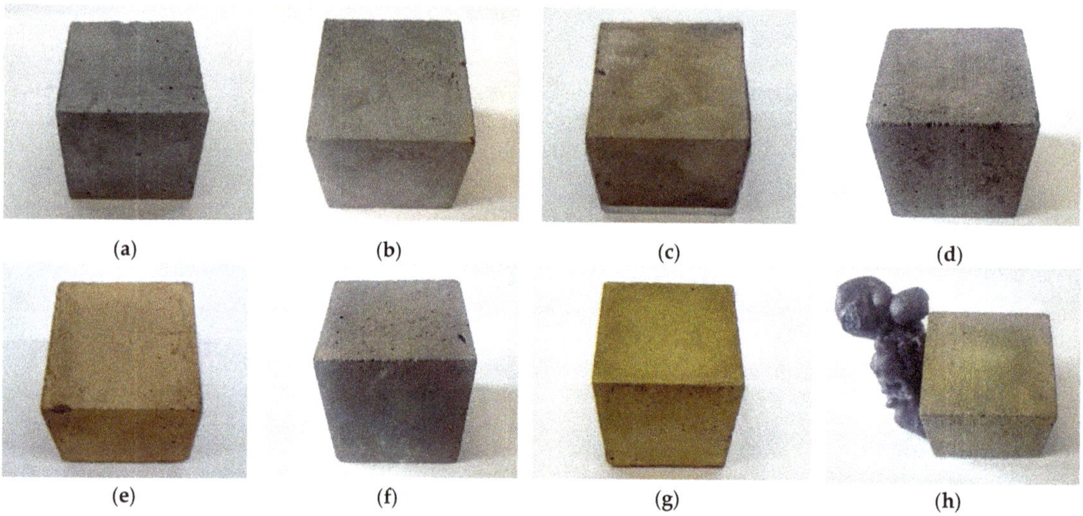

Figure 2. Alkali activated mortar specimens after 28 d of curing and thermal exposure to: (**a**) 20 °C; (**b**) 600 °C; (**c**,**d**) 750 °C with heating rates of 1 and 10 °C/min, respectively; (**e**,**f**) 900 °C with heating rates of 1 and 10 °C/min, respectively; (**g**,**h**) 1100 °C with heating rates of 1 and 10 °C/min, respectively. Mortars exposed to 105 °C are not shown here, but no significant modifications of specimens' appearance could be seen relative to the ones kept at 20 and 600 °C (**a**). At 600 °C there were no visible differences between specimens heated with distinct heating rates, with (**b**) being a representative example of both. Molten material at the surface of mortars heated up to 1100 °C at 10 °C/min is illustrated in (**h**).

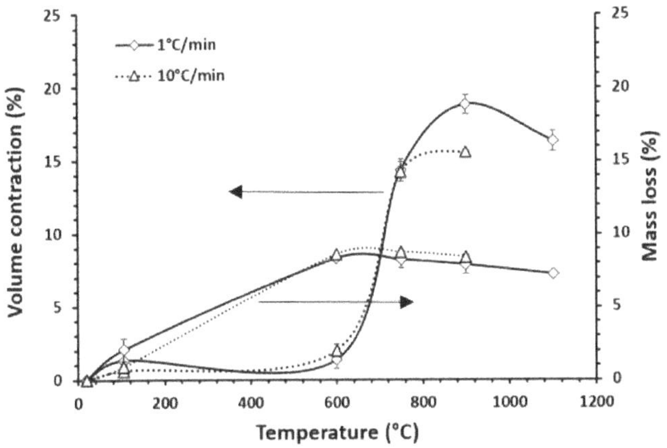

Figure 3. Mortars' volume contraction and mass loss as a function of temperature and heating rate. The curves are drawn as a guide to the eye. Due to deformation of the samples fired at 1100 °C with a heating rate of 10 °C/min data could not be collected.

When a higher heating rate was employed (10 °C/min), less distinguishable color changes were observed. The specimens heated up to 750 and 900 °C presented a less vivid brownish color relative to the ones heated at lower heating rates. Some color gradients were observed across the specimens' surfaces, especially when heated to 900 °C. The differences in color indicate that temperature gradients may have led to uneven transformations as suggested elsewhere [24]. In addition, some cracks and superficial degradation were visible on the surface of samples. The rapid increase of temperature accelerates dehydration processes, but water release occurs at higher temperatures due to lower heat and mass transfer rates (Figure 3), giving less time for steam to escape, which in turn results in excessive vapor pressure and leads to crack propagation [15]. Khan et al. [23] also reported severe cracks on Ca-rich fly ash/GGBFS mortars fired at 5 °C/min, suggesting some similarities in the effects of increased heating rates. However, all samples heated up to 900 °C maintained their structural integrity, regardless of the heating rate used. The samples heated to up 1100 °C with a heating rate of 10 °C/min present considerable molten material on their bottom surfaces (rotated in Figure 2h). Apart from the excessive vapor pressure generated, the large temperature gradients across the samples' cross-sections resulted in differential shrinkage, which increased the stresses formed and crack propagation (Figure 4), allowing for the runoff of molten material. Similar findings were reported by Fernández-Jiménez et al. [26], who observed that high heating and cooling rates form new or intensified pre-existing cracks due to the high thermal strain generated, ultimately leading to local failure. The viscosity of the melts also decreases as the temperature rises, which further facilitates their flow. Hence, the amount of molten material is expected to be significant at temperatures higher than 900 °C, as the increased flowability of the melts and the high number of voids and cracks in the mortar structures resulted in molten material coming out of the surfaces, as seen in Figures 2h and 4.

Figure 4. The surface of alkali activated mortar after thermal exposure to 1100 °C with heating rates of 10 °C/min. Molten material was removed from the sample surface for clarity.

In samples heated to 1100 °C with a lower heating rate, no molten material was visible after thermal exposure. The more gradual heating process diminished crack formation, which allowed the molten material to be retained inside the monolith, positively affecting the mortars' structural stability and high-temperature performance. Although mortars heated up to 1100 °C with heating rates of 10 °C/min retain their cubic shape and were tested for their compressive strength, they were excluded from further physical analysis (e.g., mass loss). The physical and mineralogical modifications that occurred upon heating had clear repercussions on AAMs' residual properties. From room temperature to 600 °C a slight increase of compressive strength after thermal exposure was observed, as shown by Figure 5.

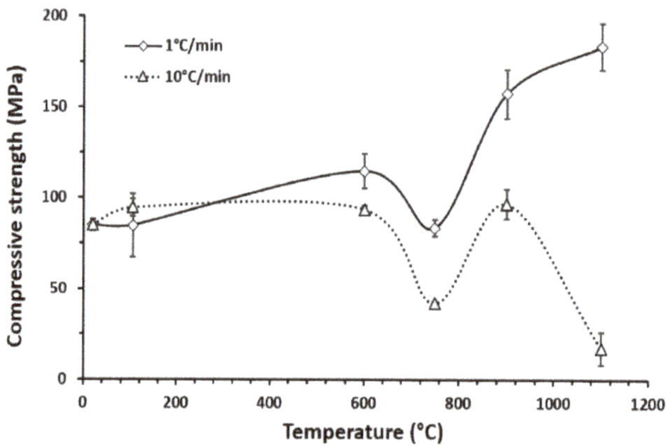

Figure 5. Mortars' compressive strength as a function of temperature and heating rate. The curves are drawn as a guide to the eye.

The improvement of compressive strength after thermal exposure to temperatures that do not exceed 600 °C is not related to melting and crystallization processes (see Section 3.3), but most likely to the increment of on-going polymerization reactions. In previous work [27,28], the authors have shown that in very similar alkali-activated systems, exothermic reactions have not completely ceased after more than 20 d of curing. Elevated temperatures stimulate these late-stage reactions and prompt the formation of more mechanically resistant binding phases. Along with the absence of significant crack formation, this beneficial effect overcomes the deleterious impact of dehydration and improves AAMs' mechanical features. The increase of compressive strength in this temperature range is often associated with densification [7,29]. However, it can be seen in Figure 6 that the apparent density slightly decreases at 600 °C. Although more resistant and probably denser polymeric chains have been formed, this decrease is the combined effect of water loss and shrinkage. As mass loss (≈8.5 wt.%) was significantly higher than the volumetric contraction up to 600 °C (<2 vol%), apparent densities decrease slightly. The increment of on-going reactions is further supported by comparing the mechanical strength of samples thermally treated up to 600 °C with different heating rates. As can be seen, lower heating rates—meaning more prolonged residence time at elevated temperatures—favor the formation of more resistant AAMs without significantly modifying the samples' mass and volume (Figure 3) or promoting the significant formation of new crystalline phases (Figure 7). At 750 °C, considerable deterioration of AAMs' mechanical properties occurred with both heating rates. This process may be related with modifications on the iron- and aluminosilicate networks occurring even before sintering [30]. The shrinkage of the aluminosilicate network induces internal cracks and increases the mean pore size, which leads to poorer mechanical performances [30]. Nonetheless, totally dehydrated charge balance cations (Na^+ or K^+) act as flux at high temperatures and reduce the temperature at which melting and crystallization start. This can imply the formation of partially sintered phases at around 750 °C [26]; however, these are not fully able to compensate for the decline in compressive strength caused by dehydration-related shrinkage.

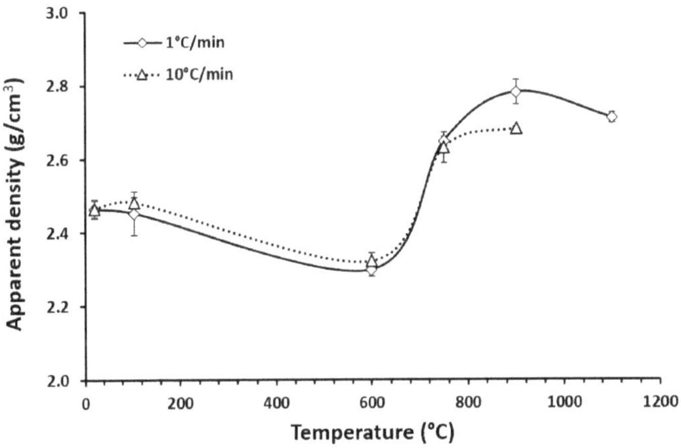

Figure 6. Mortars' apparent density as a function of temperature and heating rate. The curves are drawn as a guide to the eye. Due to deformation of the samples fired at 1100 °C with a heating rate of 10 °C/min data collection was not possible.

An additional factor that has been reported to contribute to the loss of strength at temperatures around 800 °C is the expansion of coarse aggregates [31]. Due to the type and granulometric distribution of the amorphous aggregates used in this work, however, a similar phenomenon seems unlikely to be significant. The effects of the network shrinkage could be seen at the macroscopic level by the steepest increase in volumetric contraction observed around 750 °C (Figure 3). At this temperature, increasing the heating rate was found to be detrimental to AAMs' residual strength. Along with higher crack formation due to the higher vapor pressure generated, the considerable difference in mechanical performance between mortars heated at different rates may indicate more deteriorated network structures. At 750 °C, AAMs start to lose their predominantly amorphous character, leading to the appearance of some broadened peaks of magnetite (Fe_3O_4) and esseneite ($CaAlFeSiO_6$) (Figure 7). These crystals could not be clearly identified in morphological analysis and does not seem able to compensate for the decrease of AAMs' mechanical strength; however, they may have contributed to some degree to the densification of the samples. As mass loss was mainly associated with dehydration processes occurring mostly before 300 °C, the mass remained practically unchanged at temperatures higher than 600 °C. Therefore, the apparent density followed a similar trend to the volumetric contraction. A minor increase of mass was visible from 750 °C and above (Figure 3) that can be attributed to further oxidation and crystallization of iron species into magnetite and afterwards to hematite. The reference mortar presented a density of 2.46 ± 0.02 kg/m^3 and a compressive strength of 85 ± 3 MPa after 28 d of curing. Using a heating rate of 1 °C/min the maximum apparent density was observed at 900 °C (2.78 ± 0.03 g/cm^3); however, the decrease of apparent density by 1100 °C was not detrimental to mechanical performance as the samples resulting from a heating rate of 1 °C/min presented the maximum compressive strength recorded, 184 ± 13 MPa (Figure 5).

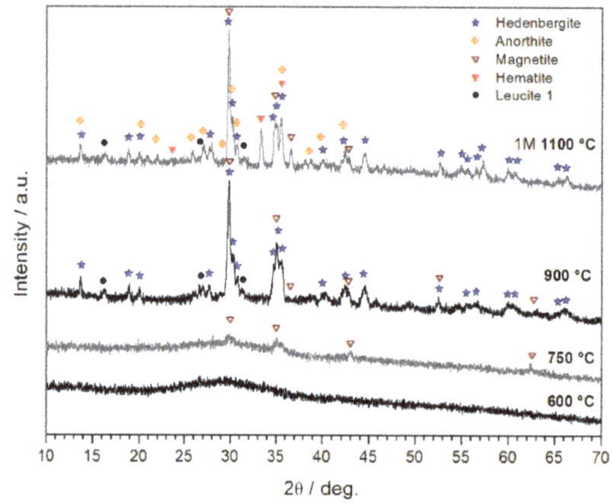

(a)

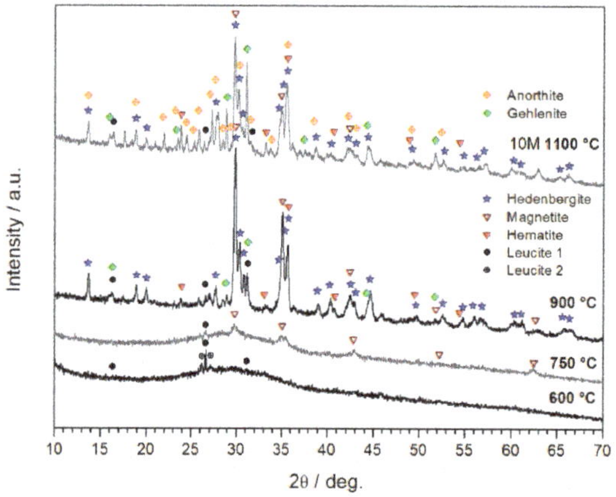

(b)

Figure 7. XRD patterns of alkali activated mortars after heating to different temperatures: (**a**) with a heating rate of 1 °C/min; (**b**) with a heating rate of 10 °C/min.

In those samples, when the temperature exceeded 750 °C and the crystallization of new phases became more intense, the compressive strength decrease was fully reversed. The compressive strength of sintered samples was progressively increased as the temperature rose, with an increase relative to the parent mortar of 85% and 115% at 900 and 1100 °C, respectively. Such strength gains contrast with the losses reported by several authors in low-iron systems, with proclivity towards lower mechanical performances as the temperature rises [23,32]. Saxena et al. [32] investigated the influence of the Si/Al ratio in the thermal performance of fly ash geopolymer mortars and reported loses in compressive strength ranging from 54–77% when firing at 1000 °C. Conversely, Villaquirán-Caicedo and de Gutiérrez [33] reported significant increases in the mechanical strength of metakaolin geopolymers, but the maximum strength reported was approximately 160 MPa and required firing at 1200 °C for 2 h.

Increasing the heating rate to 10 °C/min not only provoked a higher decrease of compressive strength at 750 °C, but also significantly impaired the performance recovery at higher temperatures. On those samples, the maximum compressive strength was halved (97 $\pm$ 8 MPa) and attained at 900 °C. Despite the recovery from the decay observed at 750 °C and the 13% increase relative to the parent mortar, much higher compressive strengths were observed in samples heated up to 900 °C with the lower heating rate (158 $\pm$ 13 MPa). At 1100 °C, the mechanical strength of samples heated with different heating rates became even more divergent. Unlike samples heated at 1 °C/min that yielded the maximum strength at this temperature, when high heating rates were imposed, a considerable degradation of the AAMs' performance occurred. As shown in Figure 3, these samples were not dimensionally stable, with some molten material coming out of the samples' structure. This explains the much lower compressive strength. The reason for material run-off in some rapid heated samples might be that crystallization is not fast enough, while the viscosity decreases due to the increasing temperature. At the same time, capillary forces generated at the solid surface may further favor some material outflow from the molten core. At 1 °C/min, there seems to be a better match between the heating and crystallization rate, as the significant residence time and formation of more liquid phases does not promote such phenomena. With the exception of samples exposed to 1100 °C with a heating rate of 10 °C/min (17 $\pm$ 9 MPa), all sintered samples presented compressive strength values higher than 40 MPa.

3.2. *Crystallinity*

The mineralogical composition and the high-temperature mineralogical transitions are of utmost importance to fully describe the thermal behavior of AAMs and to safely classify them in terms of high-temperature resistance. At room temperature, AAMs are predominantly amorphous to semi-crystalline materials, and the amount and type of crystalline phases present are mainly dependent on the parent precursors. The powdered slag (PS) and silica fume (SF) used in this work as starting materials were almost completely XRD amorphous materials. Only one crystalline peak of small intensity was detected in the XRD patterns for SF, identified as moissanite (SiC) which was an impurity (SF and PS XRD patterns available in [27]).

Not surprisingly, the AAMs made thereof exhibited an entirely amorphous structure at room temperature, with a pronounced hump between 20–40° 2θ in their XRD patterns. Heating the samples to 105 °C did not lead to any crystallization regardless of the heating rate used (not shown here for the sake of brevity). The first signs of the formation of crystalline phases were visible in the samples heated at 600 °C. Operating at the heating rate of 1 °C/min, no crystalline peak could be detected, as shown by Figure 7a; however, increasing the heating rate to 10 °C/min likely promoted the formation of some crystalline phases, visible from the peaks at 2θ~26–28° (Figure 7b). Given the weakness of the signals, a clear identification is impossible, but the positions are consistent with the main diffraction lines of leucite ($KAlSi_2O_6$) polymorphs (Leucite 1, PDF#76-2298, and Leucite 2, PDF#85-1626). Leucite is a well-known product of the thermal transformation of K-based AAMs [9].

The formation of crystalline phases was more pronounced as the temperature increased, stimulated by the separation of iron oxide phases. As with many waste-derived iron-rich glass-ceramic systems [34,35], a temperature increase to 750 °C involved the precipitation of magnetite (Fe_3O_4, PDF#89-0951), whereas a pyroxene phase (esseneite, PDF#78-1546), i.e., an Al-doped calcium-iron silicate, crystallized at 900 °C. Magnetite, embedding iron in both Fe^{2+} and Fe^{3+} forms (Fe_3O_4=FeO·Fe_2O_3), is a typical nucleating agent, separating at 650–800 °C from 'basaltic' glasses (glasses of the SiO_2-Al_2O_3-Fe_2O_3-

CaO system, including also alkali oxides), later transformed into pyroxene glass-ceramics. Magnetite may persist or be reabsorbed in the solid solution [34].

Heating to 900 °C and above promoted some oxidation, shown by the fact that both additional phases detected, i.e., $CaAlFeSiO_6$ (PDF#78-1546) and hematite (Fe_2O_3, PDF#89-0599) host only Fe^{3+} ions [34]. Finally, it can be noted that the different heating rates determined a different evolution at the maximum temperature (1100 °C), with the appearance of calcium alumino silicate phases: while the sample heated at 1 °C/min exhibited a limited precipitation of anorthite ($CaAl_2Si_2O_8$, PDF#73-0264), the sample heated at 10 °C/min featured both anorthite and gehlenite ($Ca_2Al_2SiO_7$, PDF#72-2128). Differences in the phase assemblage, with consequences for the thermo-elastic stresses developed upon cooling, could be a further explanation for the lower strength of samples heated at 10 °C/min: in particular, gehlenite is recognized for its high coefficient of thermal expansion [36].

The formation of esseneite and magnetite as the predominant phases in all samples can be seen as a natural consequence of the abundant availability of Si, Ca, and Fe species provided by the mix design. The addition of silica fume and potassium silicate solution could promote the formation of these thermally stable and refractory phases, which favor the formation of AAMs with high thermal performance [6,7,9,10,14]. The residual properties of the developed materials were enhanced with the adoption of slower heating rates. The physico-mechanical phenomena that occurred during faster heating processes, such as the higher vapor pressure and increased shrinkage and degradation of binding phases, seem to overwhelm possible positive effects from crystal formation. To further investigate such phenomena, morphological analyses were performed in samples heated up to 1100 °C, where the most significant crystallographic differences were observed.

3.3. Morphological Analysis

Figure 8 shows representative backscattered electron imaging micrographs of the reference mortar before being exposed to thermal treatment. A binder phase was formed (Figure 8a), but the existence of embedded unreacted particles confirms that complete dissolution was not achieved. The high solid-to-liquid ratio used (2.85) limited the precursors' dissolution, but the samples show a homogeneous matrix in which undissolved particles acted as small-sized aggregates, (Figure 8b). Extensive microcracks have been observed with undissolved particles acting as physical barriers that limited crack development and propagation. Similar microcracks have been reported in the literature, and their origin associated with drying shrinkage processes [28]. EDS analyses revealed a binder phase predominantly composed of Si, Al, Fe, Ca, and K, while the spectrum of undissolved particles was in line with the chemical composition of the slag used as the main precursor and as aggregates (Table 3). Some spherical pores were also observed that can be attributed to entrapped air during mixing.

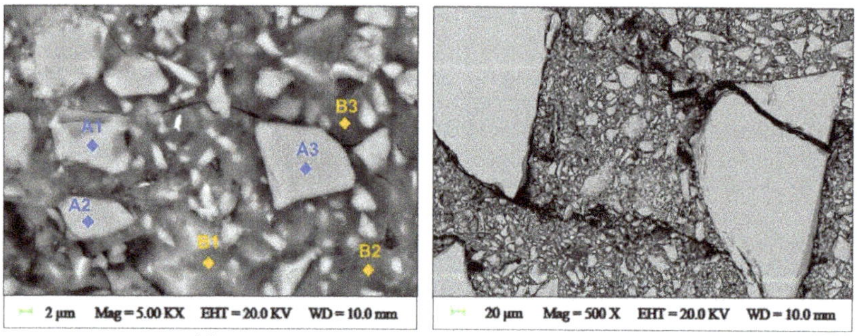

(a) (b)

Figure 8. Backscattered electron imaging micrographs of alkali activated mortar before being exposed to elevated temperatures showing: (**a**) binder phase; (**b**) microcracks.

Table 3. Representative SEM-EDS results describing the chemical composition of binder and aggregates at room temperature (B1–B3, and A1–A3, respectively) and white areas outlining the aggregates' contours after thermal exposure to 1100 °C (F1–F3). Data collection location as indicated in Figures 8a and 9b.

	SiO_2	CaO	FeO_x	Al_2O_3	MgO	TiO_2	K_2O	Na_2O	Others
B1	44.6	13.8	18.8	14.4	1.0	0.7	6.2	0.5	0.0
B2	44.9	13.7	18.4	12.7	1.0	0.5	8.8	-	0.0
B3	46.3	10.5	20.2	13.0	1.0	-	8.4	0.6	0.0
A1	34.7	22.9	23.0	16.8	1.4	0.5	0.6	-	0.1
A2	34.2	23.6	23.1	16.4	1.3	0.6	0.6	-	0.2
A3	34.2	23.5	23.2	16.4	1.4	0.7	0.6	-	0.0
F1	11.5	7.8	73.2	5.8	Nd	0.7	-	-	1.0
F2	9.2	6.2	77.6	4.8	0.7	1.4	-	-	0.1
F3	11.8	9.6	72.4	4.5	0.7	0.7	-	-	0.3

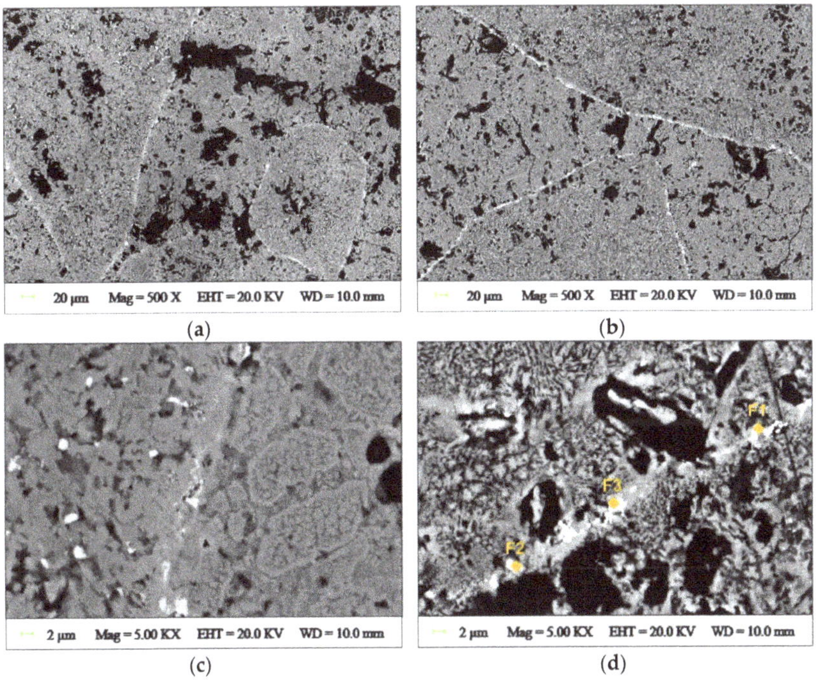

Figure 9. Backscattered electron imaging micrographs of mortars exposed to 1100 °C: (**a**,**c**) with a heating rate of 1 °C/min; (**b**,**d**) with a heating rate of 10 °C/min.

The effects of high-temperature exposure and heating rate on mortar morphology are illustrated in Figure 9. Independently of the heating rate used, a more compact microstructure with few large cracks was formed relative to the parent mortar. Partial softened aggregates have been reported to contribute to form homogenous mortars [23] and morphological analysis indicates that undissolved particles previously visible in green samples were almost completely molten (Figure 9a,b). Their previous homogeneous structure was transformed into a texture of finely exsolved phases. However, the contours of undissolved particles remain identifiable as white lines. A clearer picture of interfacial transition zones between molten aggregates and previous existing binders is better seen at higher magnification (Figure 9c,d). EDS analyses revealed the predominance of iron elements in the white areas outlining the aggregates' contours, which suggests the concentration of oxidized iron phases at these locations (Table 3 and Figure 9d). At higher magnifications, the morphological differences between samples heated with distinct heating rates become more evident.

Mortars heated at 1 °C/min exhibit a continuous interfacial transition zone with no visible cracks or microscopic pores. The progressive increase in temperature that reduces vapor pressure and cracks formed upon heating, the prolonged residence time at high temperatures that may have increased the amount of molten material, and the use of potassium-based activating solutions that assist the formation of very viscous vitreous phases [26] may have all contributed to sealing pre-existing pores and cracks. More homogeneous microstructures were formed, in agreement with the enhanced compressive strength of these mortars after being exposed to high temperatures (Figure 3).

At the highest heating rate (Figure 9d), the thermal strains generated upon heating were more difficult to absorb by the existing microstructure at room temperature. Apart from the already mentioned local failure and runoff of molten material, the formation of micropores along interfacial transition zones was also visible. A less compact and more heterogeneous structure was formed at the microscopic level, and the less effective bonding between the newly crystallized phases can explain the lower mechanical strength of mortars heated at 10 °C/min at all temperatures above 600 °C.

These differences in the mortars' morphology show that heating rate plays a vital role in the sintering process. Increasing the heating rate affects not only the phases crystallized but also their morphology and distribution, which in turn affects the dimensional stability of the mortars at high temperatures and their properties after cooling. The crystallization of new phases was particularly relevant at temperatures over 900 °C, but strength only increased if a low heating rate was employed. Nevertheless, all mortars were able to largely surpass the threshold temperature of 400 °C without any visible signs of degradation or spalling. This behavior clearly shows the superior thermal performance of the developed products relative to mainstream construction materials.

4. Conclusions

In this work, the effect of temperature and heating rate on the post-heated properties of CaO-FeO_X-Al_2O_3-SiO_2-rich AAMs was investigated. The mix design of the AAMs employed a high content of a low-value residue (85.9 wt.%). This was driven by the aim to reduce the production cost and environmental impact of the produced materials, thus providing an up-cycling solution for the large volumes of CaO-FeO_X-Al_2O_3-SiO_2-rich slags generated in thermochemical conversion processes.

The results demonstrated the superior thermal performance of AAMs produced from these residues relative to mainstream construction materials and acknowledged the vital role of both of temperature and heating rate. Lower heating rates had a beneficial effect on AAMs' mechanical performance. The maximum strength was observed when specimens were exposed to 1100 °C at 1 °C/min (184 $\pm$ 13 MPa), more than doubling the compressive strength of green mortars (85 $\pm$ 3 MPa). Increasing the heating rate from 1 to 10 °C/min was found to promote deterioration processes that resulted in extensive crack development and even the deformation of samples at 1100 °C. Along with the formation of less homogeneous microstructures, these deleterious effects were found to result in more fragile structures after thermal exposure, which limits sintering to 900 °C if high heating rates are involved.

Crystallographic analyses revealed the crystallization of thermally stable phases, indicating that the refractory and high-temperature resistant features of the developed CaO-FeO_X-Al_2O_3-SiO_2-rich AAMs could be potentially further enhanced by fine-tuning the mix design.

Comparing to benchmark materials based on OPC, no signs of spalling were visible, and the developed materials could withstand temperatures up to 900 °C with heating rates up to 10 °C/min while maintaining their structural integrity and considerable mechanical strength (>40 MPa).

Author Contributions: Conceptualization, G.A.; methodology, G.A.; validation, G.A.; formal analysis, G.A., M.S., J.V.D.S., H.R., E.B. and Y.P.; investigation, G.A.; resources, F.F., M.M., E.B. and Y.P.; data curation, G.A; writing—original draft preparation, G.A.; writing—review and editing, F.F., M.M., M.S., J.V.D.S., H.R., E.B. and Y.P.; visualization, G.A.; supervision, F.F., M.M. and Y.P.; project administration, F.F., M.M. and Y.P.; funding acquisition, F.F., M.M., E.B. and Y.P. All authors have read and agreed to the published version of the manuscript.

Funding: This research was funded by the European Union's EU Framework Programme for Research and Innovation Horizon 2020 [Grant Agreement No 721185]. G.A.'s work was also supported by the Foundation for Science and Technology (FCT)–Aveiro Research Centre for Risks and Sustainability in Construction (RISCO), Universidade de Aveiro, Portugal [FCT/UIDB/ECI/04450/2020]. This publication reflects only the author's view, exempting the Community from any liability. Project website: https://new-mine.eu/ (accessed on 6 January 2022).

Institutional Review Board Statement: Not applicable.

Informed Consent Statement: Not applicable.

Data Availability Statement: Not applicable.

Acknowledgments: The authors wish to thanks to P. Rabelo Monich for her assistance in performing the high temperature tests.

Conflicts of Interest: The authors declare no conflict of interest. The funders had no role in the design of the study; in the collection, analyses, or interpretation of data; in the writing of the manuscript, or in the decision to publish the results.

References

1. Machiels, L.; Arnout, L.; Yan, P.; Jones, P.T.; Blanpain, B.; Pontikes, Y. Transforming Enhanced Landfill Mining Derived Gasification/Vitrification Glass into Low-Carbon Inorganic Polymer Binders and Building Products. *J. Sustain. Met.* **2017**, *3*, 405–415. [CrossRef]
2. Machiels, L.; Arnout, L.; Jones, P.T.; Blanpain, B.; Pontikes, Y. Inorganic Polymer Cement from Fe-Silicate Glasses: Varying the Activating Solution to Glass Ratio. *Waste Biomass Valorization* **2014**, *5*, 411–428. [CrossRef]
3. Kriskova, L.; Pontikes, Y. Effect of activating solution on the synthesis and properties of porous Fe-Si-Ca-Rich inorganic polymers. In Proceedings of the 5th International Slag Valorisation Symposium, Leuven, Belgium, 3–5 April 2017; pp. 449–452. Available online: http://slag-valorisation-symposium.eu/2017/wp-content/uploads/downloads/Session%206/Lubica%20Kriskova%20-%20Paper%20%20Effect%20of%20activating%20solution%20on%20the%20synthesis%20and%20properties%20of%20porous%20Fe-Si-Ca-rich%20inorganic%20polymers%20%20SVS2017.pdf (accessed on 6 January 2022).
4. Hertz, K. Limits of spalling of fire-exposed concrete. *Fire Saf. J.* **2003**, *38*, 103–116. [CrossRef]
5. Rossino, C. Intrinsic Damage and Spalling Sensitivity of Concrete Subjected to High Temperature. Ph.D. Thesis, Politecnico di Milano, Milan, Italy, 2014.
6. Kong, D.; Sanjayan, J.; Sagoe-Crentsil, K. Factors affecting the performance of metakaolin geopolymers exposed to elevated temperatures. *J. Mater. Sci.* **2008**, *43*, 824–831. [CrossRef]
7. Van De Sande, J.; Pontikes, Y.; Rahier, H. Thermal Behaviour of Inorganic Polymers Synthesised from Fe-Rich Slags. In Proceedings of the Euroslag 2019 10th European Slag Conference, Thessaloniki, Greece, 8–11 October 2019; pp. 260–269.
8. Flesoura, G.; Peys, A.; Vleugels, J.; Pontikes, Y. Alkali Activation of Synthetic SiO2-CaO-FeOx-Al2O3-MgO Glass. In Proceedings of the 6th International Slag Symposium, Leuven, Belgium, 2–4 April 2019; pp. 411–414.
9. Sakkas, K.; Panias, D.; Nomikos, P.; Sofianos, A. Potassium based geopolymer for passive fire protection of concrete tunnels linings. *Tunn. Undergr. Space Technol.* **2014**, *43*, 148–156. [CrossRef]
10. Panias, D.; Balomenos, E.; Sakkas, K. The fire resistance of alkali-activated cement-based concrete binders. In *Handbook of Alkali-Activated Cements, Mortars and Concretes*; Elsevier: Amsterdam, The Netherlands, 2015; pp. 423–461.
11. White, C.; Provis, J.L.; Proffen, T.; Van Deventer, J.S.J. The Effects of Temperature on the Local Structure of Metakaolin-Based Geopolymer Binder: A Neutron Pair Distribution Function Investigation. *J. Am. Ceram. Soc.* **2010**, *93*, 3486–3492. [CrossRef]
12. Park, S.; Pour-Ghaz, M. What is the role of water in the geopolymerization of metakaolin? *Constr. Build. Mater.* **2018**, *182*, 360–370. [CrossRef]
13. Sakkas, K.; Nomikos, P.; Sofianos, A.; Panias, D. Utilisation of FeNi-Slag for the Production of Inorganic Polymeric Materials for Construction or for Passive Fire Protection. *Waste Biomass Valorization* **2014**, *5*, 403–410. [CrossRef]
14. Bell, J.L.; Driemeyer, P.E.; Kriven, W.M. Formation of Ceramics from Metakaolin-Based Geopolymers. Part II: K-Based Geopolymer. *J. Am. Ceram. Soc.* **2009**, *92*, 607–615. [CrossRef]
15. Degirmenci, F.N. Freeze-Thaw and fire resistance of geopolymer mortar based on natural and waste pozzolans. *Ceram. Silik.* **2017**, *62*, 41–49. [CrossRef]

16. Panias, D.; Giannopoulo, I.; Sakkas, K. Utilisatiion of slag for the development of fire resistant geopolymers. In Proceedings of the 6th International Slag Valorisation Symposium, Leuven, Belgium, 2–4 April 2019; pp. 387–390.
17. Close, P.; Shepherd, H.M.; Drummond, C.H. Determination of Several Valences of Iron, Arsenic and Antimony, and Selenium in Glass. *J. Am. Ceram. Soc.* **1958**, *41*, 455–460. [CrossRef]
18. Morsy, M.S.; Shebl, S.S.; Rashad, A.M. Effect of fire on microstructure and mechanical properties of blended cement pastes containing metakaolin and silica fume. *Silic. Ind.* **2009**, *74*, 59.
19. Ascensão, G.; Seabra, M.P.; Aguiar, J.B.; Labrincha, J.A. Red mud-based geopolymers with tailored alkali diffusion properties and pH buffering ability. *J. Clean. Prod.* **2017**, *148*, 23–30. [CrossRef]
20. Hager, I.; Sitarz, M.; Mróz, K. Fly-ash based geopolymer mortar for high-temperature application—Effect of slag addition. *J. Clean. Prod.* **2021**, *316*, 128168. [CrossRef]
21. Zawrah, M.; Sadek, H.; Ngida, R.E.; Sawan, S.; El-Kheshen, A. Effect of low-rate firing on physico-mechanical properties of unfoamed and foamed geopolymers prepared from waste clays. *Ceram. Int.* **2022**, *in press*. [CrossRef]
22. *EN 196-1:2016*; Methods of Testing Cement. Part 1: Determination of Strength. BSI Standards Limited: London, UK, 2016; ISBN 978-0-58084-580-2.
23. Khan, N.N.; Sarker, P.K. Effect of waste glass fine aggregate on the strength, durability and high temperature resistance of alkali-activated fly ash and GGBFS blended mortar. *Constr. Build. Mater.* **2020**, *263*, 120177. [CrossRef]
24. Rickard, W.D.A.; van Riessen, A.; Walls, P. Thermal Character of Geopolymers Synthesized from Class F Fly Ash Containing High Concentrations of Iron and α-Quartz. *Int. J. Appl. Ceram. Technol.* **2010**, *7*, 81–88. [CrossRef]
25. He, R.; Dai, N.; Wang, Z. Thermal and Mechanical Properties of Geopolymers Exposed to High Temperature: A Literature Review. *Adv. Civ. Eng.* **2020**, *2020*, 7532703. [CrossRef]
26. Fernández-Jiménez, A.; Pastor, J.Y.; Martín, A.; Palomo, A. High-Temperature Resistance in Alkali-Activated Cement. *J. Am. Ceram. Soc.* **2010**, *93*, 3411–3417. [CrossRef]
27. Ascensão, G.; Marchi, M.; Segata, M.; Faleschini, F.; Pontikes, Y. Reaction kinetics and structural analysis of alkali activated Fe-Si-Ca rich materials. *J. Clean. Prod.* **2020**, *246*, 119065. [CrossRef]
28. Ascensão, G.; Beersaerts, G.; Marchi, M.; Segata, M.; Faleschini, F.; Pontikes, Y. Shrinkage and Mitigation Strategies to Improve the Dimensional Stability of CaO-FeOx-Al2O3-SiO2 Inorganic Polymers. *Materials* **2019**, *12*, 3679. [CrossRef] [PubMed]
29. Duxson, P.; Lukey, G.C.; Van Deventer, J.S.J. Physical evolution of Na-geopolymer derived from metakaolin up to 1000 °C. *J. Mater. Sci.* **2007**, *42*, 3044–3054. [CrossRef]
30. Bakharev, T. Thermal behaviour of geopolymers prepared using class F fly ash and elevated temperature curing. *Cem. Concr. Res.* **2006**, *36*, 1134–1147. [CrossRef]
31. Kong, D.L.; Sanjayan, J.G. Effect of elevated temperatures on geopolymer paste, mortar and concrete. *Cem. Concr. Res.* **2010**, *40*, 334–339. [CrossRef]
32. Saxena, S.; Kumar, M.; Singh, N. Fire Resistant Properties of Alumino Silicate Geopolymer cement Mortars. *Mater. Today Proc.* **2017**, *4*, 5605–5612. [CrossRef]
33. Villaquirán-Caicedo, M.; de Gutiérrez, R.M. Synthesis of ceramic materials from ecofriendly geopolymer precursors. *Mater. Lett.* **2018**, *230*, 300–304. [CrossRef]
34. Höland, W.; Beall, G. *Glass-Ceramic Technology*; The American Ceramic Society: Westerville, OH, USA, 2002; p. 372.
35. Bernardo, E.; Esposito, L.; Rambaldi, E.; Tucci, A.; Pontikes, Y.; Angelopoulos, G. Sintered esseneite–wollastonite–plagioclase glass–ceramics from vitrified waste. *J. Eur. Ceram. Soc.* **2009**, *29*, 2921–2927. [CrossRef]
36. Bernardo, E.; Bonomo, E.; Dattoli, A. Optimisation of sintered glass–ceramics from an industrial waste glass. *Ceram. Int.* **2010**, *36*, 1675–1680. [CrossRef]

applied sciences

MDPI

Article

Self-Healing Glass/Metakaolin-Based Geopolymer Composite Exposed to Molten Sodium Chloride and Potassium Chloride

Patrick F. Keane [1,*], Rhys Jacob [2], Martin Belusko [3] and Frank Bruno [1]

[1] Future Industries Institute, University of South Australia, Mawson Lakes, SA 5095, Australia
[2] Forschungszentrum Jülich GmbH, Structure and Function of Materials (IEK-2), 52428 Jülich, Germany
[3] Mondial Advisory Pty Ltd., Hyde Park, SA 5061, Australia
* Correspondence: patrick.keane@mymail.unisa.edu.au

Abstract: Geopolymers (GP) are a class of X-ray amorphous, nanoporous, nanoparticulate materials that can be mixed, poured, and cured under ambient conditions. Typically, geopolymers are made using a Group 1 (G1) alkali activator such as sodium or potassium metasilicate and an aluminosilicate precursor. An analogous material to GPs is ordinary Portland cement because of the similarities in processing, however, the resulting microstructure is more similar to that of a glass. Geopolymers are more thermally stable than OPC and can therefore be used in a variety of thermal energy storage systems, as energy storage is an increasing global concern. In this study, potassium metakaolin-based geopolymer composites containing glass particles and alumina platelets were manufactured, heated in air, and exposed to molten sodium chloride or potassium chloride under an air atmosphere. Results showed the formation of an amorphous self-healing geopolymer composite (ASH-G) that could contain molten G1 chlorides for over 200 h without signs of macro or microscopic chemical degradation. The filling of cracks by glass particles in the composite after heating to 850 °C makes this material self-healing. It was found that the morphology of ASH-G composites was more affected by temperature and duration than contact with corrosive molten chlorides in air. Future works include investigating the effect of molten salt on mechanical properties during initial heating, after prolonged heating, and the material compatibility with other molten Group 1 chloride eutectics.

Keywords: geopolymer; porosity; heat treatment; alumina; glass frit; chloride; molten salt

1. Introduction

Global energy consumption in 2021 was 595 exajoules according to statistical data provided by British Petroleum [1]. Of this energy consumed, 82% came from coal, natural gas, and oil, which equates to 36.3 gigatons of carbon dioxide released in 2021 alone [1,2]. Climate change caused by the release of greenhouse gases, as well as the high cost of fossil fuels, have made renewable energies and technologies an attractive option [1,3,4]. A major drawback of renewable energy sources is the inconsistent supply compared to the grid demand. Energy storage systems coupled with renewable energy would greatly lower the cost of energy as well as reduce the amount of CO_2 released.

There are several ways to store energy such as electrochemically, in the form of batteries, mechanically, in the form of pumped hydro, or thermally, in the form of hot or cold masses. Commonly deployed batteries (e.g., lithium-ion) require rare earth elements that are expensive and in limited supply [5], while hydroelectric batteries require natural or artificial height differences to take advantage of gravity, which limits locations where they can be placed [6,7]. Alternatively, thermal energy storage systems can be made from abundant materials and placed in a variety of locations [8].

In general, thermal energy storage (TES) can be classified into three broad categories: sensible, latent, and thermochemical. It can be further characterized based on temperature, with applications ranging from sub-zero (e.g., cool rooms), to low- to medium-temperature (e.g., HVAC), or high-temperature (e.g., concentrated solar power). Of the stated categories, sensible is the most often deployed owing to its relative simplicity [9].

In sensible TES, the energy stored relies on the total mass of the storage material, the specific heat of the storage material, and the temperature difference of the storage material. A major challenge of these systems is the footprint required to house the storage material, which in many low-temperature cases is water. To reduce the footprint of thermal energy storage systems, a higher temperature difference, and therefore a higher temperature material, can be used. At higher temperatures, molten salts such as nitrates, chlorides, or carbonates can be used owing to their favorable melting points (for latent heat applications) and low cost [10,11].

Of particular interest is the use of chloride materials as they are very abundant and low-cost [12]. For example, chlorine and sodium are the third and fourth, respectively, most abundant elements in Earth's oceans behind hydrogen and oxygen which makes sodium chloride salt abundant and relatively easy to process [13]. Sodium chloride, NaCl, is solid at ambient conditions and has a relatively high thermal conductivity, specific heat, and latent heat of fusion upon melting at 801 °C. These properties are ideal for storing thermal energy [14]. The major drawbacks of using NaCl and other chlorides for thermal energy storage are corrosivity, chemical degradation, and operating temperatures. Molten salts have also been used to quench alloys, extract minerals from ore, and are found in potential nuclear reactors [15–17], making their containment an important research topic.

Container material for molten salts can be metal alloys, ceramics, or refractory composites. Common metal alloys such as stainless steel 304 and 316 have shown substantial levels of corrosion and require high-purity inert atmospheres [18–20]. Suitable alloys are not entirely corrosion-proof, use rare metals, and are expensive [21]. Ceramics require high amounts of time and energy by comparison [22]. Refractory bricks are affordable, and chemically and thermally stable but require joins and mortars which can be a major point of failure [23]. An alternative to these proposed solutions is geopolymer composites.

Geopolymers (GP) are a class of X-ray amorphous, nanoporous, nanoparticulate, inorganic covalently bonded material typically comprised of Group 1 (G1) aluminosilicates [24]. GPs are similar in processing to cements yet similar to glass in microstructure. Synthesis of GPs can be done at ambient temperatures and pressures by the addition and high shearing of aluminosilicate precursor, such as calcined kaolin (metakaolin), with an alkali activator solution. Acidic synthesis is an alternative method of producing similar phosphoric GPs, however, this route is not only relatively new compared to the alkali route but also utilizes phosphorous, which is much less common than elements used in the alkali synthesis route [25,26]. Prospect aluminosilicate precursors include various industrial waste ashes from metal processing, coal-fired power plants, and incinerators [25]. However, these tend to have high amounts of calcium, which is detrimental to geopolymer formation and maximum operating temperatures/environments [27]. This is due to the formation of calcium silicate hydrate phases, which are less thermally stable than covalently bonded aluminosilicates with minimal hydrates [27–29]. The repeatability of industrial waste ash composition is another concern, requiring attention to the individual batch formulation. Common alkali activator solutions are aqueous G1 hydroxides such as sodium and potassium and amorphous fumed silica to produce a solution of G1 metasilicate, which is also needed to fabricate geopolymers. Other alkali activating sources include G1 metasilicate, which is formed from the fusion of G1 carbonates with silica sand at high temperatures. Geopolymers made from G1 hydroxides have higher degrees of polymerization and therefore higher mechanical properties [30].

Metakaolin-based geopolymer composites can be made into complex shapes, easily incorporate functional phases, are thermally and chemically stable, and are relatively inexpensive. Additionally, ceramics can be produced by heating geopolymers to 950 °C [31].

For these reasons, metakaolin-based geopolymers could be a potentially suitable containment material for molten chlorides [32,33]. Two obstacles that need to be overcome are microcracking and significant shrinkage/warping of the geopolymer matrix after heating to 800 °C [34,35]. One method to prevent significant shrinkage during initial heating is the addition of alumina platelets [36]. The homogenous dispersion of glass-forming material in the composite has also been shown to be a successful strategy for filling microcracks [37,38]. The filling of microcracks by a uniformly distributed phase in the composite, upon heating, makes the composite self-healing [31].

Previous work on this conducted by Keane et al. (2022) showed that heat-treated, potassium metakaolin geopolymer containing homogeneously mixed alumina platelets and System 96 © glass particulates of size 10–250 μm did not chemically interact among themselves and produced a self-healing composite with uniform glaze [31]. X-ray diffraction analysis of bulk powder samples showed no new crystal formation among the phases after heating to 900 °C.

Therefore, the purpose of the current study is to investigate the compatibility and performance of a self-healing geopolymer composite exposed to either molten sodium chloride or potassium chloride to determine if any chemical degradation occurs and if this material can be used to contain molten salt at high temperatures. Potassium chloride was selected because of its similarity to sodium chloride, as well as the presence of potassium in the geopolymer composite. The aim of this research is to successfully contain molten chlorides for extended durations. If successful, the containment of molten salts by geopolymers can be used as an inexpensive solution to energy storage and fossil fuel use reduction [11].

2. Methodology

2.1. Geopolymer Composite Preparation

A metakaolin-based (MK) potassium geopolymer matrix (KGP) was prepared by the addition of MetaMAX ® from BASF and potassium metasilicate solution, also known as waterglass (KWG). MetaMAX ® is an aluminosilicate kaolin clay that has been heated to approximately 750 °C and has an average particle size of 1.3 μm [39,40]. A potassium metasilicate solution was synthesized by dissolving 85% assay potassium hydroxide flake and fumed silica, from ChemSupply Australia, into deionized water. The mixture was stirred at 300 rpm for 24 h using a magnetic stir bar, ensuring the total dissolution of the fumed silica. Metakaolin clay was then high-sheared into the KWG solution for 5 min at 2000 rpm using an IKA® RW 20 high-shear mixer and dissolver attachment to produce a homogenous KGP slurry. The KGP slurry was de-aerated for 5 min using a vibration table. Equation (1) and Equation (2) show the formation of KWG and KGP.

$$2KOH + 2SiO_2 + 10H_2O \rightarrow K_2O{\bullet}2SiO_2{\bullet}11H_2O \quad (1)$$

Equation (1): Formation of potassium metasilicate solution from SiO_2, deionized water, and KOH

$$[K_2O + 2SiO_2 + 11H_2O] + [Al_2O_3{\bullet}2SiO_2] \rightarrow [K_2O{\bullet}Al_2O_3{\bullet}4SiO_2{\bullet}11H_2O] \quad (2)$$

Equation (2): Formation of potassium geopolymer from metakaolin clay and potassium metasilicate solution

Fifty-micron alumina platelets from Micro Abrasives were then added to the KGP slurry and high-sheared for 5 min followed by 2 min on a vibration table to remove entrapped air. System 96® glass frit powder of particle size 10–250 μm from Oceanside Glass and Tile was then added to the mixture, high-sheared for 5 min, and de-aerated for another 5 min. The homogenous mixtures were then poured into plastic molds and placed in a 100% relative humidity curing chamber for five days. Plastic molds were made from modified PVC plumbing pipe and caps, as well as PVC rod. Weight percents, both reference and measured densities, and calculated volume percents of each phase are presented in Table 1. The elemental composition of glass frit powder was measured using EDS and is presented in Table 2. The glass frit powder is most similar to sodalime glass with the minor addition of impurities, which are used to prevent crystallization upon melting and cooling among other properties [41].

Table 1. Composition of Alumina-Glass-Geopolymer Composites.

Phase	Weight (%)	Theoretical Density (g/cm^3)	Measured Density (g/cm^3)	Calculated Volume (%)
KGP	50	2.0481 [24]	1.86	57.8
Glass Frit	35	2.5 [42]	2.47	33.2
Alumina Platelet	15	3.97 [43]	3.88	9.0

Table 2. Elemental Composition of Glass Frit Powder (10–250 μm).

Element	Weight (%)
O	44.0
Na	12.5
Mg	0.8
Al	1.6
Si	33.1
K	0.7
Ca	4.8
Zn	1.9
Y	0.7
Total	100.0

2.2. *Heating/Sintering*

Cured composites were demolded and acclimated to ambient humidity for 48 h before being placed into an air furnace. Composites were then heated to 900 °C for 5 h using heating and cooling rates of 2.5 or 10 °C/minute, respectively. The target temperature and ramp rate were selected as successful samples were previously produced using 900 °C and 2.5 °C/min respectively [31]. The ramp rate of 10 °C/min was selected it is a conventional heating rate for sintering ceramics [44]. Next, heat-treated geopolymer composites were filled with a known amount of sodium chloride or potassium chloride.

Isothermal container testing was conducted in an air muffle furnace for 240 h using initial heating and final cooling rates of 10 °C/min. For each Group 1 chloride, one empty control composite was placed next to a composite containing G1 chloride resulting in four composites total (two control composites [S3,5], one composite containing NaCl [S2], and one composite containing KCl [S4]). NaCl testing was conducted at 805 °C and KCl testing was conducted at 775 °C. These temperatures were selected to minimize mass loss but ensure melting.

2.3. *Group 1 Chloride Preparation*

Sodium chloride and potassium chloride were purchased from Sigma Aldrich. Both chlorides were ACS grade with a purity of >99%. Approximately 12 g of sodium chloride or potassium chloride were added to cylindrical geopolymer composite crucibles that had been heated to 900 °C. Crucibles filled with salt were then heated to 5 °C above their respective melting points and held for 240 h to indicate if a chemical reaction or degradation occurred.

2.4. Mass/Length Change, Density, and Apparent Porosity

The weight, height, and diameter of the geopolymer composites were measured using a mass balance accurate to 0.01 g and calipers accurate to 0.01 mm. Measurements were taken before and after heating, as well as after exposure to the molten Group 1 chlorides. Mass losses were determined by comparing sample weights before and after heating. Length changes were determined by comparing sample outer diameters before and after heating, similar to previous work [31]. Heat-treated sample densities and open porosities were measured before and after exposure to molten salt using the Archimedes method outlined in ASTM C20. Deionized water at 24 °C was used as the reference fluid. Samples were dried at 130 °C for 24 h and weighed, W_{Dry}. Next, samples were submerged in boiling water for at least 2 h to fill all open pores with fluid. Once cooled and still submerged, suspended samples were weighed to offset buoyant force, W_{Sub}. Finally, saturated samples were wiped of free water and weighed, W_{Sat}. Equations (3)–(7) below show how to calculate open porosity and bulk density using described sample weights.

$V_{Bulk} = \frac{(W_{Sat} - W_{Sub})}{\rho_{Water}}$	Equation (3). Bulk Volume
$V_{Open\ Pores} = \frac{(W_{Sat} - W_{Dry})}{\rho_{Water}}$	Equation (4). Open Pore Volume
$V_{Matrix} = \frac{(W_{Dry} - W_{Sub})}{\rho_{Water}}$	Equation (5). Apparent Volume
$\rho_{\ Bulk} = \frac{W_{Dry}}{V_{Bulk}} = \frac{W_{Dry} * \rho_{Water}}{(W_{Sat} - W_{Sub})}$	Equation (6). Bulk Density
$\text{Open Porosity} = \frac{V_{Open\ Pores}}{V_{Bulk}} = \frac{W_{Sat} - W_{Dry}}{W_{Sat} - W_{Sub}}$	Equation (7). Open Porosity

2.5. Microcharacterization

Samples were polished on a Struers Tegra grinding system using silicon carbide paper of 180 and then 1200 grit. Further polishing was done on a Presi Mecatech 250 polishing system down to 6 µm. A carbon coater deposited a conductive layer of approximately 20 nm thickness to prevent electron buildup during SEM/EDS analysis. Micrographs were collected using an Olympus SC50 optical microscope, a Zeiss Merlin field emission electron gun SEM, and silicon drift detector EDS. Fracture surfaces and polished surfaces were used to observe porosity, localized heterogeneity, and chloride presence. ImageJ software was used in post-processing to determine the size of features.

X-Ray diffraction patterns of samples were measured using a Panalytical Empyrean X-ray diffractometer using a step size of 0.02° from 10 to 90 2θ. The X-ray source was CuKα with a nickel filter monochromator and an approximate wavelength of 1.54 Å. Recovered salt samples were ground to less than 50 µm particulates. Sample diffraction patterns were then compared to the International Centre for Diffraction Data PDF4+ database for confirmation.

3. Results & Discussions

Geopolymer composites containing fifty-micron alumina platelets and glass particulates as seen in Figure 1 were manufactured, cured for seven days, and heated to 900 °C for five hours. The high angularity of these particles allows for a relatively higher average free space among particles [45]. This higher average free space among particles allows for the molten glassy phase to flow throughout the composite more easily at elevated temperatures once the matrix phase has uniformly shrunk [46]. This may seem counterintuitive as shrinking causes a reduction of space, however, the alumina phase remains rigid [36]. It can be theorized that capillary action and/or liquid phase sintering of the shrinking matrix phases drew in the softened, highly viscous glass particulates upon further heating [47–49].

Figure 1. Glass frit powder as received (**A**). Alumina platelets (50 µm) as received (**B**).

Figure 2 below shows crucibles before and after heating using slow (2.5 °C/min) and fast (10 °C/min) heating rates. A uniform high sheen glaze was observed among geopolymer composites that were heated at 2.5 °C/min to 900 °C for five hours; however, samples heated at 10 °C/min were nonuniformly deformed and did not form a uniform glaze. The increased volume in Figure 2B indicates an increase in porosity from the initial sample, which may have merit in other applications. In this study, however, uniform volume change after heating is a preliminary requirement for a suitable molten salt containment material. Therefore, only composites with a heating rate of 2.5 °C/min were used for salt containment testing.

Figure 2. ASH-G composites before (**A**) and after (**B**) heating to 900 °C for five hours using 2.5 °C/min (**left**) and 10 °C/min (**right**) heating and cooling rates.

3.1. *Effect of Initial Heat Treatment*

A linear shrinkage of 7.7% and mass loss of 12.0% occurred after heating cured ASH-G composites to 900 °C for five hours using heating/cooling rates of 2.5 °C/min. Additionally, a uniform glaze of approximately 200 µm thickness formed on the 2.5 °C/min heat-treated polished samples as seen in Figure 3. The densities before and after initial heating to 900 °C were 1.93 g/cc and 2.32 g/cc, respectively. Open porosities before and after initial heating to 900 °C were 19.95% and 0.73%, respectively. The results are shown in Table 3.

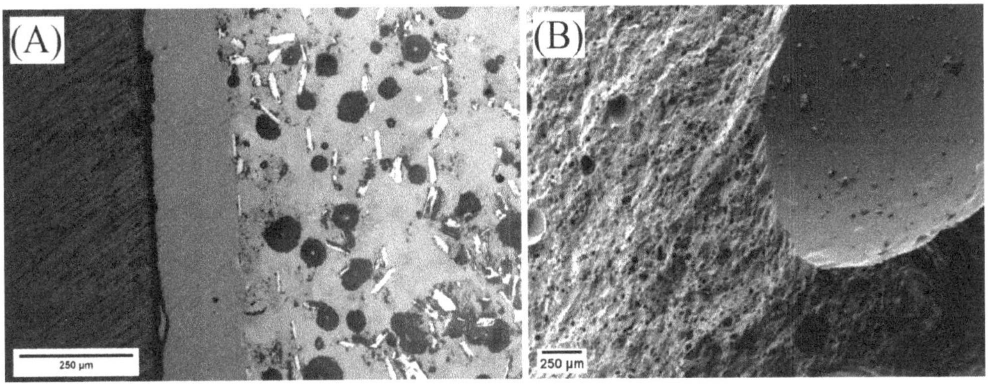

Figure 3. Micrographs of polished (**A**) and fractured (**B**) ASH-G composites after heating to 900 °C for 5 h using heating/cooling rates of 2.5 °C/min.

Table 3. Physical Changes Due to Initial Heating to 900 °C.

Sample	Linear Shrinkage (%)	Mass Loss (%)	Density (g/cc)	Open Porosity (%)
Room Temperature	0	0	1.93	19.95
Post 2.5 °C/min Heat Treatment	7.7	12.0	2.32	0.73

Previous work indicated no chemical reaction occurred among the three phases according to XRD and EDS results [31]. After heating plain KGP without additives to 900 °C a series of crystalline peaks appeared in an XRD analysis [31]. Microcracks formed during geopolymer matrix dehydration/densification were filled by molten glassy phase as indicated by little-to-no open porosity in the heated samples as expected [31,50]. Figure 4 shows how the glass phase filled much larger cracks after heat treatment. This removed the presence of discrete glass particulates and introduced oblong, form-fitting regions of glass and geopolymer matrix [31]. This phenomenon makes this material a good candidate for the containment of high-temperature liquids as the liquid cannot penetrate the glassy layer, leaving the structural material untouched and unaffected.

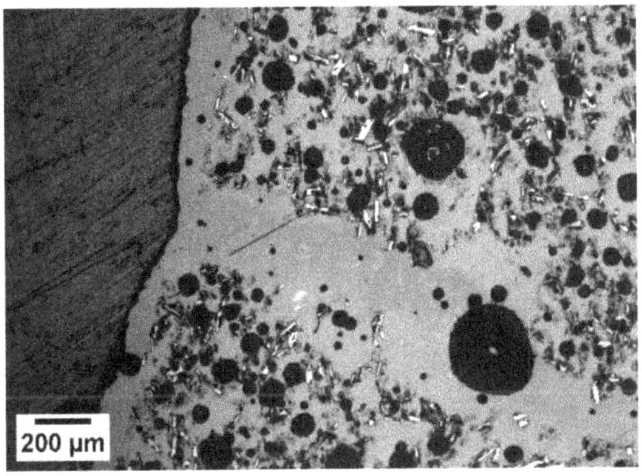

Figure 4. Micrograph of polished ASH-G composites after heating to 900 °C for 5 h using heating/cooling rates of 2.5 °C/min.

3.2. Effect of Molten Salt Exposure/Prolonged Heating

Once the composites were heated and a uniform glaze was formed, the samples were cooled, and weights measured. A known amount of either 99.9% assay sodium or potassium chloride was added to the samples and an alumina cover was placed over the opening to minimize contamination and mass loss as seen in Figure 5.

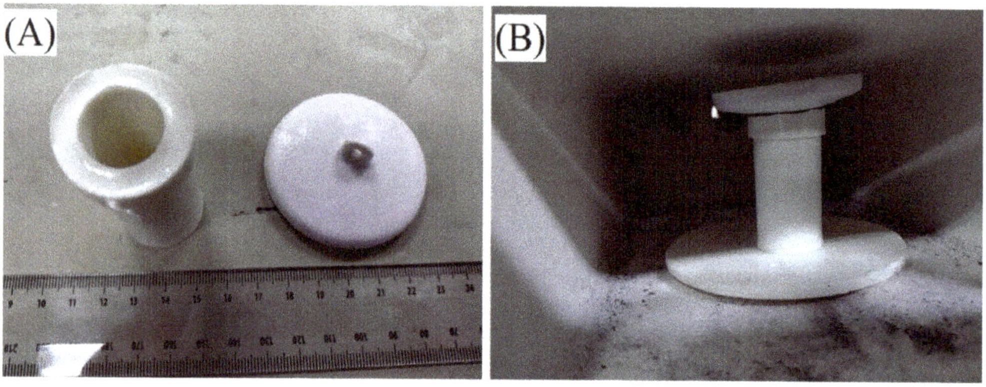

Figure 5. ASH-G composites before (**A**) and after (**B**) filling with Group 1 chloride salts.

Amorphous self-healing geopolymer containers were heated with and without Group 1 chloride to ~ 5 °C above their melting points—805 °C and 775 °C for sodium chloride and potassium chloride, respectively. Both test and control samples were held at these temperatures for 240 h. Once cooled, the ASH-G samples were measured for changes in mass and length. Solidified sodium chloride and potassium chloride samples were also taken for investigation with X-ray diffraction. ASH-G samples were then ultrasonicated in DI water for 15 min to dissolve any residual salt on the surface. Density and open porosity measurements were collected according to ASTM C20 using 24 °C de-ionized water as reference material. The results can be seen in Table 4.

Table 4. Physical Changes Due to Prolonged Exposure to Molten G1 Chlorides.

Sample	Sample Number	Linear Shrinkage (%)	Mass Loss from Cured GP (%)	Density (g/cc)	Open Porosity (%)	Salt Mass Loss (%)
Room Temperature	1	-	-	1.93	19.95	-
Post Heat Treatment	1	7.7	12.0	2.32	0.7	-
Post NaCl Exposure (805 °C)	2	8.6	11.9	2.37	2.3	4.1
Post NaCl-test Control (805 °C)	3	8.6	12.0	2.38	2.1	-
Post KCl Exposure (775 °C)	4	8.2	11.9	2.32	1.4	2.4
Post KCl-test Control (775 °C)	5	8.2	11.9	2.33	1.5	-

Data indicates that the presence of molten salt had little-to-no effect, while temperature and duration had a more significant effect. After the mass loss during initial heating, no samples experienced further mass loss due to prolonged heating. Geopolymer composites that were heated to 805 °C, containing molten sodium chloride, showed higher linear shrinkage and did not retain a surface glaze, as seen in Figure 6. In comparison, geopolymer composites heated to 775 °C, containing molten potassium chloride, did retain a reduced surface glaze of approximately 180 μm, as seen in Figure 7, and showed slightly less linear shrinkage compared to the 805 °C samples. The presence of glaze on composite containing potassium chloride in Figure 7 could also be attributed to potassium chloride being the most stable of the Group 1 chlorides, and least likely to react, according to the enthalpies of formation as well as slightly lower dwell temperature [51].

It can be hypothesized that the lack, or reduction, of the glaze, is due to some glassy phase material migrating into the bulk of the composite during reheating above the glass softening point of 573 °C, according to previous work [31]. EDS analysis of geopolymer composites in Figures 6 and 7 show the lack, or presence, of surface glaze via aluminum concentration. Areas of brighter purple indicate higher aluminum content and dimmer areas indicate lower aluminum content. Figure 8 shows lighter discrete alumina while connected irregular medium or darker phases are potassium geopolymer and glassy phase, respectively. No microcracks can be seen in the bulk of samples exposed to molten salt, indicating no chemical degradation occurred to the containment material on a microscopic level.

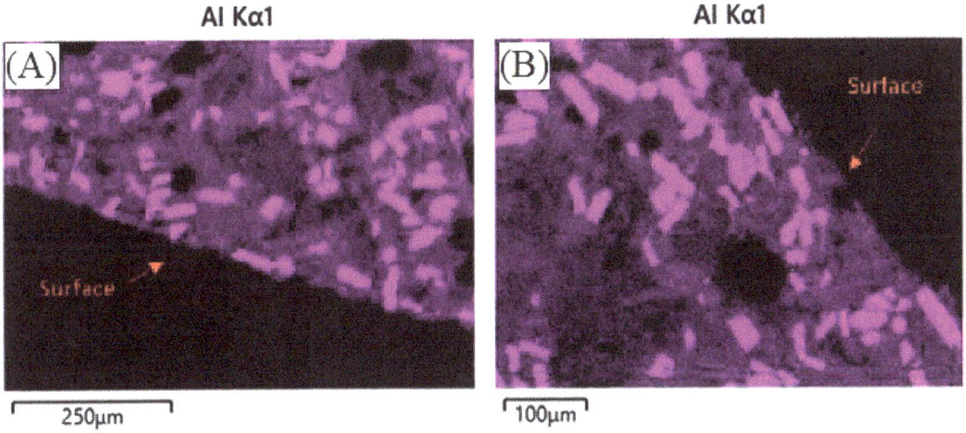

Figure 6. Surfaces of ASH-G composite after heating to 805 °C for 240 h without (**A**) and with (**B**) NaCl. No glaze is present in either case.

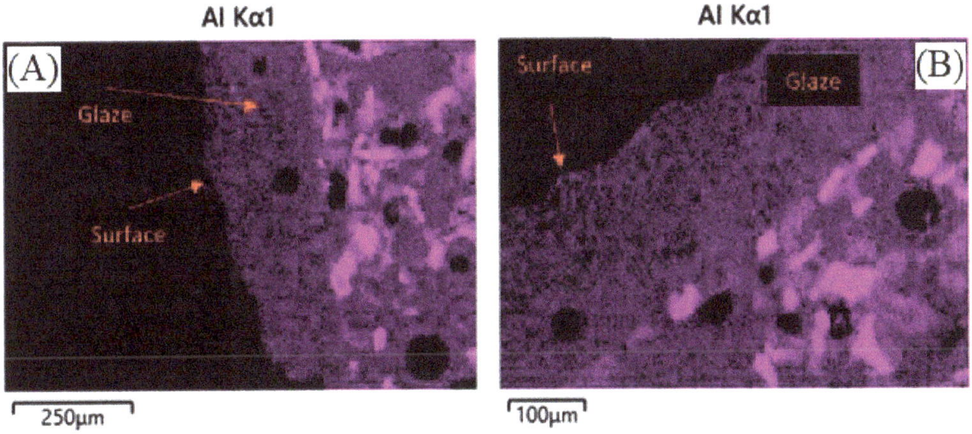

Figure 7. Surfaces of ASH-G composite after heating to 775 °C for 240 h without (**A**) and with (**B**) KCl. A reduced glaze is present in both cases.

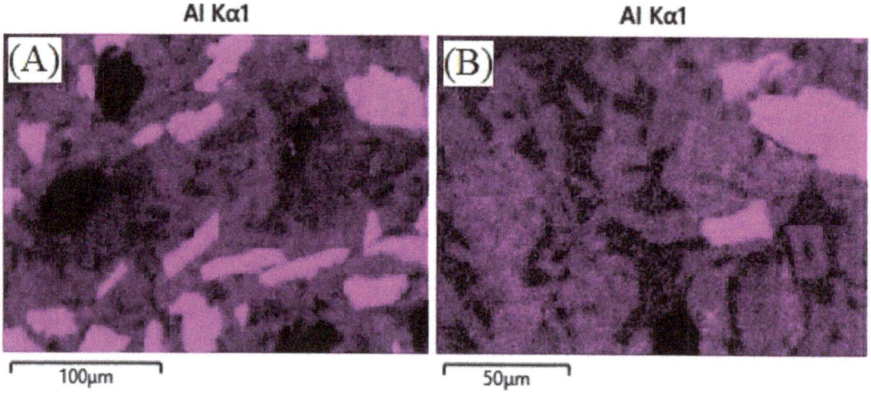

Figure 8. Bulk ASH-G composite after heating to 805 °C for 240 h with NaCl (**A**) and heating to 775 °C for 240 h with KCl (**B**). No microcracks are present.

To further check for the chemical reaction between chloride salts and geopolymer composites, bulk powder XRD analysis of ground salt samples was conducted. Representative salt samples were taken from the as-received supply and post-test crucibles. The salts were dried at 120 °C for 24 h and ground to <50 μm. The results can be seen in Figure 9 and show the presence of potassium chloride in sodium chloride samples and sodium chloride in potassium chloride samples.

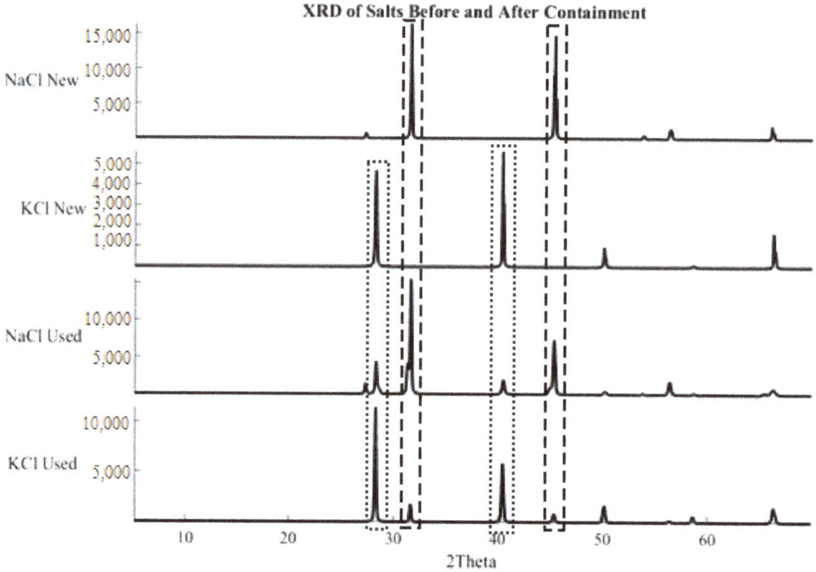

Figure 9. X-ray diffraction patterns of NaCl and KCl before and after prolonged heating in ASH-G composites. Salt samples after prolonged exposure to ASH-G show the presence of both Group 1 chlorides.

The exchange of Group 1 elements can be attributed to diffusion, and not dissolution, of the container material. Dissolution of the container material would result in microcracks, which are not present in the EDS micrographs. No other crystalline phases were observed in the recovered salt samples. If chemical degradation occurred, alumina platelets would be free to dislodge and appear in the recovered salt samples. The results indicated that the geopolymer composites did not react with air, molten salt, or chemical species generated from molten salt-air interactions to form crystalline phases.

The ability of geopolymers to incorporate additive phases of different shapes, sizes, quantities, and chemistries into stable composites is valuable and needs to be explored. There is a range of solutions for the containment of different molten salts in different geometries for different applications is a testament to the utility of engineered geopolymer composites. While no microscopic or macroscopic chemical degradation was found, how the flexural strength of geopolymer composites exposed to both molten salt and temperature changes is of particular interest. Previous work done by Tao Ai et al. found that potassium metakaolin-based geopolymer without additives exhibited superior densification and compressive strengths when heated in molten salt rather than air [46]. Initial heat treatment to 300 °C in air followed by further sintering in molten salt could not only produce stronger, more dense composites but also eliminate the contamination of molten salt over extended periods.

4. Conclusions

In this study, potassium metakaolin-based geopolymer composites containing alumina platelets and glass frit powder were formed into cylindrical crucibles. Composites were then heated to 900 °C in air for five hours using slow (2.5 °C/min) and fast (10 °C/min) heating/cooling rates. It was found that slow rates of heating and cooling caused samples to have an increased density, loss of nearly all open porosity, and develop a uniform glassy glaze of approximately 200 μm. This material can be called an amorphous self-healing geopolymer composite (ASH-G) as microcracks that occurred during heating were filled with glass phase. It is hypothesized that cracked samples can be slowly re-heated to high temperatures and the cracks healed with the molten glassy phase. Composites heated and cooled using 10 °C/min did not densify, did not produce a surface glaze, and deformed irregularly. It is speculated that sintering failure occurred during heating rather than cooling.

The as-produced amorphous self-healing geopolymer composites were able to contain molten Group 1 chlorides for 240 h. Significant findings include:

- The density and open porosity of the ASH-G samples increased and decreased, respectively.
- Molten sodium chloride and potassium chloride did not chemically react with the ASH-G container after 240 h
- A total of 4 wt% of molten salt was lost after 240 days in air
- ASH-G samples were more affected by temperature than exposure to molten salt

Analysis of the sodium and potassium chlorides indicates an ion exchange of potassium and sodium, respectively. It is shown that this occurred due to diffusion rather than the dissolution of the container material. The dissolution of any phase of the container material would result in the presence of inert crystalline alumina in the recovered salt. Future work includes mechanical evaluation to determine embrittlement or reduced flexural strength, as well as exposure to sodium/potassium chloride eutectic (T_{melt} = 665 °C) to determine salt degradation and thermocycling.

Author Contributions: Conceptualization, P.F.K., R.J., M.B. and F.B.; Methodology, P.F.K.; Formal analysis, P.F.K. and R.J.; Investigation, P.F.K.; Resources, F.B.; Data curation, P.F.K. and R.J.; Writing – original draft, P.F.K.; Writing – review & editing, P.F.K., R.J., M.B. and F.B.; Supervision, R.J., M.B. and F.B.; Project administration, M.B. and F.B.; Funding acquisition, F.B. All authors have read and agreed to the published version of the manuscript.

Funding: This research was funded by The Australian Government Research Training Development grant, The Alexander von Humboldt Foundation Early Career Post Doctorate Grant, and The Australian Solar Thermal Research Initiative P22 Grant. The APC was provided by Applied Sciences.

Data Availability Statement: The data that support the findings of this study are available from the corresponding author, P.F.K., upon reasonable request.

Acknowledgments: The authors acknowledge Microscopy Australia for the use of facilities at the Future Industries Institute at the University of South Australia. The authors thank Scott Chemical Pty Ltd. for providing the metakaolin used in this study. Rhys Jacob would gratefully like to acknowledge the Alexander von Humboldt Foundation for providing funding to support this work.

Conflicts of Interest: The authors declare no conflict of interest.

Abbreviations

ASH-G—Amorphous Self-Healing Geopolymer; DSC—Differential Scanning Calorimetry; EDS—Energy Dispersive Spectroscopy; G1—Group 1; KGP—Potassium Geopolymer; KWG—Potassium Waterglass; MK—Metakaolin; SEM- Scanning Electron Microscope; TES—Thermal Energy Storage; XRD—X-ray Diffraction.

References

1. BP. *BP Statistical Review of World Energy*; British Institute of Energy Economics: London, UK, 2022.
2. IEA. *Global Energy Review: CO_2 Emission in 2021*; IEA: Paris, France, 2021.
3. NASA. Global Climate Change: Vital Signs of the Planet. 2022. Available online: https://climate.nasa.gov/evidence/ (accessed on 12 October 2022).
4. Evans, M.; Belusko, M.; Liu, M.; Keane, P.K.; Xing, K.; Lau, T.; Bruno, F.; Taghipour, A.; Rainey, T.; Liu, A.; et al. *Electrification & Renewables to Displace Fossil Fuel Process Heating*; RACE 2030 Collaborative Research Center: Haymarket, Australia, 2021.
5. Olivetti, E.A.; Ceder, G.; Gaustad, G.G.; Fu, X. Lithium-Ion Battery Supply Chain Considerations: Analysis of Potential Bottlenecks in Critical Metals. *Joule* **2017**, *1*, 229–243. [CrossRef]
6. Görtz, J.; Aouad, M.; Wieprecht, S.; Terheiden, K. Assessment of pumped hydropower energy storage potential along rivers and shorelines. *Renew. Sustain. Energy Rev.* **2022**, *165*, 112027. [CrossRef]
7. Alisha, R.; Fernandez, S.A.B.; Reed, P.M. Operational constraints and hydrologic variability limit hydropower in supporting wind integration. *Environ. Res. Lett.* **2013**, *8*, 024037.
8. Socaciu, L. Seasonal Thermal Energy Storage Concepts. *Acta Tech. Napoc.-Ser. Appl. Math. Mech. Eng.* **2012**, *55*, 775–784.
9. IRENA. *Innovation Outlook: Thermal Energy Storage*; Institutional Renewable Energy Agency: Abu Dhabi, United Arab Emirates, 2020.
10. Jacob, R.; Liu, M.; Sun, Y.; Belusko, M.; Bruno, F. Characterisation of promising phase change materials for high temperature thermal energy storage. *J. Energy Storage* **2019**, *24*, 100801. [CrossRef]
11. Liu, M.; Omaraa, E.S.; Qi, J.; Haseli, P.; Ibrahim, J.; Sergeev, D.; Müller, M.; Bruno, F.; Majewski, P. Review and characterisation of high-temperature phase change material candidates between 500 C and 700 °C. *Renew. Sustain. Energy Rev.* **2021**, *150*, 111528. [CrossRef]
12. Myers, P.D.; Goswami, D.Y. Thermal energy storage using chloride salts and their eutectics. *Appl. Therm. Eng.* **2016**, *109*, 889–900. [CrossRef]
13. Lide, D.R. *CRC Handbook of Chemistry and Physics, 89th Edition*; Taylor & Francis: Abingdon, UK, 2008.
14. Jacob, R.; Riahi, S.; Liu, M.; Belusko, M.; Bruno, F. Technoeconomic Impacts of Storage System Design on the Viability of Concentrated Solar Power Plants. *J. Energy Storage* **2021**, *34*, 101987. [CrossRef]
15. Kriven, W.M. Geopolymers and Geopolymer-Derived Composites. In *Encyclopedia of Materials: Technical Ceramics and Glasses*; Pomeroy, M., Ed.; Elsevier: Oxford, UK, 2021; pp. 424–438.
16. Haseli, P.; Majewski, P.; Christo, F.; Raven, M.; Klose, S.; Bruno, F. Experimental Kinetic Analysis of Potassium Extraction from Ultrapotassic Syenite Using NaCl–$CaCl_2$ Salt Mixture. *ACS Omega* **2020**, *5*, 16421–16429. [CrossRef]
17. Engel, J.; Spiewak, I. Molten-salt reactors for efficient nuclear fuel utilization without plutonium separation. *Trans. Am. Nucl. Soc.* **1977**, *27*.
18. Yin, Y.; Rumman, R.; Chambers, B.A.; Liu, M.; Jacob, R.; Bruno, F.; Belusko, M.; Lewis, D.A.; Andersson, G.G. Chemical degradation in Thermally Cycled Stainless Steel 316 with High-Temperature Phase Change Material. *Sol. Energy Mater. Sol. Cells* **2021**, *230*, 111216. [CrossRef]
19. Ladkany, S.; Culbreth, W.; Loyd, N. Molten Salts and Applications II: 565 °C Molten Salt Solar Energy Storage Design, Corrosion, and Insulation. *J. Energy Power Eng.* **2018**, *12*, 517–532.
20. Jonemann, M. *Advanced Thermal Storage System with Novel Molten Salt*; Halotechnics, Inc.: Golden, Colorado, 2013.
21. LME Official Prices. *London Metal Stock Exchange*; London Metal Exchange: London, UK, 2021.
22. Duval, D.J.; Risbud, S.H.; Shackelford, J.F. Mullite. In *Ceramic and Glass Materials: Structure, Properties and Processing*; Shackelford, J.F., Doremus, R.H., Eds.; Springer: Boston, MA, USA, 2008; pp. 27–39.

23. Andreev, K.; Hoeksma, K. Using Mortar Joints to Reduce Stresses in Refractory Structures-Measurement and Modelling Experience. In Proceedings of the 49th International Colloquium on Refractories, Aachen, Germany, 7–8 November 2006.
24. Kriven, W.M. Geopolymer-Based Composites. In *Comprehensive Composite Materials II*; Elsevier: Amsterdam, The Netherlands, 2018; Volume 5, pp. 269–280.
25. Davidovits, J. *Geopolymer Chemistry and Applications*; Geopolymer Institute: Saint-Quentin, France, 2008; Volume 171.
26. Tarbuck, E.J.; Lutgens, F.K.; Tasa, D.G. *Earth Science, Global Edition*; Pearson Education Limited: Harlow, UK, 2015.
27. Sankar, K.; Sutrisno, A.; Kriven, W.M. Slag-fly ash and slag-metakaolin binders: Part II—Properties of precursors and NMR study of poorly ordered phases. *J. Am. Ceram. Soc.* **2019**, *102*, 3204–3227. [CrossRef]
28. Zhao, X.; Liu, C.; Zuo, L.; Wang, L.; Zhu, Q.; Wang, M. Investigation into the effect of calcium on the existence form of geopolymerized gel product of fly ash based geopolymers. *Cem. Concr. Compos.* **2019**, *103*, 279–292. [CrossRef]
29. Tajuelo Rodriguez, E.; Garbev, K.; Merz, D.; Black, L.; Richardson, I.G. Thermal stability of C-S-H phases and applicability of Richardson and Groves' and Richardson C-(A)-S-H(I) models to synthetic C-S-H. *Cem. Concr. Res.* **2017**, *93*, 45–56. [CrossRef]
30. Sagoe-Crentsil, K.; Weng, L. Dissolution processes, hydrolysis and condensation reactions during geopolymer synthesis: Part II. High Si/Al ratio systems. *J. Mater. Sci.* **2007**, *42*, 3007–3014. [CrossRef]
31. Keane, P.; Jacob, R.; Belusko, M.; Kriven, W.M.; Stanford, N.; Bruno, F. Microstructural evolution of amorphous self-healing geopolymer composites containing alumina and glass frit. *Int. J. Ceram. Eng. Sci.* **2022**, *4*, 327–339. [CrossRef]
32. Fang, Y.; Ahmad, M.R.; Lao, J.-C.; Qian, L.-P.; Dai, J.-G. Development of artificial geopolymer aggregates with thermal energy storage capacity. *Cem. Concr. Compos.* **2023**, *135*, 104834. [CrossRef]
33. Zhang, X.; Bai, C.; Qiao, Y.; Wang, X.; Jia, D.; Li, H.; Colombo, P. Porous geopolymer composites: A review. *Compos. Part A Appl. Sci. Manuf.* **2021**, *150*, 106629. [CrossRef]
34. Künzel, C.; Vandeperre, L.; Donatello, S.; Boccaccini, A.; Cheeseman, C.R.; Brown, P. Ambient Temperature Drying Shrinkage and Cracking in Metakaolin-Based Geopolymers. *J. Am. Ceram. Soc.* **2012**, *95*, 3270–3277. [CrossRef]
35. Duxson, P.; Lukey, G.C.; van Deventer, J.S.J. Thermal evolution of metakaolin geopolymers: Part 1—Physical evolution. *J. Non-Cryst. Solids* **2006**, *352*, 5541–5555. [CrossRef]
36. Kutyla, G.P.; Kriven, W.M. Properties and characterization of alumina platelet reinforced geopolymer composites. *J. Am. Ceram. Soc.* **2020**, *103*, 5178–5185. [CrossRef]
37. Bhuiya, A.W.; Hu, M.; Sankar, K.; Keane, P.F.; Ribero, D.; Kriven, W.M. Bone ash reinforced geopolymer composites. *J. Am. Ceram. Soc.* **2021**, *104*, 2767–2779. [CrossRef]
38. Keane, P.F.; Foltz, J.S.; Chadha, V.; Marsh, C.P.; Kriven, W.M. Amorphous self-healed, chopped basalt fiber-reinforced, geopolymer composites. *J. Am. Ceram. Soc.* **2021**, *104*, 3443–3451. [CrossRef]
39. Kuenzel, C.; Neville, T.P.; Donatello, S.; Vandeperre, L.; Boccaccini, A.R.; Cheeseman, C.R. Influence of metakaolin characteristics on the mechanical properties of geopolymers. *Appl. Clay Sci.* **2013**, *83–84*, 308–314. [CrossRef]
40. Davidovits, R.; Plelegris, C.; Davidovits, J. *Standardized Method in Testing Commercial Metakaolins for Geopolymer Formulations*; Geopolymer Institute Library: Saint-Quentin, France, 2019.
41. Martin, J.W. 4-Glasses and ceramics. In *Materials for Engineering*, 3rd ed.; Martin, J.W., Ed.; Woodhead Publishing: Sawston, UK, 2006; pp. 133–158.
42. Karazi, S.M.; Ahad, I.U.; Benyounis, K.Y. Laser Micromachining for Transparent Materials. In *Reference Module in Materials Science and Materials Engineering*; Elsevier: Amsterdam, The Netherlands, 2017.
43. Piconi, C. 1.105-Alumina. In *Comprehensive Biomaterials*; Ducheyne, P., Ed.; Elsevier: Oxford, UK, 2011.
44. Phuah, X.L.; Jian, J.; Wang, H.; Wang, X.; Zhang, X.; Wang, H. Ultra-high heating rate effects on the sintering of ceramic nanoparticles: An in situ TEM study. *Mater. Res. Lett.* **2021**, *9*, 373–381. [CrossRef]
45. Chateau, X. 6-Particle packing and the rheology of concrete. In *Understanding the Rheology of Concrete*; Roussel, N., Ed.; Woodhead Publishing: Sawston, UK, 2012.
46. Le, V.S.; Louda, P.; Tran, H.N.; Nguyen, P.D.; Bakalova, T.; Ewa Buczkowska, K.; Dufkova, I. Study on Temperature-Dependent Properties and Fire Resistance of Metakaolin-Based Geopolymer Foams. *Polymers* **2020**, *12*, 2994. [CrossRef]
47. Hrma, P.; Kruger, A.A. High-temperature viscosity of many-component glass melts. *J. Non-Cryst. Solids* **2016**, *437*, 17–25. [CrossRef]
48. Bernal, S.A.; Bejarano, J.; Garzón, C.; Mejía de Gutiérrez, R.; Delvasto, S.; Rodríguez, E.D. Performance of refractory aluminosilicate particle/fiber-reinforced geopolymer composites. *Compos. Part B Eng.* **2012**, *43*, 1919–1928. [CrossRef]
49. Liu, X.; Jiang, J.; Zhang, H.; Li, M.; Wu, Y.; Guo, L.; Wang, W.; Duan, P.; Zhang, W.; Zhang, Z. Thermal stability and microstructure of metakaolin-based geopolymer blended with rice husk ash. *Appl. Clay Sci.* **2020**, *196*, 105769. [CrossRef]
50. Adjei, S.; Elkatatny, S.; Ayranci, K. Effect of Elevated Temperature on the Microstructure of Metakaolin-Based Geopolymer. *ACS Omega* **2022**, *7*, 10268–10276. [CrossRef] [PubMed]
51. Atkins, P.W. *Shriver & Atkins' Inorganic Chemistry*, 5th ed.; Oxford University Press: Oxford, UK, 2010.

MDPI AG
Grosspeteranlage 5
4052 Basel
Switzerland
Tel.: +41 61 683 77 34

Applied Sciences Editorial Office
E-mail: applsci@mdpi.com
www.mdpi.com/journal/applsci

www.ingramcontent.com/pod-product-compliance
Lightning Source LLC
Chambersburg PA
CBHW061459210926
51850CB00008B/1762
* 9 7 8 3 7 2 5 8 5 7 6 1 6 *